MILITARY METHODS
OF THE
ART OF WAR

孫臏兵法

MILITARY METHODS

OF THE
ART OF WAR

Sun Pin

Translated, with a Historical Introduction by
Ralph D. Sawyer

MetroBooks

2002 MetroBooks

ISBN 1-5866-3608-1

Printed and bound in the United States of America

02 03 04 05 MC 10 9 8 7 6 5 4 3 2

BVG

For bulk purchases and special sales, please contact:
Friedman/Fairfax Publishers
Attention: Sales Department
230 Fifth Avenue, Suite 700-701
New York, NY 10001
212/685-6610 Fax 212/685-3916

Visit our website:
www.metrobooks.com

In memory of Fred, Ben, and Joe

Contents

Preface

The intention of the present book is to introduce and translate the reconstructed *Military Methods* attributed to Sun Pin and recovered from a Han dynasty tomb some two decades ago. As with our previous works, it is intended for a general audience rather than just for specialists, whether sinologists or military historians. Consequently, many issues that are properly the topic of academic papers, while identified and briefly considered, must be left for future works and the studies of others, including the detailed consideration of the composition of the *Military Methods* and the provenance of various statements. Furthermore, while the introduction attempts to briefly but thoroughly reconstruct the historical context and analyze the pivotal campaigns identified with Sun Pin's life, a number of complex, perhaps irresolvable problems must be deferred to our forthcoming *History of Warfare in China: The Ancient Period*.

Although we have spent years reading military texts almost exclusively, without the great work of modern scholars in reconstructing and commenting upon this text, we would never have been able to undertake the translation. Naturally we have imposed our own perspective, which is eclectic but predominately one of military history, on what is often an exclusively academic textual orientation. Moreover, many of the numerous modern Chinese editions are simply inferior rush works that either ignore or mangle the difficult portions and were designed to realize the immensely popular sales that they have attracted throughout China, Taiwan, and Japan. However, several editions are thorough, perspicacious, and detailed, especially Huo Yin-chang's *Sun Pin ch'ien-shuo* and Chang Chen-che's *Sun Pin ping-fa chiao-li*. In general, when there is severe disagreement or simply wild speculation among the commentators, we have tended to rely upon Chang's interpretations. However, there are also many cases where he seems to go dramatically astray; the most important of these are remarked upon in the Notes. Since the text is shattered and badly fragmented, we have opted to argue only those positions that clearly differ from the general position or indicate where one character rather than another has been selected. Specialists can easily refer to the relevant texts for clarification or communicate with us.

As in our previous works, we have included widely ranging notes on various aspects of potential interest to both educated readers and specialists. The general reader, in particular, is encouraged to consult the Notes, simply ignoring the more technically oriented ones. Furthermore, because of the nature of the text, we have provided extensive general commentary for each chapter, part of which is devoted to setting the *Military Methods* within the tactical thought of the period and explicating it in terms of Sun-tzu's *Art of War.* These comparisons are best made with respect to concrete cases rather than in general fashion, for the similarities and differences then become readily apparent. Consequently, the *Military Methods* translation has been conceived largely as a companion to our *Seven Military Classics,* as it should be, although it remains a fully independent volume.

The translation is based primarily upon the reconstructed text as originally published in January 1975 by Wen-wu, appropriately supplemented and revised by various book-length editions and articles that have intermittently appeared since then. We have chosen this "original edition" rather than the revised 1985 text because the former has been the basis of twenty or more modern editions, a few of which may actually be accessible apart from university holdings. Moreover, the revisions are not universally accepted, consisting mostly of reordering previously deciphered strips and making slight additions to a few chapters. Naturally the most important ones have been at least cited in the Notes, and specialists can easily consult the original where necessary.

Apart from the abstract debt we owe to the many scholars who have toiled ceaselessly on this difficult work, we wish to acknowledge Zhao Yong's ongoing assistance in locating and obtaining obscure textual materials. In addition, we have benefited greatly from wide-ranging discussions with Colonel Karl Eikenberry, C. S. Shim, Cleon Brewer, and Guy Baer here and in Asia. We particularly value and appreciate the enthusiastic efforts and expertise of Peter Kracht, senior editor, as well as the expertise and commitment of the staff at West-

view Press in making these books possible. Profound thanks are also
due to Max Gartenberg, for his optimism and skills in resolving the
difficult; Scott Lariviere, for systems consulting; and Lee T'ing-jung,
who has again honored the work with his calligraphy.

Ralph D. Sawyer

A Note on Pronunciation and Format

As our views on pronunciation and translation are unchanged, we repeat our comments from the previous works: Unfortunately, neither of the two commonly employed orthographies makes the pronunciation of romanized Chinese characters easy. Each system has its stumbling blocks, and we remain unconvinced that the Pinyin *qi* is inherently more comprehensible to unpracticed readers than the Wade-Giles *ch'i*, although it is certainly no less comprehensible than *j* for *r* in Wade-Giles. However, as many of the important terms may already be familiar and previous translations of Sun-tzu's *Art of War* have mainly used Wade-Giles, we have opted to employ it throughout our works, including the *Military Methods*. Well-known cities, names, and books—such as Peking—are retained in their common form, and books and articles published with romanized names and titles also appear in their original form.

As a crude guide to pronunciation, we offer the following notes on the significant exceptions to normally expected sounds:

t, as in *Tao:* without apostrophe, pronounced like *d*
p, as in *ping:* without apostrophe, pronounced like *b*
ch, as in *chuang:* without apostrophe, pronounced like *j*
hs, as in *hsi:* pronounced *sh*
j, as in *jen:* pronounced like *r*

Thus, the name of the famous Chou dynasty is pronounced as if written "jou" and sounds just like the English name "Joe."

Within the translation the following conventions are adopted:

. . . . . Multiple dots indicate missing portions of more than a few characters.

[entry] Words within brackets are supplemented for missing portions based on collateral sentences or texts.

[. . .] Three dots with brackets indicate one or two characters are missing.

(entry) Words within curved brackets indicate material implied or understood, supplied by translator.

Chronology of Approximate Dynastic Periods

Dynastic Period	Years
Legendary Sage Emperors	2852–2255 B.C.
Hsia	2205–1766
Shang	1766–1045
Chou:	
Western Chou	1045–770
Eastern Chou	770–256
Spring and Autumn	722–481
Warring States	403–221
Ch'in	221–207
Former Han (Western Han)	206 B.C.– 8 A.D.
Later Han (Eastern Han)	23–220
Six Dynasties	222–589
Sui	589–618
T'ang	618–907
Five Dynasties	907–959
Sung	960–1126
Southern Sung	1127–1279
Yüan (Mongol)	1279–1368
Ming	1368–1644
Ch'ing (Manchu)	1644–1911

HISTORICAL INTRODUCTION

In the centuries before the founding of China's first imperial dynasty in 221 B.C., four great military figures are commonly recognized as having been instrumental in strengthening their states and wresting power over the realm: Sun-tzu, Sun Pin, Wu Ch'i, and Shang Yang. Sun-tzu's famous book, the *Art of War*, has been transmitted through the ages and circulated around the world. Wu Ch'i is identified with the *Wu-tzu*, one of the extant *Seven Military Classics;* while many of the fragments preserved in the *Book of Lord Shang* have been attributed to Shang Yang.[1] But for two millennia Sun Pin's *Military Methods* was known only by reputation, the last concrete reference having been to a copy stored in the Han imperial library in the first century B.C. However, suddenly in 1972, after eluding scholars for many centuries, remnants of the book were dramatically recovered from a still-sealed Han dynasty tomb.[2]

For millennia every variety of antiquity has been uncovered in China. Ancient writings, often carefully stored away in tombs, walls, and caves, have been among the intriguing objects stumbled upon. Many of these artifacts were first seized upon by grave robbers, while others were accidentally unearthed during the renovation of buildings, reworking of fields, and large-scale excavations for transportation and other projects. Focused, systematic archaeological expeditions were undertaken just this century and then only after the ancient taboos against desecrating graves—which never bothered highly motivated brigands over the centuries—gradually became powerless in the context of heightened scientific interest and national pride. The Buddhist caves at Tunhuang revealed priceless texts and innumerable statues from the T'ang dynasty earlier this century, while the well-publicized tomb of China's first emperor, constructed on an almost unimaginable scale and populated by a vast terra cotta army, startled archaeologists just two decades ago. Even more important for Chinese intellectual history, in 1972 a Han military official's tomb was opened to reveal a hoard of bamboo strips that preserved numerous ancient writings on military, legal, and other subjects, including Sun Pin's *Military Methods*. Sun Pin's work had apparently vanished during the Han dynasty, causing many scholars in later centuries to insist it had never existed. Unlike the *Three Strategies of Huang Shih-kung* that reputedly had been deliberately removed from circulation to prevent its tactics from being employed by disaffected parties, only to be recovered from Chang Liang's tomb in the fourth century, it appears that these writings had been placed in the tomb either to aid the departed in the afterworld or, more likely, to deliberately preserve them for posterity—a remarkable notion, considering that such tombs would ordinarily never again have been opened.

Unfortunately, although many of the strips are perfectly preserved, over the centuries major portions of individual ones have suffered varying degrees of damage, ranging from complete physical disintegration to the partial obliteration of the characters written on them with brush and ink so long ago. Moreover, because the rolled-up strips simply collapsed into structureless piles when the cords binding them together decayed, the original organization has been significantly obscured. However, after painstaking reconstruction much of the contents have tentatively emerged, revealing incisive conceptions and integrated tactics. While some of these chapters are hopelessly fragmented and opaque, incredibly several chapters are almost perfectly preserved. However, even in this imperfect condition the *Military Methods* remains a remarkable text from the middle Warring States period, one that presumably embodies the views of the great strategist Sun Pin, who was active in Ch'i at least from 356 until 341 B.C. and perhaps lived until near the end of that century.

Sun Pin was not the first ancient Chinese military thinker to bear the surname "Sun." The man commonly referred to as Sun-tzu, the great general and strategist of the late Spring and Autumn period, whose work the *Art of War* is perhaps the earliest and certainly most famous military writing in China, preceded him. In fact, virtually every traditional source identifies Sun Pin as Sun-tzu's lineal descendant, although the actual relationship may well have been somewhat less direct. Several family trees have been offered to account for their relationship, but they are all dubious, late reconstructions that naturally ignore the possibility that Sun-tzu himself may not have existed.³ The most common view asserts that Sun Pin was Sun-tzu's grandson, but since more than a century separates their active years, "great grandson" or even "great great grandson" would be more likely. Assuming that Sun Pin would have been at least twenty-five during the unfolding of the Kuei-ling campaign in 354 to 353 B.C. yields a projected birth date of approximately 380 B.C., which is consistent with the initial statement in his biography placing him more than a century after Sun-tzu.

Even though they have a joint biography in Ssu-ma Ch'ien's famous *Shih Chi,* virtually nothing is known about either Sun-tzu's or Sun Pin's life and background. Recently numerous popular editions and even several scholars have supplied surprisingly embellished biographies for Sun Pin in their modern Chinese editions, but essentially they all derive from inferences based upon the *Shih Chi* biography and a few brief references in other writings—particularly the *Chan-kuo Ts'e*—associated with the period.⁴ Fortunately, Sun Pin's disciples, presumably based upon the strategist's own explanations and notes, compiled a description of the epoch-making battle of Kuei-ling, which now appears as the first chapter of the *Military Methods,* that largely confirms the summary presented in the *Shih Chi* biography and supplements the tactical portrait to some extent. Sun Pin's life and accomplishments, as reprised by Ssu-ma Ch'ien's *Shih Chi* some two centuries after Sun Pin's death, are preserved as follows:

About a hundred years after Sun-tzu died there was Sun Pin. Sun Pin was born between Ah and Chüan (in Ch'i), and moreover was a direct descendant of Sun-tzu. Sun Pin had once studied military strategy together with P'ang Chüan. P'ang Chüan was already serving in the state of Wei, having obtained appointment as one of King Hui's generals. Realizing that his abilities did not compare with Sun Pin's, he secretly had an emissary summon Sun Pin. When Pin arrived P'ang Chüan feared that he was a greater Worthy than himself, and envied him. By manipulating the laws he managed to have him sentenced to the punishment of having his feet amputated and his face branded, wanting to thereby keep him hidden so that he would not be seen (by the King).

An emissary from the state of Ch'i arrived at Liang (in Wei). Sun Pin, who was banished (from the court) because of his punishment, secretly had an audience with him, and exercised his persuasion on him. Ch'i's emissary found him to be remarkable and clandestinely brought him back to Ch'i in his carriage. T'ien Chi, Ch'i's commanding general, regarded him well and treated him as an honored guest.

T'ien Chi frequently gambled heavily on linked horse races with the princes. Sun Pin observed that the fleetness of his horses did not differ much from theirs. The horses had three grades—upper, middle, and lower. Thereupon Sun Pin said to T'ien Chi, "My lord should bet again for I am able to make you win." T'ien Chi trusted him and wagered a thousand gold coins with the king and princes. When they approached the time for the contest Sun Pin then said: "Put your lowest team of horses against their best; your best team against their middle one; and your middle team against their lowest one." After the three teams had raced T'ien Chi had lost one race but won two, so that in the end he gained the king's thousand gold coins. T'ien Chi then introduced Sun Pin to King Wei. King Wei questioned him about military affairs, and appointed him as a strategist.

Thereafter the state of Wei attacked Chao. Chao was sorely pressed, and requested aid from Ch'i. King Wei of Ch'i wanted to appoint Sun Pin as commanding general, but Pin respectfully declined, saying, "It is not possible for a man who has been mutilated by punishment."

Thereupon (the king) appointed T'ien Chi as commanding general, and Sun Pin as strategist. Pin traveled in a screened carriage, making plans while seated.

T'ien Chi wanted to lead the army into Chao, but Sun Pin said, "Now one who would untie confused and tangled cords does not strike at them with clenched fists. One who would disengage two combatants does not strike them with a halberd [ax]. While they stand opposed to each other you should hit their vacuities. Then as their dispositions counter each other and their strategic power is blocked, the difficulty will be resolved by itself. Now Liang (Wei) and Chao are attacking each other, so Wei's light troops and elite soldiers must certainly all be deployed outside their state, with only the old and weak remaining within it. Wouldn't it be better to lead the troops on a forced march to Ta-liang, occupying their roads and striking their newly vacuous points? They will certainly release Chao in order to rescue themselves. Thus with one move we will extricate Chao from its encirclement and reap the benefits of Wei's exhaustion." T'ien Chi followed his plan, and Wei did indeed abandon Han-tan, engaging Ch'i in battle at Kuei-ling. Ch'i virtually destroyed Wei's army.

Thirteen years later Wei and Chao attacked the state of Han. Han reported the extremity of its situation to Ch'i. Ch'i ordered T'ien Chi to take command and go forth, proceeding straight to Ta-liang. When general P'ang Chüan of Wei heard about it, he abandoned (his attack on) Han and embarked on his return. Ch'i's army had already passed by and was proceeding to the west. Sun Pin said to T'ien Chi, "The soldiers of the Three Chin (Han, Wei, and Chao) are coarse, fearless and courageous, and regard Ch'i lightly. Ch'i has been termed cowardly. One who excels in warfare relies upon his strategic power and realizes advantages from leading the enemy (where he wants). As the *Art of War* notes, 'One who races a hundred *li* in pursuit of profit will suffer the destruction of his foremost general; one who races fifty *li* in pursuit of profit will arrive with only half his army.'[5] Have our army of Ch'i, upon entering Wei's borders, light one hundred thousand cooking fires. Tomorrow make fifty thousand, and again the day after tomorrow start thirty thousand cooking fires."

P'ang Chüan, after advancing three days, greatly elated said, "Now I truly know that Ch'i's army is terrified. They have been within our borders (only) three days, but the number of officers and soldiers that have deserted exceeds half." Thereupon he abandoned his infantry, and covered double the normal day's distance with only light, elite units in pursuit of them. Sun Pin, estimating his speed, determined that he would arrive at Ma-ling at dusk.

The road through Ma-ling was narrow, and to the sides there were numerous gullies and ravines where troops could be set in ambush. There he chopped away the bark on a large tree until it showed white and wrote on it, "P'ang Chüan will die beneath this tree."

Then he ordered ten thousand skilled crossbowmen to wait in ambush on both sides, instructing them, "At dusk, when you see a fire, arise and shoot together." In the evening P'ang Chüan indeed arrived beneath the debarked tree. He saw the white trunk with the writing, struck a flint, and lit a torch. He had not finished reading the message when ten thousand crossbowmen fired en masse. Wei's army fell into chaos and disorder. P'ang Chüan knew his wisdom was exhausted and his army defeated, so he cut his own throat, saying "I have established this clod's fame!" Ch'i then took advantage of the victory to completely destroy their army, and returned home with Imperial Prince Shen of Wei as prisoner. Because of this Sun Pin's name became known throughout the realm, and generations have transmitted his *Military Methods.*[6]

This dramatic biography no doubt found strong echoes in Ssu-ma Ch'ien, for both had endured the humiliation of corporeal punishment, yet achieved historical greatness.

The tactics for the two famous battles of Kuei-ling and Ma-ling will be discussed later; here it is only necessary to comment upon certain aspects of the biography in the light of historical background. The value placed upon brilliant individuals who could provide ideas and methods critical to the state's survival is clearly apparent. Although this openness to hearing consultations from unknown, often nondescript wanderers may seem surprising, this was an age of economic, intellectual, and physical turmoil. To flourish, rulers were not only forced to welcome and heed mendicant persuaders but also compelled to depend upon and cultivate military specialists and professional officers because warfare had become too complex for either kings or otherwise capable civil bureaucrats to exercise command effectively.

That the ambitious, ruthless P'ang Chüan did not simply have Sun Pin assassinated remains puzzling; obviously P'ang Chüan enjoyed the political position and attendant power to clandestinely arrange it had he chosen to do so. No doubt many unfortunate individuals were painfully dispatched to the myriad hells depicted in Chinese thought after running afoul of the powerful; they simply disappeared from the historical stage before their acts could command interest and thus remain forever unknown. Sun Pin's own story being exceptional and his achievements outstanding, his personal travails have accordingly been preserved. Several modern writers suggest that Sun Pin may have actually gained an audience with King Hui in which he clearly eclipsed P'ang Chüan with his mastery of military affairs, but that P'ang Chüan subsequently undermined Sun Pin's influence and doomed his chances in Wei by troubling the king with doubts about Sun Pin's origin as a native of Ch'i, an avowed enemy state.[7] Furthermore, P'ang Chüan might have reminded the king that men of outstanding talent such as Sun Pin, if not employed, should be killed to prevent them from venturing to other states and subsequently wreaking havoc upon them. Perhaps, as suggested by certain scholars,[8]

P'ang Chüan was persuaded that the state of Wei could retain Sun Pin's talent and exploit his abilities on P'ang Chüan's and the king's behalf if they physically mutilated him, guaranteeing that he would be unable to encounter anyone of influence again. (People who had undergone mutilating punishments were shunned and despised, whether innocent or not, and were frequently physically impaired, thereby severely limiting their mobility.) If so, P'ang Chüan and the king underestimated his capabilities and apparently ignored the fact that the Sun family was closely related to Ch'i's ruling T'ien clan (which might explain his ability to gain an audience with Ch'i's emissary).[9] They certainly seem to have been naive in their expectation that someone subjected to such torture would willingly plan affairs for them. (Why wouldn't he simply concoct tactics that, unforeseen by them, would lead to their defeat in battle or suggest policies for the military that would debilitate it?) Clearly, P'ang Chüan had little understanding of human nature, perhaps being blinded by jealousy, fear, and insecurity, although he certainly would have orchestrated events to somehow make himself appear to be Sun Pin's savior rather than executioner. Furthermore, simply having his forehead branded, the least of the mutilating punishments, would have sufficed for their avowed purpose.

The assertion that Sun Pin studied military strategy with the semilegendary Kuei Ku-tzu also deserves note. Although a number of books on military strategy and prognostication have traditionally been attributed to the "Master of the Ghostly Valley," they are now viewed as late fabrications.[10] Historical information about Kuei Ku-tzu is nonexistent; whether he was merely a recluse or had actual experience in military affairs remains unknown. However, his name is also associated with the teaching, preservation, and transmission of other esoteric writings, and Sun Pin may well have studied with him or some reclusive expert. Popular tradition suggests that P'ang Chüan was an inferior, although presumably acceptable, student who abandoned his master early on. In contrast, Sun Pin is held to have been enthusiastic, brilliant, and determined, and he alone was therefore given Sun-tzu's *Art of War* to absorb.[11]

Insofar as the *Military Methods* is unquestionably based upon and expands the *Art of War*, it seems almost inconceivable that Sun Pin would not have received his illustrious ancestor's book through family transmission. Moreover, it would be surprising if the Sun family, known for its involvement in military affairs in Ch'i and other states over the centuries, would not have preserved a "family school" of tactics based upon Sun-tzu's thought. Certainly, given the clan background and the violent conflicts raging in Sun Pin's native area during his youth,[12] it would have been natural for him to embrace military studies. Furthermore, immersed as he was in an era of economic turmoil that witnessed the rise of the propertied class, increased mobility, and the decline of hereditary privilege, while he himself apparently lacked land and royal connections, Sun Pin might easily have found the military path inescapably appealing. Thus, while the *Shih Chi* biography states he studied with Kuei Ku-tzu, it is highly possible that he was already well versed in the *Art of War.*

The story of the competition—presumably chariot racing rather than simple horse racing[13]—encapsulates Sun Pin's cleverness, although it hardly constitutes high tactics and certainly would not recommend him to a king who had just lost significantly through a subterfuge that probably contravened competitive conventions. However, a person who can break the rules is capable of envisioning different, unexpected methods; this should be remembered in any evaluation of the possibility that Sun Pin deliberately sacrificed a portion of his army to deceive P'ang Chüan, as will be discussed later. The biography also contrasts Sun Pin's deliberate, analytical methods with T'ien Chi's direct, impulsive approach, and in the two battles Sun-tzu's emphasis upon calculation, planning, manipulation of the enemy, and deception are all embodied. As for the biography's reliability, even as essentially corroborated by the tomb text, it should be recalled that Ssu-ma Ch'ien penned it approximately two centuries after the events upon unknown materials.

Unfortunately, Sun Pin's biography ends with the battle of Maling, and he immediately vanishes from the pages of history. Since in writing his accounts for the *Shih Chi,* Ssu-ma Ch'ien, the Grand Historian, normally proceeded only upon a solid basis, the absence of concluding paragraphs suggests that no material was available. However, from brief references elsewhere in the *Shih Chi* and some few passages in the less reliable *Chan-kuo Ts'e,* it appears that the victorious T'ien Chi never returned to Ch'i, being successfully calumnized by Tsou Chi, the envious prime minister with whom he had been in constant conflict. Presumably Sun Pin, as T'ien Chi's protégé and confidant, should have accompanied him during his exile in the state of Ch'u but apparently did not.[14] The *Chan-kuo Ts'e* reprises the incident as follows:

> Tsou Chi, Lord of Ch'eng and prime minister of Ch'i, and T'ien Chi, General of the Army, were displeased with each other. Kung-sun Han addressed Tsou Chi: "Sire, why not plan an attack on Wei for the king? If we are victorious, then it will have been due to my lord's plans and you can receive the credit. If we engage in combat and are not victorious, if T'ien Chi has not advanced into battle and has not perished, you can wrangle an accusation (of cowardice) against him and have him executed." Tsou Chi thought it to be so, so he persuaded the king to have T'ien Chi attack Wei. In three engagements T'ien Chi emerged victorious three times,[15] so Tsou Chi informed Kung-sun Han. Thereupon Kung-sun Han had a man take ten gold pieces to a diviner in the marketplace and inquire: "I am T'ien Chi's retainer. T'ien Chi engaged in battle three times and was victorious three times. His fame overawes All under Heaven. If he wants to undertake great affairs, will it be equally auspicious or not?" The diviner went out and had men detain the person who had requested the divination (for T'ien Chi), and then attested to his words before the king. T'ien Chi subsequently fled.[16]

It seems incredible that Tsou Chi's simple ruse might have persuaded anyone that a brilliant and successful general such as T'ien Chi would suddenly become so profoundly stupid as to blatantly initiate such inquiries. However, several of China's rulers, and especially dynastic founders (including those of the Han and T'ang), availed themselves of—or manufactured—prophecies attesting to their "extraordinary" qualities and predestination. Had T'ien Chi actually ordered such public actions, they could well have been understood as a sort of trial balloon, a foundation for deliberately stimulating rumors and arousing public support or at least gauging the extent of potential adherents. Conversely, notwithstanding such machinations, it remains puzzling that Tsou Chi could have so easily threatened T'ien Chi except by persuading a very insecure ruler that the general's loyal, well-organized, battle-hardened army posed an imminent danger to his throne.[17]

Another passage in the *Chan-kuo Ts'e* purportedly preserves Sun Pin's perspicacious advice to T'ien Chi at this time:

> T'ien Chi, acting in his capacity as Ch'i's commanding general, had bound up Imperial Prince Shen of Liang (Wei) and captured P'ang Chüan. Sun Pin queried T'ien Chi: "Is the commanding general capable of undertaking great affairs?" T'ien Chi: "In what way?" Sun Pin said: "The commanding general should not release his troops before entering Ch'i. Have those that have become exhausted from their former efforts (against Wei) weakly mount a defense at Chu. Chu lies hard by a transport road where the axles of the wagons are constantly bumping each other as they pass by. If you mount a weak defense at Chu, one can oppose ten; ten can oppose a hundred, and a hundred can withstand a thousand. Thereafter, orient yourself with T'ai Shan at your back, the Chi river to the left, T'ien-t'ang to the right, your heavy forces at Kao-yüan, and have your light chariots and elite cavalry explosively penetrate Yung Gate (in the capital). In this manner the rule of Ch'i can be rectified and the Lord of Ch'eng (Tsou Chi) can be forced to flee. If you do not so proceed, then the commanding general will not manage to enter Ch'i." T'ien Chi did not listen and indeed did not enter Ch'i.[18]

Obviously, success in the troubled world of the Warring States period often entailed the seeds of self-destruction.

Thereafter, T'ien Chi is known to have been enfeoffed in Ch'u, perhaps with Tsou Chi's dubious assistance,[19] and apparently remained there until recalled by King Wei's successor, King Hsüan, some years later.[20] Some scholars who accept the traditional *Shih Chi* dating for King Hsüan's ascension of 342 B.C. believe that Sun Pin must have returned to Ch'i with T'ien Chi, thereby accounting for their participation in the Ma-ling campaign. However, after Ma-ling neither is ever mentioned as assuming any active role in Ch'i's military or government, in contrast to their earlier reputations, which had spread even to enemy states.[21] Because the *Military Methods* contains references to battles subsequent to Ma-ling but nothing beyond the turn of the century, there is a slight possibility (assuming such references are not simply additions by disciples) that Sun Pin may have survived until roughly 305 or 300 B.C., when he would have been nearly eighty. However, evidence is similarly lacking as to whether he played an ongoing role as a tactical adviser or simply remained in the obscurity of retirement.[22]

Historical Background

Military tactics are inescapably the product of specific environments, being founded upon historical antecedents and experience, formulated in terms of contemporary concepts, focused upon exploiting perceived capabilities, and directed toward achieving political objectives defined by national interests. Evolution produces challenges that in turn stimulate further developments; this may be clearly seen in Sun Pin's *Military Methods* in comparison with Sun-tzu's earlier *Art of War.* Accordingly, without re-creating the entire history of China, even though many events have lengthy historical antecedents, we must devote a few pages to depicting the historical setting, characterizing the weapons and armies of the time, and briefly sketching the dynamics of the volatile political situation in Sun Pin's life.[23]

The era in question, the so-called Warring States period (403–221 B.C.), is aptly named because it witnessed almost interminable strife among the surviving feudal states as they strived to strengthen themselves through internal reforms and the forcible annexation of their neighbors. Unlike the earlier Spring and Autumn period, in which a semblance of civility and vestiges of restraint had persisted under the nominal aegis of the impotent Chou king, by the time of the Warring States power in all its manifestations dictated policy and relations. Late in the Spring and Autumn period, four mighty entities had dominated the realm: Chin, Ch'u, Ch'i, and Yüeh. However, by the early Warring States period, Chin's powerful clans had rent the realm asunder, producing the three smaller, though still formidable states of Han, Wei, and Chao—also termed the *Three Chin* in recognition of their origins—plus a minor remnant briefly maintained by the old Chin royal house. Yüeh, on the eastern coast, began the period as a respected power but, doomed by internal discord, was gradually subjugated by the southern state of Ch'u, which completed its conquest by annexing Yüeh's shrunken eastern fragments in 306 B.C. To the north Yen emerged as a capable but relatively weak state, while in the west the newly risen barbarian state of Ch'in, which eventually vanquished all the major powers and exterminated the remaining minor ones to formally proclaim itself as the first imperial Chinese dynasty in 221 B.C., aggressively nurtured its power base in Chou's original territory beginning early in the fourth century.

Four of the era's so-called seven powerful states—Han, Wei, Chao, and Ch'i—were crucial participants in the events unfolding around Sun Pin, while Ch'in, ever-strengthening in the west, perpetually cast an ominous shadow over the unfolding events. However, despite (or because of) the incessant warfare, it was an era of remarkable intellectual ferment as the often beleaguered rulers vigorously struggled to unify their states, impose centralized control, and establish the material prosperity necessary to sustain prolonged military campaigns. Widely diverse viewpoints and policies were espoused, although they all essentially focused upon employing capable individuals, creating effective administration, improving agriculture, and expanding the military through better organization, training, and selection of men. Unlike in the Spring and Autumn period, when feudal rulers were still sufficiently powerful to feel unthreatened by other states, creative thinkers (such as Confucius) could no longer be ignored or merely tolerated. Now in the instability of the times a single individual, such as Shang Yang in Ch'in, sometimes played the key role in revitalizing an otherwise moribund state, often against virtually overwhelming opposition from entrenched interests. In other instances a succession of "Worthies" (moral exemplars) or a propitious conjunction of talents persuaded a ruler to forge ahead by implementing revolutionary measures. The most important figures are mentioned later in conjunction with brief characterizations of the individual states.

The interrelationship of the pivotal states in Sun Pin's personal drama, all four centered about the central plains area, may be seen from the accompanying map, which depicts approximated borders for roughly 350 B.C., a time before Ch'in swallowed Pa and Shu or Yen expanded its control over the Eastern Hu. The salient features, political objectives, and critical figures of the four states may be summarized as follows.

The State of Wei

Although history has accurately portrayed Wei as the aggressor in the twin campaigns of Kuei-ling and Ma-ling, an exclusive focus upon these ill-conceived campaigns necessarily winds up ignoring the ongoing nature of the combatants' conflicts and Wei's overall strategic objectives. At its inception as an independent state, Wei remained the

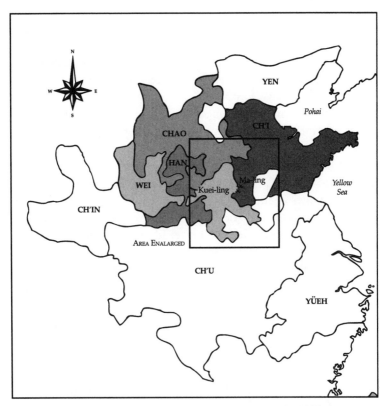

MAP 1   The seven states during the Warring States period, c. 350 B.C.

most powerful of the Three Chin because it had retained control over a significant portion of the central plains area. Although backed by mountains to the northwest, after Chin was partitioned Wei's territory consisted of two large domains of essentially flat terrain con-nected by a narrow land corridor. Consequently, Wei's strategic situ-ation was inherently flawed, its ability to mobilize and dispatch forces between the two areas easily denied by aggressive forces that could readily sever its constricted transport route. Furthermore, its initial capital of An-i on the open plains in the western portion and their subsequent capital of Ta-liang in the eastern portion were both virtu-ally indefensible without the construction of enormous defensive works because they were easily accessible from almost every direction and lacked natural obstacles to impede potential invaders. Even though Ta-liang (approximately modern Kaifeng) was situated near the Yellow River, without swift rivers or impassable mountains that could retard advancing forces or passes and ravines that might be exploited by defensive forces to defeat heavy campaign armies, Wei could not prevent massive incursions except through the development of superior numbers and the mounting of concentrated firepower.

Wei had a strong military heritage; in the Spring and Autumn period its ruler had proclaimed himself hegemon, the officially recognized sustainer of the imperial Chou ruler, in which guise he commanded the obedience (or at least grudging respect) of the other states. Furthermore, after the powerful families essentially divided Chin into three enclaves in 434 B.C., Wei was ruled by the remarkable Marquis Wen, who embarked upon a concerted campaign to strengthen the state through radical administrative reforms. The famous general and military administrator Wu Ch'i, reputed author of the *Wu-tzu,* not only defeated Ch'in in numerous battles and extended Wei's territory to the west by several hundred *li* but also proved an able reformer during Marquis Wen's reign. Wu Ch'i emphasized the welfare of the people, reformed the laws, established effective forms of rewards and punishments, and advocated a strict government still marked by benevolence and Virtue. Furthermore, as a tactician and strategist he formulated many key methods and doctrines that influenced subsequent military writings and molded generations of commanders.[24]

Marquis Wen's avowed policy of welcoming and employing capable individuals empowered the famous Li K'uei (also known as Li K'o) to dramatically transform the state's legal apparatus and its agricultural policies. He reputedly systematized the laws of Wei by arranging them into six categories and establishing criminal penalties for offenses against the social order such as robbery and brigandage. (Shang Yang evidently later employed them as a basis for his extremely successful efforts to similarly reform the state of Ch'in.)[25] Apparently because Wei's population was large in comparison to its landholdings, Li K'uei emphasized exploiting all available ground for agriculture, including utilizing yards for personal gardens and planting mulberry trees along property boundaries (to provide leaves for the silkworms and thereby stimulate sericulture). In addition, he stressed the need for plant diversity to avoid overreliance upon any single crop and the consequent risk of starvation should it be adversely affected by weather conditions or decimated by pests. Furthermore, he established regulations and a state mechanism to normalize commodity prices within certain limits, thereby stabilizing them and ensuring that the people would neither overly profit nor be unable to afford the grains of life. Coupled with Hsi-men Pao's efforts to expand irrigation and open lands for agriculture, Wei enjoyed rapidly increasing prosperity and the growth of significant cities, such as An-i and Ta-liang.

Consequently, under Marquis Wen the state expanded its territory, stabilized its borders, attained affluence, and fielded a potent military force of well-trained troops.[26] Unfortunately for Wei, in 396 B.C. he was succeeded by his son Marquis Wu, who, although generally embracing his father's visionary program, perhaps lacked his perspicacity and administrative talent. His misjudgments in court intrigues eventually resulted in the loss of the multitalented Wu Ch'i (among others), who fled to Ch'u and was subsequently instrumental in initiating that king's efforts to reform the laws and curb the intransigence of the feudal nobility.[27]

In 370 B.C. the man eventually known as King Hui (after he belligerently accorded himself the title between the battles of Kuei-ling and Ma-ling) ascended the throne and immediately attempted to capitalize upon Wei's military heritage by mounting expansive campaigns in nearly every direction. However, despite Wei being the most powerful state in China at the inception of his reign, one marked by resources, military efficiency, and bureaucratic competence, it could never be equal to the strategic demands of permanently confronting enemies in four directions or fighting successive battles on multiple fronts. Several potential courses were open to King Hui, including the relatively static, defensively oriented one of focusing upon enriching the state and strengthening the army, establishing improved border defenses and erecting walls, and thereby overawing the other states into abject submission. Concurrently, he could have maintained friendly relations with the other members of the Three Chin—

Han and Chao—which, although evolving into somewhat distinct entities, still shared much in terms of cultural background and could have been compelled by their comparative weakness to stand with, rather than against, Wei. Thus effectively allied, King Hui could have concentrated upon blunting Ch'in's growing threat and probably deterred any incursions by Ch'u or Ch'i. (Ch'i, a state located on China's eastern coast somewhat distant from Wei's capital of An-i, presented less immediate danger than Ch'u, which, although vast and blessed with bountiful resources, actually suffered from disunity in its court and weakness in its administrative apparatus.) Alternatively, a policy that focused upon subjugating and annexing just one of the other two Chin states would have been logistically reasonable, but King Hui was foolishly willing to successively engage virtually everyone—except when he needed a recovery period—and thus made constant enemies of them all. Consequently, after the debilitating losses at Kuei-ling and then at Ma-ling, Wei found its capital of An-i in the west indefensible against Ch'in's aggressive onslaughts. After suffering a severe defeat at Shang Yang's hands, Wei was therefore forced to designate Ta-liang as its capital in 340 B.C. and also cede 700 *li* of previously hard-won territory to Ch'in in the west as compensation and a peace offering to gain a temporary respite.[28]

The State of Ch'i

Physically situated in the east, Ch'i's territory extended from the plains area to the Yellow and Po seas on the Pacific coast and included inland borders contiguous with most of the important states other than Ch'in in the far west. Rich in natural resources, for centuries Ch'i had benefited from enlightened policies and economic development as well as the constant exploitation of its many mountains and the nearby seas. Initially ruled by the Chiang clan, beginning with the establishment of the Chou dynasty, it presumably enjoyed a long heritage of military expertise because the founding king was T'ai Kung, traditionally viewed as China's first famous commander and the key strategist for Kings Wen and Wu in the Chou's ascension to power. Among the *Seven Military Classics*, the *Six Secret Teachings* is attributed to his name and may possibly contain some remnants of his thought, but the extant version is clearly a product of the middle to late Warring States period. His remarkable personal history and policies, which emphasized the welfare of the people as well as hardnosed military tactics, are thoroughly discussed in our other works and need not be repeated here.[29]

Another major figure in Ch'i's history was Duke Huan, the first hegemon (in 680 B.C.) and an administrative exemplar who not only nurtured and governed the most prosperous state in the realm but also extensively employed such advisers as Kuan Chung, the famous Kuan-tzu associated with the eclectic book of the same name. While the extant *Kuan-tzu* is obviously a compilation of chapters from several hands and no doubt evolved over time, the policies advocated in the oldest strata may have been formulated by Kuan Chung and implemented by Duke Huan. Generally their measures sought to foster the welfare of the people by concretely encouraging agriculture and sericulture, not interfering with seasonal activities, imposing moral government by talented individuals, advancing good values and discouraging extravagance, effecting a system of rewards and punishments to deter miscreants, and generally nurturing an environment conducive to the work ethic.[30] Moreover, they developed an organizational hierarchy that grouped hamlets and villages by units of five and ten, thereby integrating them administratively and militarily. This organization and superimposed command system produced familiarity and unity, apparently accounting for the esprit de corps and superlative performance of Ch'i's army in battle. Finally, they reputedly created salt and iron monopolies and developed an effective tax system based upon the integrated administration of the populace, ensuring the ability to accumulate the resources necessary for military activities, including weapons. With his foundation secure, Duke Huan concluded alliances with nearby states, which were cowed into submission before Ch'i's Virtue and military might, and conducted numerous military expeditions against both recalcitrant states (presumably for the sorts of offenses listed in the *Ssu-ma Fa*)[31] and various nomadic peoples such as the Jung, who were just then establishing the state of Ch'in. Although little respected by later Confucians, Duke Huan became famous for convening the feudal lords nine times, always in the clearly formulated and avowed role of a hegemon respectfully sustaining the Chou royal house. During his rule and thereafter, Ch'i enjoyed great power and prestige, growth in its economy and cities, and essential stability. Naturally in the ensuing centuries there were fluctuations in its fortunes and even aggressive incursions by other states, but generally the major cities and the interior were untouched, while Lin-tzu, the capital, remained an unthreatened bastion for more than six hundred years.

By early in the Warring States period, the political situation in Ch'i and the other major states had changed dramatically, with much of the power shifting from the original ruling families to newly empowered clans. In Ch'i's case, after several generations of serving the ruler in the capacity of prime minister and continually expanding influence and control, the T'ien clan finally became formally recognized among the feudal lords. Shortly thereafter it deposed the last marquis in order to establish itself as the new ruling family in 386 B.C. T'ien Ho, the first family member to rule as sovereign, embarked upon military campaigns in several directions during the next few years, attacking Wei, Lu, Sung, and Yen. Eventually his grandson (who subsequently accorded himself the title of King Wei) succeeded him, although the exact year remains a matter of disagreement.[32] Apparently a vigorous, dynamic ruler, he employed Tsou Chi (who is sometimes portrayed in the *Shih Chi* and *Chan-kuo Ts'e* as venal, jealous, and opaque) and welcomed talented individuals from throughout the realm; accepted criticism and remonstrance; implemented policies designed to ensure good order, including severe rewards and punishments; revised the taxation system; exerted control over the state monopolies; and generally fostered agriculture and nurtured the state's material welfare.[33] Under his rule Ch'i was also militarily successful, its own might and topography sufficient to deter attacks by other states, while it retook land previously lost to Wei (which was then forced to cede additional territory) and Chao, including a portion of a long wall Ch'i had previously constructed for defensive purposes.[34]

Something of King Wei's emphasis upon individual talent may be seen from the following anecdote preserved in the *Shih Chi:*

> In his twenty-fourth year King Wei of Ch'i went hunting in the suburbs with King Hui of Wei. The king of Wei inquired: "Do you also have some treasures?" King Wei said: "I do not." The king of Wei said: "Even a state as small as mine still has ten pearls one inch in diameter

that cast radiance over twelve carriages in front and behind them. So how is it that a state of ten thousand chariots lacks treasures?"

King Wei said: "What I take to be treasure is different from what your majesty does. Among my ministers there is T'an-tzu. When I deputed him to defend the southern cities Ch'u's troops did not dare make plundering incursions in the east and the twelve lords from Ssu-shang region all came to court. Among my ministers is P'ang-tzu. When I deputed him to defend Kao-t'ang, the people of Chao did not dare go east to fish in the Yellow River. Among my officials there is Ch'ien-fu. When I deputed him to defend Hsü-chou the people of Yen offered sacrifice at the northern gate while the people of Chao offered sacrifice at the western gate.[35] Those that followed him and moved there were more than seven thousand families. Among my ministers there is Chung-shou. After I had him make preparations against robbers and brigands the people would not pick up things left on the road. These ministers illuminate a thousand *li*, not merely twelve carriages!"

The King of Wei flushed and unhappily departed.[36]

Additional stories and anecdotes are preserved in the ancient historical writings; however, as their reliability is questionable, they are of limited value.

The States of Han and Chao

Although both states have chapters in the *Shih Chi* and *Chan-kuo Ts'e*, the historical records for Han and Chao are less detailed in comparison with the mighty powers of the highly active states of Wei, Ch'i, and Ch'in. However, the information preserved, while sometimes contradictory, is sufficient to reconstruct the important events and numerous battles of the period. Moreover, ongoing archaeological discoveries, such as of the existence of the capital cities of Han-tan and Cheng, each with impressive outer wall dimensions of 10,000 to 17,000 feet per side, attest to the basic validity of the accounts.[37] Since Han and Chao's individual histories continued to be closely intertwined even after Chin's division, they may be discussed together.

Geographically, Han and Chao initially occupied the southern and northern portions of the ancient state of Chin, respectively. By 355 B.C., just before the battle of Kuei-ling and after both states had annexed considerable additional territory from the lesser states of Wey, Cheng, and Sung, Han's borders extended upward between Wei's two territories to form a mushroomlike incursion into Chao's otherwise solid hold over the northern area. Han also shared an eastern frontier with Wei, a northwestern border with Wei's western section, a small but critical frontier with Ch'in to the west, and a very long border with Ch'u to the south. Sometimes referred to as the "throat of China," Han (like Wei) was a linch pin in the control of the pivotal central plains area. Chao benefited from its northwestern portion falling in the high plains, while it shared a southern border with Han that also encompassed a rising plains area. To the north Chao fronted barbarian territory (which it eventually annexed under King Wuling), to the northeast it bordered the state of Yen, and to the east it had to contend with Ch'i and tolerate the newly reestablished state of Chungshan.

Neither state's territory had been particularly well developed economically by the division of Chin, but both were fertile enough to offer excellent prospects for intensively promoting agriculture and thereby increasing their wealth. Unfortunately for Han, it was (in Sun-tzu's terminology) a "focal" territory, one confronted by potential or actual enemies on all four sides, whereas Chao benefited to a considerable degree from mountainous terrain and constricted roads that could be exploited by limited numbers of troops to vanquish even large invading forces. Chao was generally regarded by Su Ch'in and the other strategists as a state of great potential, one easily capable of controlling the central plains and thus the entire realm.[38] Although Chao never realized this much-proclaimed potential, Su Ch'in's evaluation was subsequently verified by historical events: Once Han and Chao perished, Ch'in was rapidly able to subjugate Yen, Wei, Ch'i and Ch'u.

The rulers of Han and Chao failed to implement the sort of government reforms sweeping the other states in the early Warring States period, although both states made limited efforts to recruit talented individuals and effect a more thoroughgoing rule of law under central government authority, thereby diminishing the influence of both the old noble clans and the newly risen great families. At Chao's initial establishment in 403 B.C. at the start of the Warring States period, Marquis Lieh employed Kung-chung Lien in his service. He in turn introduced three capable individuals to reform the laws, advance the worthy, nurture the people's welfare, and impose accountability through such systems as rewards and punishments. However, it appears that their efforts met with limited success because Chao lacked the sort of dynamic ruler needed to effect them until King Wu-ling ascended the throne late in the fourth century.[39] Accordingly, in the intervening years Chao waxed relatively powerful only after the battle of Kuei-ling, in corresponding measure with Wei's decline. Although Chao clashed militarily with many other states from its inception, including Ch'i—which would eventually become its savior at the battle of Kuei-ling—Chao's main success was defeating (in alliance with Han) the minor state of Wey and annexing a significant portion of its territory, including the city of Han-tan, which Chao then designated as its capital in 386 B.C. (Somewhat reduced in size, Wey continued to survive as a client state of Wei, which had itself annexed some territory from Wey to expand eastward. This relationship was to prove a factor in instigating Wei's attack on Han-tan roughly three decades thereafter.) Somewhat later, in 368 B.C., again in conjunction with Han, Chao divided the remnant Chou state into eastern and western portions.

The state of Han largely had its origin in Han Ch'üeh's military activities as one of Chin's six ministers prior to its breakup, the Han and Wei clans being the only ones to survive the internecine battles that followed. Early in the fourth century B.C., Han, Chao, and Wei had jointly divided the small but warlike state of Cheng into three parts, and in about 375 B.C. Han exterminated Cheng and then moved its own capital to the city of Cheng in the east—a curious strategy because it thereby shifted Han's economic, military, and administrative focus from the relatively secure mountains to the less defensible plains. About the time Wei besieged Chao's capital city of Han-tan, Shen Pu-hai gained employment in Han and for the next fifteen years attempted to resolve the numerous conflicts between the old laws and practices and the newly formulated measures of central authority. Although only fragments of Shen Pu-hai's writings remain and much controversy surrounds his teachings, they essentially focus upon developing methods for the ruler to exert complete control over the officials necessary to govern the realm and administer its policies. In particular, the doctrine of accountability known as "name and function," whereby an official should perform only those duties precisely prescribed by his office and be evaluated solely upon that performance, is closely identified with Shen.[40] His efforts were apparently successful, for Han's reputation grew and the state was largely untroubled by external assaults until Wei besieged the capital roughly a decade later. Moreover, Han's industries seem to have developed well; archaeological excavations of recent decades have uncovered extensive cities, immense bronze and iron workshops, and pottery production centers.[41] In particular, Han became known for manufacturing the superior crossbows that provided the means to multiply the combat power of its comparatively smaller army.[42]

Once the Three Chin states of Han, Wei, and Chao had bloodily emerged from Chin's dismemberment, their history naturally continued to be entangled as they formed temporary alliances as well as nourished old factional enmities and desires for aggrandizement. The pivotal event that cemented King Hui of Wei's hatred toward Han and Chao developed with his father's death in 371 B.C., for he and his brother each received factional support as they contended to succeed their father. Han and Chao seized the opportunity created by this internal strife to mount a united attack against King Hui and the Wei forces, defeating them at Chuo-che and then surrounding An-i, Wei's original capital. Chao's ruler wanted to continue on with the campaign, seizing and presumably annexing Wei's territory (to be divided with Han) and leaving King Hui's opponent nominally on the throne of a much reduced state, whereas Han sought to divide the state and permanently weaken it by installing both princes on separate thrones. In this manner Han probably sought to maintain a buffer state between itself and Ch'in to the west, eliminate Wei as an enemy, gain an ally from at least one of the factions, and prevent Chao from growing stronger. Remaining at loggerheads, Han and Chao both eventually withdrew (only to engage Wei again three more times in the immediately forthcoming years) without resolving the issue. King Hui eventually consolidated his rule and thereafter continued to harbor a hatred for them that was only partially assuaged by his separate victories over each of them the next year. However, Chao still aided Wei when Ch'in attacked its city of Shao-liang in 363 and even assisted Wei in attacking Ch'i in 360, thus illustrating how fluid and temporary alliances were in the Warring States period.

Military Organization, Weapons, and Comparative Strength of the Major Warring States

Military organization strengthened considerably in the Warring States period as both campaign and standing armies expanded.[43] However, it was not just augmented numbers, but rather a dramatic increase in the basic qualifications and expertise of the soldiers that had the greatest impact. Infantry development peaked and the cavalry was introduced, although the latter was not a factor—if it even existed—in the battles of Kuei-ling and Ma-ling. Wu Ch'i is generally credited with being the first to assert the need for men to be physically qualified for military service, and the extant *Wu-tzu* further emphasizes the necessity for creating elite units.[44] Beginning with Wu Ch'i's reforms and thereafter, Wei's soldiers enjoyed an apparently well-deserved reputation for strength and agility, and the other states rapidly imitated Wei's efforts.

Another major difference from the Spring and Autumn period was the imposition of year-round service and fixed periods (such as three years) of duty, rather than just assembling semiskilled forces to supplement the warrior nobility during appropriate seasonal mobilizations and training. The early military writings frequently contain injunctions against violating the seasons, and many historic campaigns were in fact initiated in the fall, the season of death, but when prolonged engagements began to require from six months to a year, seasonal concerns obviously became moot. Rigorous training and discipline invariably became the norm in the Warring States, with the soldiers being taught not only weapons skills but also a variety of unit formations such as are described in the *Military Methods*. With troops acquiring articulation, segmentation, and maneuver capabilities, generals could configure their deployments to the terrain and the enemy's disposition (as will be discussed in the section on concepts and tactics).

From the late Spring and Autumn period, military organization was almost exclusively founded upon the squad of five, each squad having its designated leader and its members being inexorably bound together. Many armies also combined two squads into a functional, decimally based double squad, while ten squads made up a platoon (where they existed), and either 100 or 125 men (twenty-five squads, following the rule of five) constituted a company. Battalions usually numbered 500 men; regiments, 2,500; and armies, 12,500. The degrees of hierarchy actually imposed varied by state, with the units of 500, 2,500, and 12,500 very common early on, but as the Warring States period progressed and campaign armies began to approach 100,000 or more, the size of the individual "army" ballooned to 20,000 or more. Most states also thoroughly implemented the sort of mutual responsibility system—both horizontally and vertically—seen in the *Wei Liao-tzu,* a text probably composed about this time or within fifty years thereafter. Rewards and punishments were severe, and defeat, the loss of a comrade from the basic unit (as well as the death of any officer), and cowardice were normally punished by death. Naturally, the soldiers' responsiveness to orders largely depended upon the difficulty of the task confronting them coupled with the strictness of their commander. However, it is unlikely that discipline was lax in any army during the period because defeat normally meant death at the hands of the enemy. Capturing prisoners was no longer a priority; destroying enemy forces had become paramount.[45]

By the Warring States period, the infantry had become the primary force, long having displaced the chariot and outmoded old-style engagements between chariot-mounted nobles. However, chariot-centered units remained significant, and on reasonably level terrain they could provide a formidable penetrating force, especially when deployed against unprotected infantry arrays. The most common form of chariot-based organization saw anywhere from 75 to 100 men attached to each chariot, often grouped into three or four platoons of 25 men each. In addition to the 3 warriors, usually officers, riding in the chariot itself—the driver, archer, and shock-weapon warrior—according to some sources another 25 support personnel were attached, responsible for supplies, equipment, and perhaps one provisions wagon. The combat platoons would advance with the chariots, providing close support and protecting them from flank attacks by enemy infantrymen when they were not engaged in moving, chariot-to-chariot engagements. No doubt these infantry units could also be detached to act as an independent ground force and supplement any infantry regiments. Within the limits of terrain-imposed constraints, the commander could select the appropriate mix of chariot, infantry, and (eventually) cavalry forces to be employed. Mobility is historically identified with the cavalry, but one thousand chariots could rapidly deploy at least as many bowmen, as well as 1,000 warriors armed with shock weapons, assuming the driver remained with the team. In both of Sun Pin's famous battles, P'ang Chüan's disorganized forces rushed onto the battlefield, presumably led by their chariots and accompanied by fleet infantry running on foot, to confront an established, well-entrenched enemy with prepared defenses. From the *Military Methods* it appears that Sun Pin did not employ either the chariots or the cavalry as mobile forces in the actual conflict, deploying them instead prior to battle as fixed elements. Especially on the more open terrain of Kuei-ling, he may well have used the chariots as war wagons (perhaps like John Zizka) or constructed temporary fortifications from them.[46]

The weapons of the period were primarily still bronze, but iron weapons, especially for the infantrymen who were drawn from the common people rather than the nobility, had appeared and were becoming common. The technology of the former still produced far superior blades that were not susceptible to rusting, but iron was more common and readily available and could be used to arm a larger number of men far more economically. However, irrespective of the actual material the main weapons were the hand *chi*, a sort of shortened halberd with a spear point affixed to allow thrusting, piercing movements as well as the halberd's normal mode of slashing and hooking, and spears for fighting at middle distances, supplemented with daggers and true swords, which had by now appeared and were becoming popular for close combat. However, these early swords were still designed for thrusting and piercing rather than slashing; long swords capable of slashing did not evolve and become significant until very late in the Warring States period. Hand-wielded axes were also common, and weapons use was specialized and apparently mixed even within squads. Naturally there was basic armor, primarily a lamellar leather tunic for the infantry, for both the men and chariots, as well as wooden shields and bronze helmets. As the state's investment in its soldiers increased, so presumably did their protection. However, it should be remembered that the soldiers were primarily pressed into service through a levy that was implemented as part of an extensive centralized tax system that required each family to provide either men or material support according to its landholdings and similar factors. The troops no doubt appropriated some share of the plunder and more was granted to them as rewards, but the expenses to the state were also considerable. Since different districts within the same state, as well as different states, varied greatly in their levels of material prosperity, a certain lack of uniformity among component forces of field armies obviously resulted.

The campaigns in question involved the central area of China and basically distances of only a few hundred miles. Consequently, although the massive number of troops imposed a significant logistical burden, given the fertile area being crossed, the support (willing or otherwise) of allies, and the ability of the men to subsist on simple rations supplemented by foraging, this burden was not as overwhelming as it would be less than a century later, when many times more men would be fielded. Ox-drawn wagons would have provided the basic means of transport, horses being expensive and their numbers always insufficient throughout the history of central states China. Depending upon the season, water supplies should have not occasioned any great difficulty within the central area because of the presence of numerous ponds, lakes, wetlands, streams, and great rivers. Wei's army would have also had the additional burden of conveying specialized forces and transporting heavy equipment for undertaking siege warfare. The art of siegecraft had advanced greatly by this time, and siege engines, such as battering rams, overlook towers, and large mobile protective shields, would all have been required to break down defensive walls and perimeter fortifications 20 or 30 feet thick and more than 25 feet high, assuming the outer moats and ditches could be successfully crossed.

Estimates of force strength for the various states in Sun Pin's era present a rather incomplete and possibly questionable picture because the main sources remain Su Ch'in's famous passages dating from at least some decades thereafter found in the *Chan-kuo Ts'e*. As part of his tactical persuasions, Su Ch'in analyzed the numerical strength of various states, and historians have employed the figures ever since. Whether his numbers were mere guesses, calculated estimates, or actual figures remains an important but unanswered question. Presumably each state kept reasonably updated counts of its soldiers; alternatively, an astute external observer, if unable to employ spies to ferret out the actual information, should have been able to roughly estimate the numbers with basic information about unit size and the overall number of armies. Generally speaking, each of the stronger states apparently could have drawn upon several hundred thousand troops at this time, although how many actually constituted standing forces needs to be determined. Obviously Ch'i and Wei each maintained at least double the troops they committed to their battles, although King Hui of Wei seems to have been rather easily perturbed by threats to Ta-liang. In a discussion with King Min of Ch'i, Su Ch'in asserted that at the time of the battle of Ma-ling Wei had 360,000 troops.[47] However, the *Shih Chi* biographical account indicates King Hui mobilized the entire state, yet dispatched only another 100,000 men, suggesting that Su Ch'in overestimated Wei's strength. Based upon Chang I's and Su Ch'in's persuasions in the *Chan-kuo Ts'e*, the maximum figures for the mid-fourth century might have been as follows:[48]

> Wei: 360,000 trained, mailed soldiers plus perhaps another 100,000 assigned to border defense; 1,000 chariots (possibly a serious underestimate); and 10,000 cavalry
> Chao: 100,000 trained, mailed troops; 1,000 chariots; 10,000 cavalry
> Han: 200,000 trained, mailed troops; 100,000 additional on border duty
> Ch'in: 100,000 trained, mailed troops (perhaps a serious underestimate, certainly increasing extremely rapidly); 1,000 chariots; 10,000 cavalry
> Ch'i: several hundred thousand trained, mailed troops
> Ch'u: 100,000 trained, mailed troops; 1,000 chariots

The Campaigns

The twin campaigns of Kuei-ling and Ma-ling made Sun Pin famous, creating his image as a brilliant but tragic strategist. Although his own writings were subsequently lost, the accounts recorded in the ancient *Chan Kuo-ts'e* and *Shih Chi* secured his reputation. However, even af-

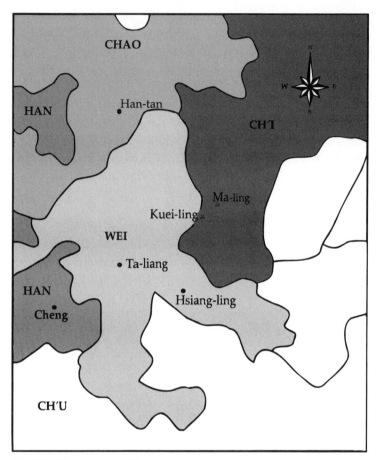

MAP 2   The area of conflict

ter we supplement these works with other early writings, including
the *Mencius,* the *Bamboo Annals,* and the *Lü-shih Ch'un-ch'iu,* a
number of significant discrepancies regarding the dates and sequences of events in this period remain, many impossible to defini-
tively resolve. Among the major questions are the date Wei moved its
capital to Ta-liang; who the speakers in various court conferences may
have been; whether P'ang Chüan commanded at the battle of Ma-
ling as well as at Kuei-ling (where he had supposedly been captured);
when T'ien Chi, Sun Pin's mentor in Ch'i, was driven out of Ch'i

through Tsou Chi's machinations; who was king of Ch'i at the time of the second campaign; whether attacks were ever mounted against Hsiang-ling and P'ing-ling; and whether Imperial Prince Shen died at Ma-ling or not. The depictions that follow attempt to reconcile the apparent facts as comprehensively as possible consistent with the capabilities of ancient armies, the known political situation, the types of tactics generally practiced, and the dominant personalities involved. Without additional discoveries such as commemorative bronzes or bamboo writings, more decisive conclusions remain unattainable. However, most of these issues, as well as the campaigns themselves, are extensively discussed here and also in our *History of Warfare in China: The Ancient Period.*

The Battle of Kuei-Ling

As already noted, the state of Wei was highly experienced in military matters and could be expected to have fielded a well-trained, mobile, and effective fighting force. After a pre-emptive strike at Ch'ih-ch'iu, its primary objective was Chao's capital of Han-tan, for Wei hoped to subjugate the government by seizing Chao's most valuable asset with a single, bold thrust employing 80,000 troops in 354 B.C. (See Maps 3A and 3B.) Clearly warfare had evolved considerably since Sun-tzu's era for one state to be immediately targeting another's most significant city for assault rather than avoiding or perhaps isolating it until conditions became more conducive. However, instead of a quick victory, roughly one year had to be expended for siege operations before Wei apparently emerged victorious, no doubt only after repeated clashes and mutually significant losses. While nothing is now known about the course of events during this prolonged siege, Chao's forces would certainly have resisted by utilizing every possible defensive measure, mounting harassing fire from the battlements, and sallying forth at intervals to strike Wei's encamped army. Otherwise, Sun Pin's basic premise that the two forces were decimating each other would be invalid and his strategy fundamentally flawed.

In striking Chao, Wei intended to block Chao's expansion toward the southwest and systematically reannex territory that had once been part of the mighty state of Chin. If successful, after the additional step of vanquishing Han, it could forge a new power base out of an area of

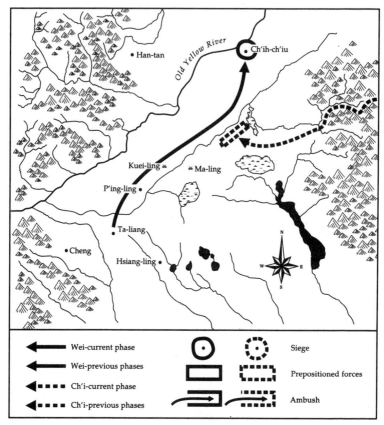

MAP 3A   *The battle of Kuei-ling, first phase. Wei advances to besiege Ch'ih-ch'iu. Ch'i dispatches a defensive army to the border area.*

once-common but ever-diverging cultures. Alone, although still powerful, Wei could probably not withstand continued onslaughts from Ch'in's growing strength in the west or repeated offensives from Ch'u and Ch'i. Conversely, by thus reintegrating all the territory of the former Three Chin, Wei might prevail over almost any force or alliance, just as Chin had in the glorious days of its undisputed hegemony. Moreover, insofar as their borders were complexly entangled, Chao's and Han's proximity naturally made them priority targets. Because King Hui was greedy and ambitious but lacked the wisdom of his forefathers, he could not envision any alternative to

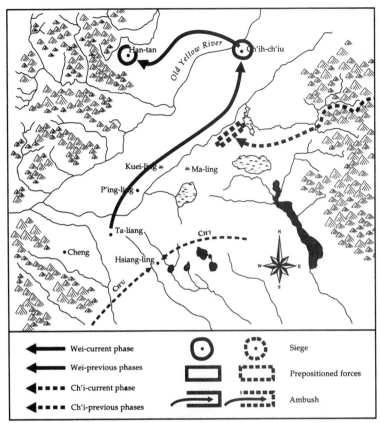

MAP 3B   The battle of Kuei-ling, second phase. Wei advances to besiege Han-tan. Ch'i and Ch'u launch a limited strike on Hsiang-ling.

brutal military conquest and therefore failed to effectively bind Han and Chao through close, mutually supportive alliances.

Although a rancorous heritage of conflict between Wei and Chao dated back several decades, more immediate causes decisively provoked King Hui into launching his ill-considered attack on Han-tan in 354 B.C. In addition to the long-festering political enmity caused by the incidents accompanying his own turbulent ascension in Wei in 370 and Marquis Ching's successful defeat of Prince Ch'ao in Chao in 386, Chao's aggressiveness toward the smaller state of Wey precipitated King Hui's response to a newly perceived political threat. In

356 Wei had intimidated the minor states of Lu, Sung, Wey, and Cheng into attending a court conference intended to visibly reaffirm its awesomeness in the central region. Chao obstreperously declined to attend, opting instead to conclude an alliance that same year with Ch'i that apparently included provisions for mutual aid in the event of an attack by outside forces. This naturally infuriated King Hui, who must have felt that any actions that further increased Chao's independence would correspondingly diminish his own authority and security, and probably stimulated him to further alliance making in anticipation of initiating military actions against Chao.[49]

Early in 354 B.C. Wei, in conjunction with Han, Chao, and Wey, mounted successful attacks against perimeter states and annexed captured territory. However, Chao, which had expended considerable military effort but whose borders were isolated and distant, alone among the four states failed to benefit by similarly increasing its territory. Chao therefore felt compelled to proportionately augment its own domain just to keep pace with its potential enemies Wei and Han and accordingly attacked the small state of Wey in the southeast. Wey had actually been a favorite, oft-exploited target for Chao over the years. In 383 Marquis Ching had attacked Wey, which managed to survive only through Wei's forceful intercession. In 372 Chao invaded Wey, seizing and retaining some seventy cities; Wei again dispatched the troops necessary to defeat Chao. Finally, early in 354 Chao launched a major offensive against Wey, vanquishing its army, annexing cities, constructing defensive works, and forcing Wey to pay homage at Chao's court. These developments enraged King Hui of Wei because a subject state had suddenly been roughly wrenched from his grasp as Chao aggressively expanded its power in a threatening manner.

Ch'i for its part did not enjoy any particularly beneficial relationship with Chao (or even Han later on) but was severely troubled by the specter of a newly invigorated Wei commanding the allegiance of an increasingly huge populace in a vast, fertile area. For Ch'i to respond was merely a matter of self-interest, as Chao certainly knew and adroitly exploited when framing its appeal for aid. However, shunning an immediate response would be advantageous for Ch'i because its armies would ultimately confront a significantly diminished and enervated Wei force, while Chao would also have been weakened. If effectively implemented, this strategy would considerably reduce both the immediate and long-range potential dangers each state might present to Ch'i in the ever-changing swirl of Warring States alliances and events. The *Chan-kuo Ts'e* contains a passage in which Tuan-han Lun, an otherwise unknown figure, provides a summary of this stratagem, which is usually identified with Sun Pin:

> When the difficulty at Han-tan developed, Chao beseeched Ch'i to rescue them. King Wei summoned his high ministers to make plans and asked: "Shall we rescue Chao or not?" Tsou Chi said: "It would be better not to rescue them." Tuan-han Lun said: "If we do not rescue them it will not be advantageous to us." Marquis T'ien said: "Why?" (Tuan-han Lun replied:) "If Wei annexes Han-tan, of what advantage will it be to Ch'i?" King Wei said: "Assuredly!" Thereupon he mobilized the army, directing them to encamp at the suburbs of Han-tan. Tuan-han Lun said: "When I sought to determine whether it would be advantageous or not, it was not with reference to this. Now if we rescue Han-tan by encamping in its suburbs, Chao's capital will not have been seized, while Wei's (army) will still be intact. Thus it would be better to go south to attack Hsiang-ling in order to exhaust Wei. When Han-tan has been taken and we exploit Wei's exhaustion, Chao will have been destroyed and Wei weakened." King Wei said: "Excellent." Then he mobilized the army to go south to attack Hsiang-ling. In the seventh month Han-tan succumbed; Ch'i then exploited Wei's exhaustion to severely destroy (their army) at Kuei-ling.[50]

This account and some others refer to an attack on Hsiang-ling that is otherwise unmentioned in the biographical versions in the *Shih Chi* and *Military Methods*.[51] Some scholars have therefore concluded that "Hsiang-ling" is an error for "P'ing-ling," where Sun Pin deliberately sacrificed two Ch'i commanders. However, more likely this initial attack constituted a minimal show of force mounted in conjunction with paltry units coerced from Sung and Wey as an immediate response.[52] (See Map 3B.) While opportunistic gains would not be foregone, the fundamental objectives could have been five: stiffen Chao's resolve to resist,[53] establish a holding force to constrain a sizable number of Wei troops, threaten Wei territory, display a certain ineptness in military matters to encourage P'ang Chüan to largely ignore Ch'i's activities in favor of simply concentrating upon the siege of Han-tan, and maintain Ch'i's reputation as a righteous state by fulfilling the requirements of its covenants with Chao.[54] Actually, as targets both Hsiang-ling and P'ing-ling would have been reasonably well chosen not only because they lay pivotally in the area of Wei's supply lines but also because the region could have been plundered for materials, food, and other resources. Furthermore, striking into the heart of Wei's staging area for its forward deployment might even have resulted in unexpected gains that could not have been disdained. However, the main purpose may well have been luring Wei into complacently maintaining intense pressure against Han-tan while stimulating Chao to exert maximum effort in its fight for survival. Some accounts state that Ch'i's attack on Hsiang-ling turned into a prolonged siege that was lifted only after Wei defeated the troops still deployed there the year following its own defeat at Kuei-ling. If accurate, this minimal force obviously accomplished significant achievements and remained an annoying threat to Wei throughout the confrontation.

Several works indicate that Han-tan actually fell about one year after Wei initiated its onslaught, although others raise minor doubts on whether the city ever succumbed.[55] While tactically there would have been considerable differences, for Ch'i's purposes they probably would have been equally opportune. If the city fell, which appears more likely, Wei's army would no doubt have expended great effort in a final assault, perhaps even with the men "swarming over the walls like ants" (after much preliminary battering, mining, and assaulting) as described in the *Art of War*.[56] This would have resulted in the elation of victory, the plundering of the city, and then the psychological letdown that inevitably follows. As the *Wu-tzu* indicates, an army burdened with the spoils of war loses its mobility and also its inclination to fight.[57] Therefore, after having suffered the normal attrition expected during a prolonged siege from illness and injury, from lengthy periods of intense heat and freezing cold, and from combat casualties, the substantially weakened Wei army would have been thrust into a psychological maelstrom if suddenly faced with the need to resurrect its spirits and race out to confront a fresh army threatening its homeland. Of course, the Wei army might have been enraged, piqued to the frenzied energy that P'ang Chüan himself displayed, but generally speaking armies that have endured campaigns of more than a half year tend to be worn out and dispirited, their equipment having grown moldy and the men decrepit. Even if the city never fell, after a year in the field they would physically have been in essentially the same condition, whereas mentally they might have been tougher. However, if not moved to bravado and defiance by this new threat, according to the psychology of *ch'i* (morale) found in the military classics, they might easily have become terrified at being forced to fight on two fronts, frightened at being threatened from behind.[58]

In either case, after about one year and the commitment of limited forces to Hsiang-ling, Ch'i's strategists apparently determined the appropriate moment had arrived. Although the campaign's essential features as recorded in Sun Pin's *Shih Chi* biography and referred to in other *Shih Chi* annals (despite some apparent errors) largely cohere with their depiction in Sun Pin's recently recovered *Military Methods*, the latter expands certain aspects and provides additional insights. In particular, the first chapter, entitled "Capture of P'ang Chüan," reveals a complexity to Ch'i's strategy not previously preserved by the historical records. (It is this absence that causes many analysts to conclude that references to Hsiang-ling are errors for P'ing-ling, or P'ing-ling for Hsiang-ling, and that the new account therefore merely expands an already known factual matter.)[59] In this account, while the army is actually en route toward Ta-liang, Sun Pin is clearly seen advocating a second ruse to deceive Wei's commander into believing Ch'i's leaders have blundered and Ch'i's forces need not be a cause for concern. In essence Sun Pin was willing to suffer casualties in a detached force under inept commanders to create this impression.

P'ing-ling, the chosen target, was located in Wei's southeastern corner, a fortified military area clearly being employed to support the attack on Chao, for the roads were filled with soldiers. In addition, it had vast resources and a large population and probably maintained troops in strength even in normal Warring States times. Since any thrust toward it would require Ch'i to project itself across the enemy's border, penetrating a corridor between potentially unfriendly states, their supply routes could easily be severed, stranding them

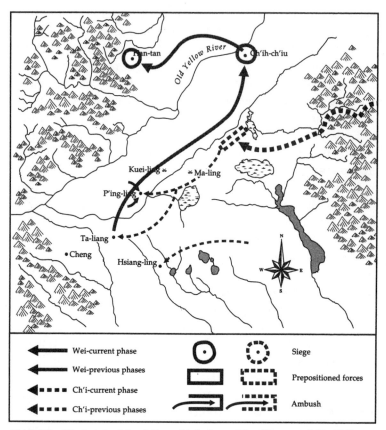

MAP 3C   The battle of Kuei-ling, third phase. After Han-tan falls, Ch'i dispatches two small, detached forces to attack P'ing-ling. After defeat at P'ing-ling, Ch'i sends a light, limited force to visibly threatened Ta-liang.

with inadequate resources for an assault on a fortified town like P'ingling. (See Map 3C.) As Sun Pin states in the first chapter of the *Military Methods,* "I suggest that we go south to attack P'ing-ling. The town of P'ing-ling is small but the district is large; the population is numerous; and its mailed soldiers abundant. It is a military town in (Wei's) Tung-yang region, difficult to attack. We would thereby display something dubious to them. When we attack P'ing-ling, Sung will be to the south, Wey to the north, and Shih-ch'iu will lie along our route. Accordingly, since our supply route will be cut off, we will show them we do not understand military affairs." Without doubt Sun Pin exemplified the dictum that "that those who excel in employing the army can bear to kill half their officers and soldiers" to achieve their ultimate objective.[60] While some analysts claim that the preceding quotation does not preserve Sun Pin's own words but rather those he hypothetically attributes to P'ang Chüan analyzing the situation, the entire chapter is coherent, explicit, and tactically reasonable.

Once the segmented thrust at P'ing-ling resulted in the intended, highly visible defeat, Ch'i mounted another apparently ill-conceived reactive response by turning toward Wei's eastern capital of Ta-liang to threaten its political and economic center.[61] From P'ang Chüan's perspective, this could only be an impulsive, desperate, and badly planned venture easily disposed of—exactly the misconception that Sun Pin intended to foster. Consequently, P'ang Chüan, compelled to respond to King Hui's urgent command to return and deflect this new threat while confident in his army's capabilities, rashly abandoned his heavy supply train (and probably his heavy forces as well) to quick march back toward Ta-liang. (See Map 3D.) The "Capture of P'ang Chüan" further indicates that Sun Pin deliberately dispatched only light or chariot forces toward Ta-liang, visibly followed by only a portion of his strength "to show that they were few." This was designed to accomplish two objectives: to prod P'ang Chüan into a rapid response to deal with highly mobile troops and to convey an impression of limited numbers that might easily be countered. In this light P'ang Chüan's own use of mobile elements reflects a decision made with respect to tactical requirements. He did not fatally blunder forth simply out of anger or arrogance, although they were probably contributing factors that occluded his own estimation of the situation and doubtlessly were consciously exploited by Sun Pin's tactics.

Since the roads available for swiftly returning to Ta-liang were limited, Sun Pin could easily foresee which route P'ang Chüan would choose. Then, following Sun-tzu's dictum and his own tactical advice, Sun Pin mobilized his main armies, which had remained in reserve near the border; arrived at the battlefield first; encamped and improved their position to fully exploit the terrain; and finally confronted P'ang Chüan at Kuei-ling. Deployed in depth and probably

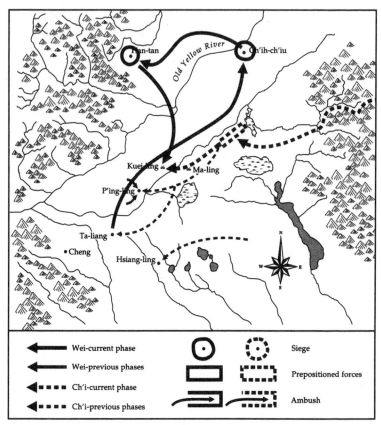

MAP 3D   *The battle of Kuei-ling, final phase. Wei rushes mobile and light forces from Han-tan toward Ta-liang and is attacked by Ch'i's well-rested army at Kuei-ling.*

with part of the forces concealed in ambush, Sun Pin's well-rested troops easily defeated a harried, exhausted Wei army surprised by Ch'i's intense attack mounted with the benefit of overwhelming tactical advantages.

In this campaign Sun Pin's basic principle of dividing his forces—often into three—and keeping a large portion in reserve is clearly seen. Those segmented off to P'ing-ling and Hsiang-ling realized valuable tactical objectives as well as acting as holding forces in themselves. Sun Pin's main army proceeded unobserved in an orderly fashion behind those racing to Ta-liang; he thus extended apparent forms to P'ang Chüan and manipulated him by obscuring his actual capabilities and intentions within these forms. Therefore, it might well be said that Sun Pin concealed the invisible in the visible, reaching the height of the formless and the pinnacle of the Sun family strategy.

Tactical Evaluation and Consequences. (Tactics will also be further identified and commented upon in the last section and in the individual chapter commentaries, which should be read in conjunction with this section.) As already mentioned, King Hui's eagerness to directly attack the enemy's major city marked a dramatic shift in warfare doctrine from the Spring and Autumn period to that of the Warring States. In the former the equipment for conducting siegecraft operations would have been lacking, and any force unprepared to endure a long siege operation designed to starve out the defenders could only impetuously mount feverish attacks against well-entrenched defenders enjoying innumerable advantages. However, by the middle Warring States period the advantage had perhaps swung to the attackers: They had unfettered access to resources and equipment, whereas the defenders, while entrenched and fortified, suffered ever-dwindling supplies of food, fuel, and even water. Without external aid, the defenders would eventually reach a critical point when their strength would be inadequate to resist an enemy's onslaught and the citadel would fall.

The two large cities in question, Han-tan and Ta-liang, and later, in the Ma-ling campaign, Han's capital of Cheng had all attained significant size by the middle of the fourth century B.C. Excavations conducted at their ancient sites have revealed walled enclosures incorporating gates oriented to the four cardinal points of the compass that range from roughly 10,000 to 17,000 feet per side, with the walls themselves averaging up to 25 feet high, generally with protective outer dike works or moats.[62] Portions of Cheng's walls soared 55 feet in height above the surrounding terrain, with proportional dimensions for the base and top. Each of these cities was well organized in a rectangular pattern and no doubt strictly administered. Considerable wealth in addition to that possessed by the ruling clan and the government could be expected as plunder, as well as weapons and tools that would add immeasurably to the hoard from any vanquished city. Therefore, in addition to the primary military objective of defeating the enemy by killing his soldiers and capturing the ruler, now that cities themselves had become much wealthier economic centers, any tangential material gains would opportunely offset some of the costs incurred by undertaking lengthy sieges.

Han-tan and Ta-liang were both easy targets because they lay on focal, traversable terrain accessible from three sides in the former case and four sides in the latter.[63] Han-tan, the site of modern Handan in Hopei province, was backed by a mountain range to the west but was easily accessible to Wei's armies venturing north from Ta-liang or coming across from An-i in the constricted corridor that connected the two portions. Ta-liang in Wei was similarly open to incursions from Chao to the north, Ch'i to the east, and Ch'u to the south, as well as annoying skirmishes with minor border states. From Ch'i, which was largely protected by oceanic and mountainous borders and essentially unthreatened by Yen to the north, the major access routes to the central plains areas would be down toward the southwest through one of two western passes. Thus, its natural corridor of movement would tend to orient it toward cities such as Hsiang-ling and Ta-liang, inescapably making them optimal targets.[64]

Ch'i's tactical predisposition (potentially exploitable by an astute enemy) immediately raises doubts about King Hui's wisdom in making Ta-liang his eastern, and eventually the final, capital. Positioned at a crossroads near the Yellow River, it had no doubt developed militarily and economically long before being selected for an augmented administrative role. When Wei was strong, King Hui must have felt confident that if properly fortified and manned, despite being situated in an exposed position, Ta-liang would still be an ideal site from which to control the region. However, having been established on the alluvial Yellow River plains, it could easily be attacked in force even by chariot-based armies and remained vulnerable despite the construction of artificial fortifications and long external walls to defend it. Kuei-ling, the site of the first encounter, was located in the ancient state of Ts'ao (which had been absorbed by Ch'i) and was similarly accessible from every direction, facilitating Ch'i's troop movements but perhaps deliberately making it too easy for P'ang

Chüan to rush forward across flat terrain with just his more mobile elements.

Over the ages orthodox Chinese thinkers have emotionally argued from political and military perspectives about Ch'i's wisdom in responding to Chao's request. In fact, a balanced evaluation is necessary because at the time of Wei's attack on Han-tan, King Wei of Ch'i had only just embarked upon his reform programs. Although the state had ample natural resources and a long military heritage, it was not generally considered a match for Wei by outsiders as well as members of its own government.[65] Thus, when Ch'i's limited battlefield successes against Wei were weighed against its earlier defeats, Ch'i realized it had to exercise caution in directly engaging Wei's experienced troops outside the protective confines of its own borders. Consequently, Sun Pin's argument that Wei and Chao should be allowed to diminish each other fully accords not only with the concept of attacking when the enemy's *ch'i* has deflated but also with the realities of tactical estimation. It was a Sun family dictum that one should fight only when assured of victory;[66] Sun Pin ensured that such conditions would be realized by first exploiting the existent situation and then further manipulating the enemy to achieve additional gains. He compelled the enemy into movement—destabilizing, wearying, and debilitating him—and then simply applied the impulse of a well-organized and well-entrenched force onto a disorganized army unexpectedly encountering its foe. Moreover, by choosing a battlefield in Ch'i, he kept his supply lines short and his requirements for cartage and support personnel to a minimum and avoided exhausting the army en route to the engagement. Under the primitive conditions of the tamped roads in the central plains, the army that traveled farthest invariably suffered the most breakdowns in equipment and morale. Just the easily raised dust that befouled equipment and exhausted the men and horses would have been a significant factor affecting the comparative balance of power.

Politically Ch'i had to respond to Chao's request in order to maintain its honor, image, and prestige among the other states. No doubt the youthful King Wei initially welcomed the alliance with Chao as a means to strengthen the latter's independence and thereby keep the central states fragmented and comparatively weak. However, apart from any strategic considerations, if Ch'i failed to respond to Chao's distress every state within its expanding sphere of influence would regard it as untrustworthy and unrighteous, neither a potential ally nor a qualified hegemon. Accordingly, Sun-tzu said, "The highest realization of warfare is to attack the enemy's plans; next is to attack their alliances; next to attack their army; and the lowest is to attack their fortified cities."[67] Moreover, relying upon the unreliable could easily doom a weaker state to extinction. Therefore, in order to survive, the smaller states had to cultivate *reliable* allies among the major powers, whereas the major powers (of Ch'i, Wei, Ch'in, and Ch'u) needed the support of the smaller states to counter other alliances and the constant threats posed by enemies. Political considerations thus loomed large in Ch'i's deliberations, causing one speaker to state unequivocally that "it would be unrighteous" if Ch'i did not aid Chao because of their alliance.[68]

The minor states of Sung and Wey were supposed to be client states of Wei, and Wey itself had been attacked by Chao. However, Ch'i also exerted significant power over them, while Chao had forced Wey to pay homage, so the records indicate somewhat contradictory roles. Sung was initially pressured by Wei to participate in the assault on Chao but merely created the facade of an attack to appease King Hui of Wei (who must have either been incredibly dumb or lacked a competent intelligence service to have been so easily fooled).[69] Later, Ch'i seems to have successfully coerced Sung and Wey into dispatching a minimal number of troops to join the attack on Hsiang-ling. If true, this would vividly illustrate the political complexities in the Warring States period. In Sun Pin's argument for attacking P'ing-ling, he also characterized Sung and Wey as potentially dangerous states, although he was probably speaking from the perspective of how Wei would evaluate the situation because King Hui would no doubt feel confident in his overarching power and ability to control them.

Unfortunately, there is a major contradiction in the records as to when Ch'i attacked Hsiang-ling, with some accounts even noting that the action occurred the year after Kuei-ling. However, assuming that the *Ku-pen chu-shu* records indicating that Wei (with Han's support) attacked the feudal lords (of Ch'i, Sung, and Wey) at Hsiang-ling allows most of the difficulties to be reconciled and confirms that

Wei's power had not been significantly reduced by the defeat at Kuei-ling.[70] This also suggests that the troops initially came from Ta-liang rather than An-i, as is often asserted, thereby rendering it temporarily vulnerable, and that a large reserve of troops existed in Wei's western portion.

Wei supposedly committed roughly 80,000 troops to the siege at Han-tan, and Ch'i reportedly then dispatched an equal number. At the battle of Kuei-ling Ch'i's forces probably engaged far fewer than the initial 80,000 Wei troops due to the need for P'ang Chüan to assign part of his original number as a policing force at Han-tan coupled with the large portion presumably left behind to facilitate his hasty advance. Even if Wei had lost 50,000 troops in this engagement, or perhaps even all 80,000 between its earlier casualties at Han-tan and then Kuei-ling, in comparison with a standing core army of perhaps 360,000 the battle would have proved costly but not devastating. Apart from men and horses perishing, more important in an age of limited industrial resources would have been the extensive loss of weapons and the disaster of forfeiting expensive, difficult-to-replace equipment such as chariots and wagons. Moreover, to the extent that Sun Pin's armies were able to seize them intact, Ch'i thereby benefited, thus augmenting its own resources, offsetting its expenditures, and compensating for minimal casualties. However, because Wei had implemented an intensive training program decades earlier under Wu Ch'i to select the best men and prepare them thoroughly, perhaps (as Sun Pin indicates) the best troops—at least from the east—were deployed outside the state. If so, their loss would have had a disproportionately greater impact.

Although one or two modern Chinese analysts have created fairly detailed tactical battle diagrams for the encounter at Kuei-ling, the presently available historical sources do not provide any basis for such reconstructions. It can be assumed that each side employed three operational groups, somewhat along the lines of the ancient three-army system (since this is how Sun Pin formulates his tactics), but that their troops would have been divided into six or more armies since the ancient system of just three armies would have been incapable of imposing basic command and control measures on such large numbers. Both sides certainly would have had the ability to segment and reunite, to deploy in various formations, to change the shape of their deployments, and to utilize weapons groups appropriately. The *chi*, a sort of halberd with a spear tip, was probably the main infantry weapon, with detachments of bowmen (and perhaps crossbowmen) providing long-range fire support. For close combat the sword was beginning to appear in numbers but remained inferior in massed confrontations to the *chi* and spears, except in mixed usage. The role of chariots had become more constrained than in the Spring and Autumn period, although they still provided considerable mobility on the flat plains and could act as firing and command platforms. However, their role would probably have been supplemental, perhaps as described in the *Six Secret Teachings*.

As depicted in the earlier *Mo-tzu* and books from the middle to late Warring States periods such as the *Six Secret Teachings* and the *Wei Liao-tzu*, techniques for mounting effective sieges and tactical methods for breaking them had been formulated and become commonplace.[71] The employment of a rescue army conjoined with a defending force presented Ch'i with a strong option—this was apparently T'ien Chi's intention—but was dismissed by Sun Pin as entailing too much risk and requiring too great a force expenditure in comparison with indirect assault methods. While unstated, logistically the burden of dispatching an 80,000-man army over the considerably greater distance to Han-tan, of projecting power northwest rather than down the most accessible corridor toward Ta-liang only to probably encounter a well-entrenched Wei force long forewarned about the army's coming, would have been immeasurably greater and fraught with uncertainty. Even when weighed with the degree of support to be expected from Chao's encircled troops, the possibilities for success from a two-pronged inner and outer combined strike would have been much reduced by the enervated condition of the defenders and the exhausted state of the onrushers.

Wei's precipitous movement on Han-tan correspondingly exposed it to strong threats from Ch'u and Ch'in, both of which exploited the opportunity to launch border incursions. After Wei's defeat at Kuei-ling, its first major loss in the Warring States period, Ch'in became particularly aggressive on the western border, threatening An-i and other major cities on Wei's perimeter and exacting heavy losses. Accordingly, Wei found itself at a disadvantage on virtually every front, compelling King Hui to focus on one enemy at a time and cultivate more friendly—although hardly completely pacific—relations with his neighbors for the next decade.

The Battle of Ma-ling

The Ma-ling campaign, which unfolded more than a decade after Wei's embarrassment at Kuei-ling, again began with a provocative move by King Hui of Wei to seize the capital of another state.[72] In this case he sought to vanquish Han, the third of the three states that had resulted when the major clans divided powerful Chin into Han, Wei, and Chao. Thus, despite years of bitter experience and the counsel of many wise individuals, his logic had remained unchanged: He still hoped to annex territory and absorb people by military force from states that might otherwise have been retained as effective allies. Again there was a more proximate cause: Han's unwillingness to submissively attend the convocation of states instigated under King Hui's leadership (ostensibly to honor the Son of Heaven) at Fengche in 344 B.C.[73] Han's geographic location in the south, mushrooming up to essentially divide Wei into two unintegratable territories, virtually a lance in Wei's body politic, also cried out to be excised. Moreover, through the implementation of Shen Pu-hai's administrative measures, Han continued to strengthen, posing an ever-greater threat, and had taken advantage of the debacle at Kuei-ling to annex territory from Eastern Chou while consolidating its position in the center of the plains area.

In preparation for his assault on the capital city of Cheng, King Hui had improved Wei's diplomatic relations with Ch'in and Chao to preclude predatory actions while his main force was preoccupied outside the state. Because of Han's relatively isolated position in the south, once Wei launched its invasion Han's strategic options were limited. (See Map 4A.) Even though it could request military assistance from many states, only the three in closest proximity could mount an immediate, effective response: Ch'i, Ch'in, and Ch'u. Realistically, both Ch'in and Ch'u might exploit the situation to occupy Han themselves after effecting a much-vaunted rescue. Moreover, Ch'in and the northern state of Chao, which could have mounted a coercive at-

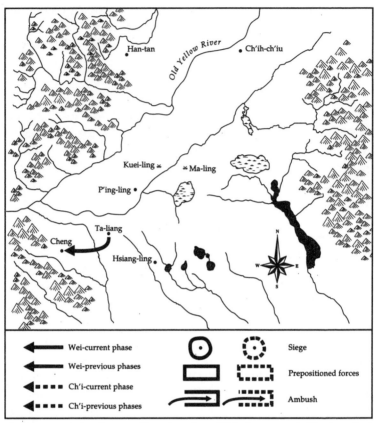

MAP 4A   The battle of Ma-ling, first phase. Wei's forces attack Chao's capital of Cheng.

tack on Wei's rear if so inclined, were supposedly on friendly terms with Wei. Therefore, following Chao's earlier example, Han's ruler dispatched emissaries to beseech Ch'i to provide the military forces vital to its survival.

According to the *Shih Chi*, Han's urgent plea for assistance stimulated another debate in King Wei of Ch'i's court, one argued along lines similar to Chao's identical request some twelve years earlier.[74] Tsou Chi, who had prominently opposed undertaking any action in the case of Wei attacking Han-tan, again selfishly advocated a policy of inaction, while T'ien Chi and many others believed that they should rescue Han to prevent its irreversible annexation by Wei.[75] Eventually their camp prevailed against the naysayers as well as the proponents of an immediate response because Sun Pin cited the need for the antagonists to diminish each other before Ch'i should venture forth to engage Wei in battle:

> King Wei[76] summoned T'ien Chi to resume his former position.[77] Han requested rescue from Ch'i. King Wei summoned his high ministers to make plans, saying: "Should we rescue them early on or rescue them later?" Tsou Chi said: "It would be best not to rescue them." T'ien Chi said: "If we do not rescue them then Han will quickly be broken and absorbed into Wei. It would be better to rescue them early on." Sun Pin said: "Now if we rescue them before the armies of Han and Wei have been exhausted, then we will be replacing Han to receive (the brunt) of Wei's forces and conversely obeying Han's commands. Moreover, Wei is determined to destroy their state. When Han sees it is about to perish, it will certainly face east and advise us. Then by relying on our close alliance with Han and our late taking advantage of Wei's exhaustion, we can doubly profit and gain an honored name." King Wei said: "Excellent!" Then he secretly informed Han's emissary and sent him forth. Han, relying upon Ch'i, then fought five battles in which it was not victorious, and then came east to entrust their state to Ch'i. Ch'i then mobilized its army, commissioning T'ien Chi and T'ien Ying as generals and Sun Pin as strategist to rescue Han. Ch'i[78] then attacked Wei and severely defeated it at Ma-ling, killing its general P'ang Chüan and taking Imperial Prince Shen of Wei prisoner. Thereafter the kings of the Three Chin, in accord with T'ien Ying's (suggestion), paid court to Ch'i at Po-wang, departing after concluding an alliance.[79]

The *Chan-kuo Ts'e*, which may have been the basis of the *Shih Chi* account, contains a similar passage but with the identity of the disputants significantly changed. This has stimulated much scholarly discussion, including articles that assail Ssu-ma Ch'ien for mangling the original version so that it would cohere with his other biographical material. For the purpose of understanding the basic issues, the identity of the speakers is hardly critical, although Sun Pin's stature is somewhat enhanced by having the essential strategic principles attributed to him alone.[80]

Events proceeded pretty much as predicted because Han, encouraged by the prospect of imminent aid from Ch'i (and probably secure

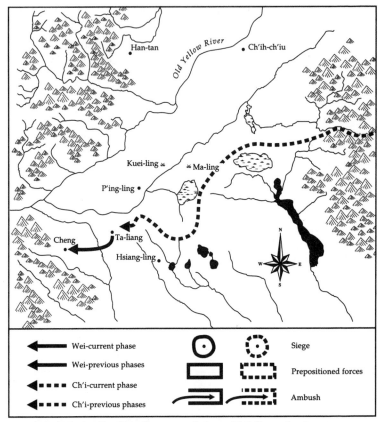

MAP 4B   *The battle of Ma-ling, second phase. Ch'i dispatches forces to threaten Ta-liang.*

in the knowledge that Ch'i had prevailed in past situations, although remarkably ignorant in failing to realize it could be exploited in exactly the same fashion as Chao), engaged the enemy five times before acknowledging the hopelessness of the situation by submitting to Ch'i in a last-ditch effort to retain some semblance of an independent existence. Ch'i then knew the moment had arrived to dispatch the forces it had no doubt been preparing for many months. Once again its visible strategic principle was to launch a strike at Wei's heartland in an apparent effort to seize Wei's critical point. (See Map 4B.) Naturally a terrified King Hui hastened to respond, ordering the Heir-apparent, Prince Shen, to personally command all the troops mobi-

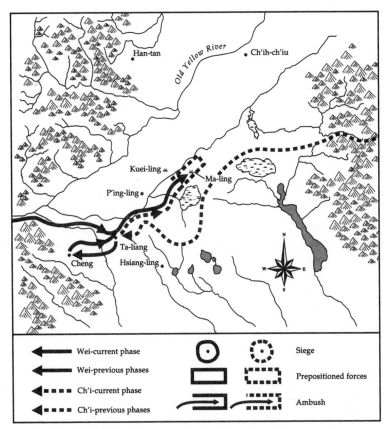

MAP 4C   *The battle of Ma-ling, final phase. Wei mobilizes a second army;
P'ang Chüan rushes mobile forces back from Chao; Ch'i visibly withdraws to
establish ambush at Ma-ling.*

lized to blunt the threat and unite with P'ang Chüan's forces, which,
already in the field, would turn back to create a combined force. (See
Map 4C.) Although the historical records disagree as to whether
P'ang Chüan actually participated in this campaign or not—some indicating he had been captured at the battle of Kuei-ling—from the
Shih Chi account there seems to be little doubt of his dramatic involvement.[81]

When going forth, Prince Shen was himself warned by a man
named Hsü-tzu in Sung not to participate in the conflict[82] and was
also advised that Ch'i's two generals, T'ien Pan and Sun Pin, were

hoary and experienced.[83] In fact, one passage in the *Shih Chi* describes Sun Pin's army by saying, "They eat people and cook the bones, the officers have no thought to turn outside—these are the troops of Sun Pin."[84] Yet the prince was already trapped by the surge of events, committed to a precipitous course of action in which he could not control his own destiny. Perhaps even though he theoretically wielded overall command, he felt compelled to defer to P'ang Chüan's assessment of Ch'i's condition, or perhaps he was simply overjoyed at the prospect of an easy victory. The well-known *Shih Chi* account speaks of P'ang Chüan, not Prince Shen, charging forth and arriving at the predestined tree, suggesting the latter was farther back and less impulsive. P'ang Chüan paid for his misperception, for allowing his estimations to accord with his wishes rather than probing the actualities of the situation, with his death. The Heir-apparent was then either captured (according to the *Shih Chi*) or killed (based on other versions, including the *Mencius*).[85]

Sun Pin's strategy for the Ma-ling campaign exploited tactical principles already witnessed in the Kuei-ling debacle: Manipulate the enemy, play upon his expectations, tire him, disorganize him, realize the advantages of terrain, and concentrate forces for the attack. Rather than choosing an open field, direct confrontation of heavy forces certain to incur heavy casualties on both sides, Sun Pin actively structured the situation to bring maximum firepower (offered by the massed crossbows) from protective cover upon a totally disordered and disoriented enemy hurtling onto constricted terrain where no maneuver or effective counterdeployment of troops would be possible. Furthermore, in the darkness of night Wei's troops had no idea of the terrain's configuration, no knowledge of the defensive or offensive contours, and thus no ability to mount even an intuitive response in the panic of the moment. In short, they had rushed into a trap much like a darkened pit and had no hope for survival.

Tactical Evaluation and Consequences. Although the surviving accounts provide a comprehensive, if compressed, portrait of the personalities and events in this conflict, some aspects bear emphasizing within the context of tactical thought. Traditional commentators focus on King Hui's anger at finding himself once again opposed by

Ch'i to account for his fevered mobilization and commitment of another 100,000 men to the campaign.[86] However, while not denying he must have felt great annoyance at Ch'i's resurgent threat, it certainly should not have been unexpected because Han had few other potential allies, was known to have requested assistance from Ch'i, and had been promised such aid. Moreover, Ch'i had proven itself willing to intercede in order to sustain its allies among the central states, while Wei's bitter experience at Kuei-ling only twelve years before should have been much in memory. It is more likely, being again threatened by the specter of a massive army directed toward Ta-liang, that in the light of the previous debacle King Hui decided to ensure certain victory by mobilizing the major portion of the troops held in the eastern portion of Wei. His errors were thus chiefly strategic, not emotional: He failed to secure an alliance with Ch'i beforehand, whether through bribes or gestures of feigned goodwill, and he undertook the campaign without the active support of the other strong states of Ch'u and Ch'in. Moreover, perhaps to ensure some semblance of loyalty and royal command, he appointed the inexperienced Heir-apparent as commander in chief. This was a serious blunder because the youthful general could not possibly control or countermand P'ang Chüan's authority, even though the latter was apparently subordinate to him.

Wei's problems were compounded by P'ang Chüan himself, for rather than profiting from his expensive historical lesson at Kuei-ling, he succumbed to the dangerous error of evaluating military intelligence and analyzing behavior through the matrix of his own expectations and desires. Naturally this is a complex issue, for raw, unevaluated information may easily prove worse than useless, leading to completely skewed actions. However, assessments based upon false premises, which detect in enemy behavior the confirmation of these premises, are equally fatal. In this Sun Pin's genius for "knowing the enemy" proved vastly superior. While P'ang Chüan's reaction may be interpreted as simply anger and arrogance, in fact the more fundamental issue might best be termed *fostered misperception*, Sun Pin having exploited a tendency he recognized in P'ang Chüan by structuring events to sustain and nurture it. This allowed him to manipulate the enemy into making an advance instead of, for example, forcing Ch'i to attack their entrenched forces or employing a pincer movement that would have taken advantage of Wei's overwhelming numerical superiority.

The story of the debarked tree and P'ang Chüan's ignominious death contained in Sun Pin's *Shih Chi* biography presented at the beginning of this introduction may be partly based on an actual historical event or may be completely apocryphal. In either case it illuminates the tactics that ensured his defeat because the story suggests that P'ang Chüan rushed forward without accurate intelligence, an effective battle plan, or any sense of the many lessons taught by the earlier military strategists, such as Wu-tzu, about pace, measure, and conservation of strength. The fact that an exhausted army has little chance against a prepared, rested foe was axiomatic and hardly needs repeating. However, note what Sun-tzu said about such haste over 100 *li*, just about the distance that Wei's army covered on its forced advance:

> Combat between armies is advantageous; combat between masses is dangerous. If the entire army contends for advantage, you will not arrive in time. If you reduce the army's size to contend for advantage, your baggage and heavy equipment will suffer loss. For this reason if you abandon your armor and heavy equipment to race forward day and night without encamping, covering two days normal distance at a time, marching forward a hundred *li* to contend for gain, the Three Armies' generals will be captured. The strong will be the first to arrive, while the exhausted will follow. With such tactics only one in ten will reach the battle site. Accordingly, if the army does not have baggage and heavy equipment it will be lost; if it does not have provisions it will be lost; if it does not have stores it will be lost.[87]

If this admonition was not well known to P'ang Chüan—although it was probably axiomatic in any event—Wu Ch'i's instructions should have been:

> In general the Way to command an army on the march is not to contravene the proper measure of advancing and stopping; not miss the appropriate times for eating and drinking; and not completely exhaust the strength of the men and horses. These three are the means by which the troops can undertake the orders of their superiors. When the orders of superiors are followed, control is produced. If advancing and resting are

not measured; if drinking and eating are not timely and appropriate; and if, when the horses are tired and the men weary, they are not allowed to relax in the encampment, then they will be unable to put the commanders' orders into effect. When the commander's orders are thus disobeyed, when encamped they will be in turmoil, and in battle they will be defeated.[88]

Moreover, the *Ssu-ma Fa*, which probably took form about this time but integrated portions from a much earlier date, emphasized the dangers of following retreats, especially feigned retreats: "In antiquity they did not pursue a fleeing enemy too far or follow a retreating army too closely. By not pursuing them too far, it was difficult to draw them into a trap; by not pursuing so closely as to catch up, it was hard to ambush them."[89] Even Sun-tzu had advised, "The strategy for employing the military: Do not approach high mountains; do not confront those who have hills behind them. Do not pursue feigned retreats. Do not attack animated troops. Do not swallow an army acting as bait. Do not obstruct an army retreating homeward. If you besiege an army you must leave an outlet. Do not press an exhausted invader. These are the strategies for employing the military."[90] With the military thinkers having expressed such intense concern about these issues, only P'ang Chüan's overwhelming desire to exploit a perceived opportunity can possibly account for his precipitous actions. After all, even Sun-tzu had condemned any general who failed to capitalize upon and exploit advantage.[91]

While dissenting voices are heard, the site Sun Pin chose for the battle was probably Ma-ling within the southwestern portion of Ch'i. History records other battles at Ma-ling, including among the same combatants, but the locations differed greatly, while the name was apparently common.[92] Engaging the enemy on home territory would obviously have been to Ch'i's advantage because the populace could have readily been impressed to provide ancillary services, such as building fortifications, and would presumably have been loyal. Moreover, Ch'i's commanders would have had ample opportunity to carefully explore any unfamiliar terrain and improve their positions, although this is an advantage any army arriving on its chosen battlefield well in advance of the enemy would enjoy. In this case the terrain about Ma-ling Mountain (from which the site derives its name) was apparently constricted, with the road passing through a moderate valley marked by ravines and wooded hillsides.[93] Clearly the trees and heavy vegetation provided concealment for the frequently mentioned 10,000 crossbowmen, if not all the troops. However, deploying 100,000 men along even the widest valley would not have been a simple task, and it is likely many troops were held in reserve near the mouth of the valley to close the trap behind P'ang Chüan's army once it found itself engaged by the crossbowmen lying in wait deeper within the defile. In accord with tactical principles advocated in the *Military Methods*, still more troops may have been deployed at the far end of the valley as a reserve to exploit developments. That they would all have waited in undetected silence is not impossible, for ancient troops were highly disciplined, effectively controlled by a severe system of rewards and punishments and the fear of being mutually implicated by their comrades' transgressions. Furthermore, they would probably have been uneasy about the size and frenzy of the approaching enemy but confident in Sun Pin's strategy, while China also had the tradition of the gagged mouth to ensure silence.[94]

Probably the key issue in any reconstruction of this battle is whether the events portrayed in Chapter 4 of the *Military Methods* specifically describe the encounter at Ma-ling or merely represent a general response to T'ien Chi's query about fortifications. While this is also discussed in our chapter commentary, in essence the issue is whether T'ien Chi participated in the campaign and battle but was unclear exactly how Sun Pin attained the victory, or whether he did not, and whether Sun Pin is generally reprising such situations or the particulars of Ma-ling. If the latter, then the *Shih Chi* and the other records that indicate Ch'i's attack was a complete surprise would be incorrect and would need to replaced with a portrait of Wei's army advancing into a well-prepared defensive position that no doubt exploited the local cover and darkness of the night but also relied upon caltraps and other hindrances to entangle Wei's forces once they entered the killing zone targeted by the massed bowmen. Moreover, Ch'i's troops would then have been clearly visible since they would have employed the chariots in stationary, defensive positions, creating temporary fortifications well integrated across the terrain that Wei would have been compelled to assault. (This would not cohere with Sun Pin's advocacy of circular formations in constricted terrain. However, the context for the latter is not clear.)[95] Clearly the Ma-ling area satisfied all the criteria for "fatal terrain"; the question is whether the traditional account of their complete surprise is accurate or must be melded with the *Military Methods* sequence. This is not impossible, as such fortification could well (and should have been) prepared, being undiscovered until the assault was sprung and Wei's soldiers tried to escape by breaking through to the end of the valley, ascending its slopes, or turning about. The caltraps previously deployed by Ch'i would have prevented Wei's units from maneuvering off the main road, confining them to an easily targeted zone.

P'ang Chüan's death at the head of his troops is perhaps lent credence by the oft-voiced advocacy in this period of generals "leading in person" when commanding troops. P'ang Chüan's own impulsiveness was also undoubtedly a factor. If the enemy had been unexpected, P'ang would have been exercising leadership in preference to command functions by encouraging his exhausted troops to press forward rather than let enemy troops, their certain victims, escape. The role of Prince Shen is somewhat less clear, for he apparently led his troops close on the heels of P'ang Chüan's army, and his masses and heavier equipment may well have prevented the advance units from being able to wheel about and escape the bottleneck. According to the records, P'ang's troops should have been highly mobile and therefore primarily chariots and cavalry (if the latter existed at this time), whereas Prince Shen's force should have been a fully complemented one and thus more weighty and ponderous. Whether Prince Shen's force was just joining up with P'ang or had already been integrated as the rear of his advancing column is unclear. However, given the likelihood that they were burdened with a baggage train and therefore suffered much-reduced mobility, if Sun Pin's army had partially deployed outside the valley and could have contained Prince Shen's force by springing an attack from behind, in the confusion that ensued 100,000 men from Wei could easily have been killed and captured. Naturally most of them would have been from the prince's army as P'ang would presumably have rushed forward with only about 20,000 troops, having left his own baggage and apparently even his heavy infantry behind. (If he in fact acted in this pre-

cipitous fashion, as the historical accounts assert, he thereby contravened every orthodox tactical principle of the day, including the
belief that an assault of this type would require at least double the enemy's strength to be effective.) Prince Shen's fate at Ma-ling remains
unresolved, with some sources stating he was captured, others that he
perished, as already noted.

The combined size of P'ang Chüan's and Prince Shen's armies certainly dictated a cautious plan of action on Ch'i's part. In theory, by
mobilizing only a 100,000-man force, Ch'i could well have found itself outnumbered two to one by reasonably experienced troops already in the field. Why Ch'i elected to employ such a limited number
when its total standing force was approximately 400,000 and there
were only a few miles of exposed border to protect is puzzling. Perhaps the access corridor naturally limited transport possibilities; perhaps Sun Pin planned the encounter well in advance and realized he
could not successfully mobilize and deploy more than this number on
Ch'i's constricted southwestern terrain. Even more likely, a
100,000-man force may well have been disproportionately composed of fighting men since they could expect to be supplied from advance depots in their own state, obtain water easily from mountain
streams, and retreat if necessary behind prepared defenses farther in
the interior. Conversely, Wei's army would have had to carry supplies
for a several days march, including food and water that would have
required numerous supply wagons. To the extent that a general out-
stripped or left them behind, his army, as Sun-tzu noted, would suffer
accordingly.

After the massacre at Ma-ling, Wei was forced to acknowledge
Ch'i's hegemony. The latter, now relatively free from threats to its
borders, flourished briefly before inexorably declining due to incom-
petent rulership and internal strife until the debacle known as the Battle of the Fire-oxen several decades later. Wei itself never regained its
former power or position of dominance and in fact became the target
of major attacks by Ch'in, the first coming under Shang Yang's personal direction the following year. After Shang Yang severely defeated Wei through subterfuge, it was forced to yield all the territory
west of the Yellow River and, to escape Ch'in's relentless pressure and
hostile surrounding forces, move the capital from An-i to Ta-liang.[96]

With Wei no longer posing a serious obstacle, Ch'in's future expansion became easy, and the central states grew increasingly weaker. Thus, the collapse at Ma-ling marked a turning point in China's history and made possible Ch'in's growth and eventual accession to empire.

Problematic Aspects of Kuei-ling and Ma-ling

Although the Kuei-ling and Ma-ling campaigns appear to be well documented in the historical records, close inspection reveals several irresolvable problems. The battle accounts here provided represent a probable reconstruction based upon the available evidence; however, there is certain to be disagreement about various aspects of these depictions. For the convenience of those interested in these issues, even though they have been extensively covered in the Notes, the basic problems may be summarized as follows.

Movement of Wei's Capital to Ta-liang. Several sources and modern historians indicate that Wei moved its capital to Ta-liang in 361 B.C., seven years before it attacked Chao's new capital of Han-tan. However, others (as indicated in the Notes) suggest King Hui moved the capital in 340 only after being resoundingly defeated at the battle of Ma-ling and suffering an invasion by Ch'in under the command of the legendary Shang Yang. In either case the king's motive would have been to shift the capital farther away from Ch'in, although strategically Ta-liang could hardly be considered more advantageous due to its easily accessible location amid level terrain.

P'ang Chüan's Command Role. A problem arises because the basic records indicate he was captured at the battle of Kuei-ling. Therefore, how could he have survived and again been entrusted with command? If Ch'i's forces had failed to kill him, King Hui would normally have been expected to execute P'ang for bringing about such a horrendous defeat. However, perhaps he was ransomed, for the vividness of the Ma-ling account, while no doubt amplified and embellished, argues greatly for his perishing there. Alternatively, the character meaning "captured" might simply refer in a general sense to the army under his command being taken.

T'ien Chi's Command Role: The question revolves around when he succumbed to Tsou Chi's machinations. Was it after the battle of Kuei-ling or not until after Ma-ling, as the historical records seem to imply? Moreover, when did King Wei die? If King Hsüan assumed the throne between the two campaigns, he could (as some sources note) have recalled T'ien Chi to assume an active role at Ma-ling. However, if King Wei ruled an additional number of years (as other historians assert and our analyses assume), the date of T'ien Chi's banishment becomes uncertain. Some records indicate that other members of the T'ien clan commanded the expedition to Ma-ling, suggesting T'ien Chi may have already been forced to the sidelines.

The Attack on Hsiang-ling. It appears from the combined accounts plus the references in the books of the various states that separate attacks were indeed launched against Hsiang-ling and P'ing-ling. Moreover, it appears that the siege of the former proved unsuccessful and Wei forced its release the year after Kuei-ling. The details preserved in the *Military Methods* about the tactics and underlying rationale suggest that P'ing-ling was a deliberate sacrificial ruse. (The text's custom of referring to Han-tan as the capital of Wey rather than Chao [which had annexed it] lends further credence to claims about the *Military Methods'* archaic nature and reliability.)

Subjugation of Han-tan. The *Shih Chi* account records that Han-tan fell before Sun Pin moved to rescue Chao, but a few other writings state that it never fell or that it managed to hold out for three years. While Sun Pin's tactics became famous as "rescuing Chao by besieging Wei," from the various accounts it is apparent that the "rescue" was only a secondary consideration because Ch'i astutely made its own self-interest paramount. However, there are dissenting interpretations, as previously noted.

The Death of Imperial Prince Shen. In the *Shih Chi's* various accounts he is captured rather than killed; however, some passages in the *Chan-kuo Ts'e* indicate that he died there. Moreover, the *Mencius,* a fairly reliable text dating from roughly this period, preserves a conversation (already translated in the Notes) in which King

Hui informs Mencius that his son died at Ma-ling. Perhaps he was slain in the battle's confusion; perhaps he committed suicide thereafter. Certainly there are enough suggestions in Wei's writings to indicate that the prince was doomed, supporting the view that he did not survive the engagement.

Main Concepts in the *Military Methods*

Because the *Military Methods* is badly fragmented and the recovered passages tend to focus on concrete issues rather than general principles, a brief overview of the main concepts and fundamental tactics may prove beneficial. Although disagreement continues about whether the two portions of the text should be considered a single work or not, the analysis that follows treats all thirty-one chapters as an integral work, the product of Sun Pin's school of thought, if not Sun Pin himself. (References to the historical Sun Pin will be appropriately qualified or clear from the discussion's context.) This position is adopted because much of the most interesting and extensive material appears in the second half; the latter introduces new topics as well as explicates brief passages from the first portion. While major differences between the two halves will be appropriately noted, since this is a work for the general reader no attempt has been made to distinguish Sun Pin's pristine concepts amid the various layers and possible voices. Moreover, even though he clearly founded his tactics upon Sun-tzu's *Art of War*, the analysis focuses upon Sun Pin's own thought; comparative material will be found in the chapter commentaries, where his concrete statements can be examined in detail.

Sun Pin, just as Sun-tzu before him, felt that warfare is of paramount importance to the state, critical to its survival in a dangerous, troubled world. However, the *Military Methods* also indicates that warfare should not be undertaken except when unavoidable and moreover should not be pursued for pleasure or—contrary to Lord Shang—profit. In fact, warfare invariably became and continued to be "unavoidable" because Virtue, while fundamental and essential, even in the period of the legendary Sage Emperors, China's incomparable paragons of Virtue, proved both inadequate and ineffective in controlling evil. However, the issue is complex, and the parameters for determining what constitutes the "unavoidable" are lacking in the extant text. One principle is certain: Frequent battles exhaust the army and debilitate the state, resulting in substantial losses and inevitable disaster.

Warfare cannot be undertaken without a suitable basis. Accordingly, Sun Pin advised King Wei that the first policy had to be to enrich the state; next the army had to be strengthened. Naturally victory in combat is one way to achieve both, as was thought during the Warring States period, but this was not a policy Sun Pin advocated. The ruler (and commanders as well) must cultivate Virtue, righteousness, and the other aspects of a true king and thereby attain *Te*, the encompassing personal power necessary to overawe the realm. However, these and other issues of righteous qualification are mentioned only briefly, never focused upon or as fully developed as in many other writings.

Tactical Principles

Sun Pin clearly absorbed many of Sun-tzu's fundamental concepts; apart from being fragmentary, the *Military Methods* also implicitly assumes certain principles. However, among those actually incorporated into Sun Pin's thought, the most important is this: Manipulate the enemy to create weakness, and then aggressively exploit that weakness. As many modern commentators have pointed out, Sun Pin's work is little concerned with defensive warfare, although the function and establishment of fortifications are mentioned. Prior to embarking on a military campaign, preparations must be made, the enemy evaluated, and a comprehensive plan formulated. (Among the greatest errors any general or state can make is to engage in battle ignorant and unprepared, lacking minimal military intelligence.) Thereafter the army's efforts should be directed to exploiting any weaknesses already present and creating opportunities where none exists. The methods, all familiar ones found in the *Art of War,* include being deceptive, luring the enemy onto fatal terrain, enticing him into movement in order to destabilize and then attack him; and realizing the advantages of terrain (as discussed later).

Whenever one's own forces confront a strong enemy, they should be divided into three or more operational groups (not necessarily identical with the formal organization along three or more armies). One group should then be employed to engage the enemy, the others kept in reserve and or deployed in ambush to take advantage of the enemy's movements and contain any unexpected developments. (Sun Pin is thus the first theoretician East or West to advance and employ the concept of segmenting the troops to create a strategic reserve.)[97] Similarly, the enemy should be coerced into splitting into disjointed groups, for then it can be easily engaged piecemeal by locally superior numbers. Naturally the commander must seek out and exploit every possible weakness, deliberately exacerbating those inherently marking the enemy. Among those Sun Pin identified are character flaws in the commanding general, such as easily becoming angered or, as in P'ang Chüan's case, being arrogant. Armies that are tired and weary make easy targets, just as do those on constricted terrain where their movements and potential responses are severely limited. The confused, doubtful, unprepared, and weak can all be summarily attacked and defeated. Whether the enemy appears strong or not, tactics should always focus upon acting unpredictably; going forth where unexpected; attacking where the enemy is unprepared; striking their weak points, their vacuities, and undefended areas; assaulting their flanks; and especially encircling the rear. Emotional factors should also be considered: Fear, doubt, and confusion should be induced in the enemy through every means possible, including feigned retreats and sudden, unfathomable movements. In every instance the army's actions must be timely and confident, appropriate to the overall situation, and directed toward ultimate success.[98]

Sun Pin also provides categories for classifying enemy armies and suggests methods for response that are similar to Wu Ch'i's series in the *Wu-tzu*.[99] Furthermore, Sun Pin describes the behavior that should characterize an invading army moving onto enemy territory and concretely analyzes cities in terms of whether they may successfully be attacked within the overall perspective of targeting them as economic, political, and military centers. However, most of this material, for which he has justifiably become famous, appears in the second half of the work. In fact, the famous principle attributed to him • in the various biographical accounts of the twin campaigns—manipulate the enemy by seizing what it loves—only appears there.[100]

Strategic Configuration of Power. The origination and explication of the concept of *shih*, the "strategic configuration of power," have generally been identified with Sun-tzu because the concept receives prominent treatment in the *Art of War.* In essence *shih* is a measure of the relative power an army derives from positional advantage combined with its overall combat strength.[101] Naturally the astute commander seeks to fully exploit whatever advantages of terrain his numbers, firepower, morale, superior provisions, and other force multipliers will make possible. When this objective is attained, the soldiers will realize that their army enjoys an advantage in strategic power and will enthusiastically engage the enemy; when not attained, it will prove difficult to wrest victory with a force of reluctant warriors. For Sun Pin, whose era witnessed the introduction and first widespread use of the crossbow, it was the crossbow itself that could be taken as a model. The bowman acts at a distance, unseen and unknown, just like the general, but the arrow flies forth to inflict great damage. The release of *shih*, strategic power, should be like this, just as Sun-tzu imagized.[102] An army's strategic power varies with many factors, all of which must be considered, including terrain and formations.

Configurations of Terrain. Classifying and exploiting various configurations of terrain constitute a significant topic in the *Art of War,* and Sun Pin clearly incorporates most of his predecessor's categories in the *Military Methods.* The initial "deadly terrains" that Sun Pin enumerates are all familiar, having being drawn from the former work: Heaven's Well, Heaven's Jail, Heaven's Net, Heaven's Fissure, and Heaven's Pit. To these he adds a number of generally recognized "entrapping terrains" that can retard an army's progress and convert even the most aggressive army into a vulnerable target: gorges with streams, valleys, river areas, marshes, wetlands, and salt flats. Obviously water hazards are particularly troublesome, and Sun Pin provides further injunctions about going contrary to the current's flow or being caught fording rivers. Moreover, he characterizes the viability of flowing water in terms of *yin* and *yang* (based upon its direction), thereby warning against relying upon certain types of rivers to sustain the army in the odd chapter entitled "Treasures of Terrain."[103] *Yin* and *yang* classifiers are also applied to mountains, formations, and seasonal indicators, while five-phase categorizations are similarly employed, generally phrased in terms of conquest relationships, for such objectives as soil classification.[104] Any ground that cannot sustain life—for example, incinerated areas—should be avoided whatever the season or location.

More important is the general principle that the commanding general must investigate the terrain, become thoroughly familiar with it, and actively exploit the topography, the aspects of Earth, to emplace his troops and defeat the enemy.[105] When the advantages of terrain are realized, the troops will naturally be inclined to fight. The enemy should be targeted on deadly ground (and of course manipulated or forced into entering it), while easy terrain should be exploited only when the commander enjoys a decisive superiority in numbers or mobile elements. Correspondingly, constricted terrain—warned against by all the military writers—and especially ravines should be fully utilized to control and vanquish the enemy. They provide not only the means for the few to attack the many but also the ground for exploiting a height advantage and for concealing troops in ambush. With appropriate fortifications erected across the mouth of a ravine, including interconnected chariots deployed with shields to fill the voids, they become strongholds not easily assaulted.

Relative Strength and Appropriate Tactics. One of the main concerns of all generals is an enemy's strength. Selection of appropriate tactics depends upon relative strength (or, more astutely, strategic power, which would be realized as a tactical imbalance of power [*ch'üan*] in any particular situation).[106] The *Military Methods* is no exception, although much of the concrete discussion appears in the second half of the book. Three possibilities for appropriate tactics exist: the few/weak against the many/strong, equal strength, and the many/strong against the few/weak. Confronting a vastly superior foe is of course every commander's nightmare, but the imbalance also presents the greatest possibilities for glory if victory can somehow be achieved. Appropriately, there are more concrete tactical suggestions for this situation than all the others combined, with the main advice being to employ persisting, rather than direct assault, tactics; to attack where the enemy is unprepared, a general principle in any case to conserve strength and forces; to divide and strike with "death warriors" whose commitment might wrest a telling advantage; to avoid easy terrain and exploit the possibilities of constricted ground; to segment the enemy so that he will be ignorant of each group's actions and can therefore be struck in relatively localized strength; and to stretch the enemy out, always avoiding direct confrontations.

Situations of equal strength can employ many of the same tactics, but the measures need not be so urgently dictated or so limited in possibility. Sun Pin particularly advises that in attacking strength with strength, one should still choose a complimentary configuration or a deployment that will prove effective in the specific situation rather than engage in a direct confrontation. Naturally dividing the enemy continues to be the single most powerful tactic, presumably a preparatory measure in most circumstances.

The third situation, one of overwhelming superiority, apparently proved problematic in ancient times because tacticians were constantly surprised when rulers inquired about it. No doubt many kings and commanders fell into the pitfall of underestimating their enemies and suffered accordingly. Sun Pin suggests that to avoid this blunder the commander should feign disorder and entice the enemy into movement, tricking him into coming forth where he can be engaged and overcome by superior numbers. Alternatively, weak troops can be employed to foster the false sense of elation that will inevitably result when the enemy scores apparently easy gains, thereby drawing them out for an expanded engagement across a more extensive front. Surprisingly, many of the manipulatory tactics used against overwhelming odds, such as feeding the enemy's arrogance and spreading the enemy out, may also provide a key. Finally, where the enemy is contained or trapped, an outlet should be left (so as to prevent a sudden determination to fight to the death, which may result on fatal ground without escape, as Sun-tzu noted).[107] Conversely, if trapped oneself, elite troops, acting as if deranged, can be employed to disrupt the enemy and create an opening.[108]

The Unorthodox and Orthodox. The pivotal concepts of *ch'i*, the unorthodox, and *cheng*, the orthodox, apparently originated with

Sun-tzu, for whom they are the very essence of transcendent strategy and mark only the most astute commanders. Fundamentally, things that are orthodox—whether they be forces, tactics, or strategies—obey the basic rules that govern general situations, whereas the unorthodox deliberately (but not simply or naively) runs contrary to normal expectation and thus entails the element of surprise. It is the latter that accounts for the surpassing effectiveness of unorthodox tactics.

However, the ability to envision and employ appropriate unorthodox tactics within particular contexts requires genius, as will be seen from the lengthy commentary appended to the intriguing chapter entitled the "Unorthodox and Orthodox." While opposites, such as rest and motion, may be characterized by the orthodox/unorthodox pairing, the successful creation of overwhelming tactics is far more complex, taking the practitioner into the realm of employing the formless against the formed, exploiting strength through weakness, and subverting superiority through complimentary deficiencies. Although Sun Pin only focuses upon this doctrine in a single chapter, the principles may be seen in several earlier chapters as well in his choice of tactics for defeating various formations: segmenting the troops, mounting flank and encircling attacks, and generally manipulating the enemy to render it vulnerable to unexpected measures. However, his lengthy exposition, which synthesizes the original concepts with a Taoist understanding of the formed and formless, appears in the second half of the book and may well represent the integrated product of his disciples' analytic work.

Formations. In the first half of the *Military Methods* Sun Pin discusses the nature of formations, while in the second half the book provides an integrated analysis of ten different formations, with their advantages and countermeasures. Numerous individual formations are also briefly correlated with their applicable situations in many chapters throughout the entire work. However, of particular interest is Sun Pin's use of the sword (and later the arrow) as an analogy for the nature of an effective formation. Essentially, there has to be a substantial structure or basis coupled with a sharp, piercing edge (since swords were primarily thrusting, rather than slashing, instruments in China). In the chapter entitled "Eight Formations," Sun Pin asserts that formations must be suited to the topography and that their employment should generally adhere to the principle of dividing one's forces into three operational groups, with one to execute the formation, the others to be reserve units and unorthodox forces. Unfortunately, he never elucidates the eight formations or indicates in what ways they are individually suited to various configurations of terrain. However, other chapters specifically mention several formations: The Awl, Wild Geese, Fierce Wind, and Cloud Array appear in the first half, with some basic characteristics (such as the Cloud Array being designed for arrow warfare). Chapter 16, which begins the second half of the book, enumerates ten different deployments or formations. (The Chinese character is the same and the meaning not clearly segregatable, but a distinction should be made between formations as theoretical organizational constructs and as actual deployments.) The Wild Geese and Awl formations reappear among these ten, but the others are primarily a question of shape or array, such as circular, hooked, dense, and sparse. Additional ones scattered about the book in various chapters include an "extended horizontal array," a "basketlike" deployment, a "sharp hooking array," and the "full" and "vacuous." Preferred modes of employment are indicated for some of them—such as the "square" formation being the means to solidify control and the "circular" to facilitate turning movements—but generally the discussion of them is terse and lacks any real explanation of their shape or mode of action. Later military compilations provide some odd renditions for them, but extensive research will be required before even minimally accurate reconstructions become possible.

Based upon these distinctive formations, it is obvious that organization, segmentation, and articulated deployment are all essential elements in any commander's tactics. From other writings the basic outlines of organization, such as by squads of five and companies of 100, are clear and to be expected in the state of Ch'i as well. If the training techniques of Sun Pin's era were at all like those preserved in the *Six Secret Teachings* and similar writings,[109] the troops would have been well drilled and easily capable of responding on signal to deploy in any particular formation and subsequently change into a different one. The commander's responsibility is to choose appropriately, providing himself with advantageous forms while denying the enemy an opportunity to attack or find an easy opening. Some forms, such as the diffuse formation, are obviously designed to allow the enemy to penetrate and become trapped, particularly as Sun Pin otherwise advises that the middle should be kept open or void. Moreover, based on the arrow analogy, normally the general should place his best troops at the front so that their combat power is greatest there, even though large numbers of soldiers are kept in reserve.

Command and Control

The Commander's Qualifications

Insofar as the commanding general controlled the power that could ultimately decide a kingdom's fate in the Warring States period and had to formulate strategy as well as execute complex tactics during lengthy campaigns such as Kuei-ling and Ma-ling, every ruler had to exercise great care in evaluating and appointing qualified men to the position.[110] No doubt the candidate's loyalty was paramount in the ruler's mind: Once deployed in the field and consequently in command of increasingly loyal troops, an ambitious or disaffected general could easily topple the civil authorities. Moreover, the king would be doubly endangered to the extent that the best troops had naturally been committed to combat. Conversely, having been appointed, the general should (as all the military writers emphasized) not only enjoy the ruler's complete confidence but also be empowered with the sole authority to conduct military affairs as his judgment might dictate. After formally accepting his position and taking active command, he could not permit any interference in field operations by the ruler or other high-ranking government officials.[111]

Basically, the commanding general (and virtually any officer exercising authority over military units) should be qualified for his position by his personal characteristics, intelligence, knowledge, and command skills. He should be a man of Virtue[112] in every sense: benevolent, courageous, righteous, incorruptible, and caring. Moreover, he must not only manifest positive characteristics but must also be free from the innumerable character flaws that can doom campaigns and easily be exploited by astute enemies: arrogance, greed, frivolity, cowardice, indecisiveness, laziness, slowness, brutality, selfishness, argumentativeness, carelessness, doubt, irascibility, and dejection.

Intelligence, knowledge (gained from study and experience), and the wisdom to make appropriate evaluations are also minimal requirements. Furthermore, the general has to exercise command and control skills so that the army will be united, disciplined, submissive, and spirited. He must be awesome (to evoke respect and obedience) but not brutal; treat the men well, evincing concern but without forfeiting their martial spirit; elicit great effort without exhausting the army, thereby inevitably dooming it to defeat; confident so that people will trust his orders; and decisive, rarely changing his commands or directions.

Selecting, Training, and Controlling the Troops

The era of basing qualifications for military duty upon increasingly severe standards is frequently identified with Wu Ch'i in Wei about the end of the fifth century, even though he was perhaps only articulating a trend already visible in the stronger states. Sun Pin apparently assumes that the men will be at least minimally qualified for military duty, although he also emphasizes using elite units composed of selected men for the critical tasks of spearheading assaults and penetrating formations. Moreover, he seems to disparage the role of large numbers alone, deemphasizing the importance of the masses because they cannot be relied upon for success. However, his view is not unequivocal, for in other places (in the first half of the book) he stresses that the masses must be gained and that they can provide the means to victory.[113] Apparently the overriding issue is whether they can be effectively controlled, motivated, and molded into a unified force in which all the units cooperate and sustain one another. However, the task of wresting victory apparently falls to selected, spirited forces.

Apart from the hierarchical organization imposed on the troops together with whatever form of mutual responsibility Ch'i may have employed, the main control method was the strict implementation of a system of rewards and punishments. Neither Sun Pin nor any of the other military writers could refrain from commenting upon the twin handles of rewards and punishments, but only the edges of a fully formulated doctrine are apparent in the several statements included in the *Military Methods,* primarily in the first half. He astounded King Wei by suggesting that rewards are not the most critical issue; however, he clearly assigned a major role to them in motivating men, causing them to forget death and willingly enter into battle. Moreover, he apparently subscribed to the idea that an excess of material wealth in society would blunt the effect of such rewards, although his economic vision is unclear since he also stressed enriching the state as a fundamental doctrine. Conversely, punishments make it possible to instill order and compel men to obey odious commands because they fear their superiors for their punitive powers. Clearly the punishments he envisioned were severe, although he does not discuss their psychology extensively. In all cases the actions to be required must be reasonable so that the men will be motivated to perform them and thereby earn the rewards, while the implementation of both rewards and punishments must be thorough, clear, and just.

Sun Pin's motivational psychology is clearly based upon the concept of *ch'i,* the essential breath of life that might, in human manifestation, be identified with martial spirit or morale, although the theory of vital breath and intentionality is rather more complex than a simple identification of the two.[114] Rewards were the primary method for stimulating men in the Warring States period, but Sun Pin also devotes an entire chapter to describing the motivational stages required as the army prepares to engage in battle. This remarkable material remains unique; other texts discuss *ch'i,* and a psychology of *ch'i* can be constructed from several of them, but Sun Pin's multistage analysis is not otherwise duplicated.

Weapons and Their Employment

Weapons are little discussed in the *Military Methods*—in contrast to, for example, the *Six Secret Teachings*—and then largely only as analogies for explaining the nature of military formations or deployments. In an intriguing passage Sun Pin attributes the invention of original weapons to the legendary Sage Emperors and separately also mentions the combined use of weapons to implement field tactics. However, throughout the book he mainly speaks about bows and crossbows; references to the latter are among the first found in any ancient text, while their employment at Ma-ling is the first engagement for which they are recorded. This perhaps reflects their recent appearance and the probable trend in his era to enthusiastically exploit their increased range and power despite a slower rate of fire. Of particular note is their combined use with the caltraps: An enemy ensnared in the latter, possibly foundering but certainly suffering severely reduced mobility, presents an easy target of opportunity.

Other Concepts and Principles

Scattered throughout the book are a number of concrete admonitions, such as not attacking an enemy backed by hills, similar to those found in the other military writings. For most of these the basic reasoning is apparent; others remain somewhat opaque and await the recovery of further material to understand how they might be integrated into an overall tactical science. (Empirically derived techniques and critical experience-based observations are of course more valuable than mere theory, however reasonable and convenient.)

The necessity for gathering military intelligence is mentioned but mainly as a reminder rather than as the burning issue it is in the *Art of War.* While in the field, lookout posts should be established; good communications should be effected and maintained. Fortifications provide the basis for defense, and temporary ones can be constructed from chariots, shields, caltraps, and similar equipment at hand. While they are not enough for victory, they may ensure survival in difficult circumstances, particularly if the army can occupy a ravine or other difficult terrain and certainly if the army is on fatal ground.

The component forces receive very little attention in the book, unlike, for example, the *Six Secret Teachings,* which ponders the nature of infantry, chariots, and cavalry and their appropriate modes of employment. Sun Pin mentions chariots—primarily with regard to their limitation to easy terrain—and the cavalry, but focuses upon the latter's application on more difficult terrain. In both cases they are to be divided into three operational forces in accord with his general principle, but he does not discuss how they might be employed in conjunction with the other arms or the various formations. The infantry is mentioned only in the context of exploiting ravines. In short, force components were either well understood and therefore little discussed, or Sun Pin simply tended to focus on more concrete issues— such as the detailed situations and suggested tactics found in Chapter 14—always assuming the missing portions of the *Military Methods* are similar in content and orientation to the extant portions of the text.

The Text and Its Author

While there are far fewer questions about Sun Pin and his relationship to the *Military Methods* than haunt Sun-tzu and the *Art of War*, certain issues deserve brief consideration. The translated work that follows was diligently pieced together from somewhat disordered piles of bamboo strips discovered in a previously unopened Han dynasty tomb in 1972.[115] While the bundled rolls presumably were perfectly ordered when stored away, particularly as they may have been deliberately entombed for future discovery by an unenvisioned posterity,[116] during the intervening twenty-one centuries the bindings disintegrated, leaving them in somewhat jumbled piles. Naturally the sequence of the layers provided some information, but even after numerous *pien* (or chapters) had been reconstructed, exactly how many should be considered integral portions of the *Military Methods* remains a subject of divisive debate. The committee of scholars entrusted with reconstructing the various texts originally identified thirty chapters as belonging to the *Military Methods* and divided them equally into upper and lower sections. They conveniently included the fifteen *pien* that have identifiable speakers, such as Sun Pin or T'ien Chi, in the upper portion and consigned those that lack any designated spokesman and tend to be more like monologues or discourses upon military affairs into the lower section. Consequently, some analysts felt that the second part, although discussing essentially the same topics in a consistent manner but being more expansive and detailed, did not properly belong with the first half. Accordingly, several modern Chinese and Japanese editions include only the first fifteen chapters found in our translation and completely ignore the second half, even though it contains important, even remarkable material. Conversely, those that include the second half do so with certain caveats, particularly as one or two chapters seem rather discordant.

Partly in response to this issue and to the development of additional insights with years of study, the committee issued revised, distinct volumes of the two parts in 1985.[117] In addition to including a new chapter entitled "Five Teachings" in the "true" Sun Pin portion and rearranging the strip order slightly, they consigned all the other material to a separately titled version, thereby indelibly characterizing it as extraneous. However, as the second portion of the text is clearly important and probably stems from Sun Pin's hand or from his school of thought, we have opted to translate the work in its entirety. Moreover, since virtually all the readily available modern Chinese editions (such as Chang Chen-che's outstanding *Sun Pin ping-fa chiao-li*) are based upon the original text, we have retained it here, supplementing appropriately only where significant differences have resulted and providing the new "Five Instructions" chapter at the end. Most of these issues are of interest solely to specialists, who may consult the Notes and the original, although not easily obtained, publications.

Physically, the present book has been reconstructed from some four hundred bamboo strips comprising about eleven thousand characters.[118] However, many of the strips are badly damaged, while others are clearly missing; therefore, given the fragmentary nature of the text, it is difficult to discern which chapters should be subsumed by Sun Pin's book and which ones belong to other works, known or unknown. Even though the character count of the reconstructed book is already almost twice the length of Sun-tzu's classic *Art of War* (of some six thousand words), the *Military Methods* probably contained more than thirty chapters. In fact, many heavily damaged chapters with only a few characters preserved contain indications of the entire *pien's* character count on the last strip. From these it is immediately obvious that the heavily fragmented portions have lost upward of several hundred words, leading to the conclusion that the original may have totaled over thirty thousand characters.

The present, lengthy text of thirty-one chapters still does not accord with the only dynastic reference to Sun Pin's *Military Methods*.

The entry contained in the "Treatise on Literature" in the *Han Shu* (History of the Han) indicates a book by "Sun-tzu of Ch'i in eighty-nine *pien* with four *chüan* [rolls] of maps." Without becoming buried in the morass of textual transmission issues, we must observe that there is no evidence whatsoever for determining whether the eighty-nine-*pien* book and the present *Military Methods* are the same work or not. The recently recovered text may constitute a portion of the eighty-nine chapters, or it may be a separate, entirely unrelated work that was preserved by descendants of the Sun family in Ch'i and then into the Han.[119] There are minimal indications that the *Military Methods* was slightly known in the Han dynasty since it is sometimes referred to and phrases from it appear in other works.[120] However, Ts'ao Ts'ao, the first commentator for the *Art of War,* made himself thoroughly conversant with every available military work in his quest to rule the realm, yet he never mentioned the *Military Methods* or cited any of its teachings, suggesting that any and all texts had disappeared from public provenance by the third century A.D.[121]

While disagreement continues about many aspects of the text, most scholars who have pondered its origins agree on certain conclusions. With regard to the physically recovered copy, the style of characters written by brush on the bamboo strips was popular from the late Warring States through the Ch'in; therefore, it either predates the Han or was written very early in the Han by an older person still employing this style.[122] From an almanac placed in the tomb, perhaps deliberately to indicate the date it was closed, the work must have been composed before 134 B.C.[123] Furthermore, based upon avoiding the taboo name of the emperor, some analysts have concluded that this particular copy was written during Han Kao-tsu's founding reign (between 206 and 194 B.C.)[124]

As to the date of the original *Military Methods* and to what degree this copy represents an embellished or otherwise altered version, it appears that the book is based upon Sun Pin's thought and teachings but was compiled and edited by his disciples. Sun Pin himself, perhaps wishing to emulate his famous predecessor, may have composed a prototext or developed a kernel of fixed teachings, and then his disciples or family members reworked it into the present form. As already noted, the first fifteen chapters are cast in the dialogue form common to other early writings (such as the *Mencius*) and always indicate the speaker as "Sun-tzu said" rather than simply launching into the subject or prefacing it with "Your subject says," as was common. Assuming that Sun Pin would not refer to himself in the third person, this format clearly implies these chapters were edited, if not actually compiled, by his disciples.

The second part of the book may have originally comprised extended discussions on concrete topics (such as found in the critical chapter on the unorthodox and orthodox) that Sun Pin had not put into dialogue form or that simply had not been formatted in the same way by his students. If these discussions were originally composed or preserved separately, even though by the same hand, the missing "Sun Pin said" would not be problematic. However, neither the absence nor presence of his name in this format constitutes a final, or even adequate, criterion for decisively judging the authorship and authenticity of the two sections.

Sun Pin obviously acquired disciples during his lifetime, for they are mentioned in the text questioning him about his discussions with King Wei and T'ien Chi.[125] Moreover, it is unlikely that he would have written the passage praising the Tao of the Sun family, although Sun Pin may have deliberately proceeded in unorthodox ways in order to establish the credibility of his book with posterity.[126] This suggests that his disciples may have finished the *Military Methods* at about the end of his life or compiled it shortly thereafter from memory to preserve the master's teaching.

A few references to major historical events embedded in one or two chapters within the book itself provide internal evidence from which a last date of composition can be deduced. The battles cited in the chapter entitled "Strengthening the Army," even if simply appended by a later hand, had all erupted by the end of the fourth century. The tactical conceptualizations and strategic priorities reflect prominent developments characterizing the middle Warring States period when Wei's hegemony was declining and Ch'i was ascending through the exercise of its military power. Sun Pin lived during a period of unremitting struggle, when states invariably committed themselves to the quest for survival by centralizing their administrative power and implementing radical policy reforms, as discussed previously. The *Mili-*

tary Methods mirrors these changes and the new emphasis upon enriching the state and strengthening the army. Moreover, as several scholars have asserted, the text advocates an aggressiveness in warfare not previously seen, focusing upon taking the initiative and targeting cities for attack.[127] Naturally these tactical principles were not created ex nihilo by Sun Pin himself; they continued themes already present in Sun-tzu's *Art of War*. However, the shifts in emphasis and aggressiveness accord with developments in the middle Warring States period. Sun Pin's tactics at Kuei-ling and Ma-ling consciously exploited the army's increased mobility (and possibly cavalry) and the greater strength suddenly provided by weapons such as the crossbow. Only the single disjointed chapter entitled "Treasures of Terrain" contains passages reflecting obviously later thought, indicating they were probably inserted by his disciples perhaps a few decades after his death.[128] When we consider these aspects in conjunction with the internal evidence and the tactics employed in the two victorious campaigns, it appears that the *Military Methods* probably assumed present form about the end of the fourth century or perhaps in the first decades of the third century.[129]

TEXT AND COMMENTARY

1
CAPTURE OF
P'ANG CHÜAN

Formerly (King Hui), Lord of Liang,[1] being about to attack (Chao's capital of) Han-tan, had General of the Army P'ang Chüan lead eighty thousand mailed troops to Ch'ih-ch'iu.[2] (King Wei), Lord of Ch'i, hearing about it, had General of the Army T'ien Chi lead eighty thousand mailed troops to . . . . . on their border.[3]

P'ang Chüan attacked Wey's city of [. . .].[4] General of the Army T'ien Chi [addressed Sun Pin: "P'ang Chüan has taken] Wey's city of [. . .]. Should we rescue them or not?"

[Sun Pin replied: "We should not."

T'ien Chi] said: "If we do not rescue Wey, what should we do?"

Sun Pin said: "I suggest that we go south to attack P'ing-ling. The town of P'ing-ling is small but the district is large; the population is numerous; and its mailed soldiers abundant. It is a military town in (Wei's) Tung-yang region, difficult to attack. We would thereby display something dubious to them.[5] When we attack P'ing-ling, Sung will be to the south, Wey to the north, and Shih-ch'iu will lie along our route.[6] Accordingly, since our supply route will be cut off, we will show them we do not understand military affairs." Thereupon they broke camp and rushed to P'ing-ling.

[Approaching P'ing]-ling, T'ien Chi summoned Sun Pin and asked: "How will this affair be managed?"

Sun Pin said: "Among the *ta-fu* of (our nearby border) cities, which ones do not understand military affairs?"[7]

T'ien Chi said: "The (*ta-fu*) of Ch'i-cheng and Kao-t'ang."[8]

Sun Pin said: "I suggest that we take the place where . . . . . stored ·. . . . . the two *ta-fu* (then) will . . . . . at (Wei's) cities of Heng and Chüan.[9] The area is crossed by regional roads in all directions, and is one where the (cities of) Heng and Chüan can [easily] deploy (their forces).[10] These wide roads are (already) occupied by chariots and soldiers. If our vanguard[11] remains stalwart, and our main force remains intact, (their forces) will move along the roads to attack and destroy the rear (of our two independent forces) and our two *ta-fu* may be killed."[12]

Thereupon (T'ien Chi) segmented off (the forces of) Ch'i-cheng and Kao-t'ang into two, and had them launch a flurried assault on P'ing-ling.[13] (Wei's forces from Heng and Chüan came forth) along the regional roads in a continuous wave to mount a (pincer) attack upon their rear. (The *ta-fu*) of Ch'i-cheng and Kao-t'ang fell prey to these tactics and were severely defeated.

General T'ien Chi summoned Sun Pin and inquired: "We unsuccessfully attacked and lost (our forces from) Ch'i-cheng and Kao-t'ang, which fell prey to (our) tactics and were defeated. How will affairs (now) be managed?"

Sun Pin said: "I suggest that you dispatch light chariots to the west to race to the suburbs of Liang in order to infuriate them. Divide up the troops (and dispatch a portion) to follow them, to show that we are few."

Thereupon T'ien Chi did it. As expected P'ang Chüan abandoned his supply wagons and arrived after a forced march at double pace. Without allowing P'ang Chüan's army any rest, Sun Pin attacked and captured him at Kuei-ling.[14] Thus it is said that Sun Pin fully realized (the Tao of the military).

Commentary

Since this chapter is extensively analyzed in the introduction, it need only be noted that it surprisingly focuses on Wei's attack on the minor state of Wey; in fact, the actual assault against Han-tan and that city's fate are never discussed. Taken in isolation, it would indicate that Ch'i's rescue operation was directed toward releasing the pressure on Wey's city rather than Chao's capital of Han-tan, as recorded in numerous historical sources. (No doubt the attack was a preliminary action to secure the corridor of approach to Han-tan, as discussed in the Historical Introduction.)

2
[AUDIENCE WITH KING WEI]

Sun Pin, in his audience with King Wei, said: "Now the military does not rely on an unvarying strategic configuration of power.[1] This is the Tao transmitted from the Former Kings. Victory in warfare is the means by which to preserve vanquished states and continue severed generations. Not being victorious in warfare is the means by which to diminish territory and endanger the altars of state. For this reason military affairs cannot but be investigated. Yet one who takes pleasure in the military will perish, and one who finds profit in victory will be insulted. The military is not something to take pleasure in, victory not something through which to profit.

"Move only after all affairs have been prepared. Thus one whose walled city is small but defense solid has accumulated resources.[2] One whose troops are few but army is strong has righteousness.[3] Now mounting a defense without anything to rely upon, or engaging in battle without righteousness, no one under Heaven would be able to be solid and strong.

"At the time when Yao possessed All under Heaven there were seven (tribes) who dishonored the king's edicts and did not put them into effect.[4] There were the two Yi (in the east), and four in the central states.[5] . . . . . (It was not possible for Yao) to be at ease and attain the profit (of governing All under Heaven). He was victorious in battle and his strength was established; therefore, All under Heaven submitted.

"In antiquity Shen Nung did battle with the Fu and Sui;[6] the Yellow Emperor did battle (with Ch'ih Yu) at Shu-lü;[7] Yao attacked Kung Kung;[8] Shun attacked Ch'e[9] . . . . . and drove off the Three Miao[10] . . . . . Kuan;[11] T'ang deposed Chieh;[12] King Wu attacked Chou;[13] and the Duke of Chou obliterated (the remnant state of) Shang-yen when it rebelled.[14]

"Thus it is said that if someone's virtue is not like that of the Five Emperors;[15] his ability does not reach that of the Three Kings;[16] nor his wisdom match that of the Duke of Chou, (and yet he) says, 'I want to accumulate benevolence and righteousness, practice the rites and music, and wear flowing robes and thereby prevent conflict and seizure,' it is not that Yao and Shun did not want this, but they could not attain it. Therefore they mobilized the military to constrain (the evil)."

Commentary

This chapter, the first substantive one on tactics in the book, opens just as Sun-tzu did in the *Art of War*—with a statement emphasizing the critical, life-and-death nature of warfare. Sun-tzu had previously said, "Warfare is the greatest affair of state, the basis of life and death, the Tao to survival or extinction. It must be thoroughly pondered and analyzed."[17] The initial paragraph similarly summarizes Sun Pin's attitude toward military affairs: Insofar as evil or threats to security remain in the world, the military and warfare are both necessary and unavoidable. The state's very survival depends upon understanding the principles of warfare, undertaking military preparations, and acting when necessary with commitment and resolve. From a more humanitarian perspective, military forces provide the only means to eradicate the great scourges of mankind, to act on behalf of others to eliminate evil and repression, just as Ch'i (in a very self-interested manner) did in rescuing Chao and Han from King Hui's forces. In this regard Sun Pin's sentiments generally accorded with the thoughts of the other military writers of the Warring States period, although there were sometimes significant variations in their perspectives.

The main thrust of the chapter is justified with citations of the historical character and inevitability of weapons and warfare (a theme reiterated in Chapter 9 where Sun Pin stresses that warfare is inherent to mankind). As indicated in the Historical Introduction, when the evil encroached upon the good, and especially upon the Sage rulers of antiquity, it was painfully discovered that only force could constrain them. Consequently, even such great paragons of Virtue as Yao and Shun were compelled to create weapons[18] and evolve tactics as they proceeded to mount military actions to extirpate the evil. Sun Pin cites numerous examples of these ancient, semilegendary conflicts (among tribes and totems) to support his argument, all of which are considered in our *History of Warfare in China: The Ancient Period.* Sun Pin thus directly contradicted the Confucians, who were led by the able but pedantic Mencius, who vociferously claimed that antiquity was an ideal period of civilization when Virtue alone held sway over civilization and rulers cultivated such pristine purity that even incorrigibles were shamed into submissive obedience.[19] As Sun Pin pointedly concludes, Yao and Shun wished to govern with benevolence and righteousness, but Virtue simply proved inadequate to the daunting task of contending with force and brutality.

At the same time Sun Pin warns, as Sun-tzu and many other military writers do, against the danger of becoming enthralled with warfare, of being seduced by the apparent profits, and of thereby dooming the state to extinction.[20] Although only explicitly raised once more,[21] the belief that frequent battles debilitate a state and that numerous victories can lead to ruin clearly underlies the entire *Military Methods.*

One final point not previously discussed: Apart from being physically prepared, the soldiers must embrace a moral cause, must fight out of and for righteousness. Only those properly motivated by Virtue (in addition to the immediate stimulus of rewards and fear of punishments) prove committed and effective in combat. Although Sun Pin does not again mention the importance of righteousness for the troops, he does stress the need for righteousness in the commander[22] and asserts that individual warriors will fail to qualify for assignment to the chariots if they lack a constellation of virtues.[23] Among the other military writings, Wu Ch'i spoke the most explicitly about righteousness: "In general to govern the state and order the army, you must instruct them with the forms of propriety, stimulate them with righteousness, and cause them to have a sense of shame."[24] Naturally the authors of such other works as the *Six Secret Teachings* and the *Ssu-ma Fa* assumed that the ruler would personally exemplify surpassing righteousness and that his cause would be pure: to rectify the evil, deliver the people from evil and terror, and bring peace and order to the realm.[25]

3
THE QUESTIONS OF
KING WEI

King Wei of Ch'i, inquiring about employing the military, said to Sun Pin: "If two armies confront each other, their two generals looking across at each other, with both of them being solid and secure so that neither side dares to move first, what should be done?"[1]

Sun Pin replied: "Employ some light troops to test them, commanded by some lowly but courageous (officer). Focus on fleeing. Do not strive for victory.[2] Deploy your forces in concealment in order to abruptly assault their flanks. This is termed the 'Great Attainment.'"

King Wei asked: "Is there a method (Tao) for employing the many and the few?"[3]

Sun Pin said, "There is."

King Wei said: "If we are strong while the enemy is weak, if we are numerous while the enemy is few, how should we employ them?"

Sun Pin bowed twice and said: "This is the question of an enlightened King! To be numerous and moreover strong, yet still inquire about employing them is the Way (Tao) to make the state secure. (The method) is called 'Inducing the Army.'[4] Disrupt your companies and disorder your ranks in order to accord with the enemy's desires. Then the enemy will certainly engage you in battle."

King Wei asked: "If the enemy is numerous while we are few, if the enemy is strong while we are weak, how should we employ them?"[5]

Sun Pin said: "The strategy is termed 'Yielding to Awesomeness.'[6] You must conceal the army's tail to ensure that (the army) will be able to withdraw. Long weapons should be in front, short ones to the [rear].[7] Establish roving crossbow (units) in order to provide support in exigencies. [Your main force] should not move in order to wait for the enemy (to manifest his) capabilities."[8]

King Wei said: "Suppose we go forth and the enemy comes forth. We still do not know whether they are many or few. How should we employ (the army)?"

Sun Pin said: "The method is called 'Dangerous Completion.'[9] If the enemy is well ordered,[10] deploy into three formations. One [should confront the enemy; two can][11] provide mutual assistance. When they can halt, they should halt; when they can move, they should move. Do not seek [a quick victory]."[12]

King Wei asked: "How do we attack exhausted invaders?"

Sun Pin said: " . . . . . You can make plans while waiting for them to find (a route) to life."[13]

King Wei asked: "How do we attack someone of equal (strength)?"

Sun Pin said: "Confuse them so that they disperse (their forces),[14] then unite our troops and strike them, do not let the enemy know about it. But if they don't disperse, then secure your position and halt. Do not attack in any situation that appears suspicious."

King Wei said: "Is there a Way (Tao) for one to attack ten?"[15]

Sun Pin said: "There is. 'Attack where they are unprepared; go forth where they will not expect it.'"[16]

King Wei said: "If the ground is level and the troops well ordered, but after engaging in battle they retreat, what does it mean?"

Sun Pin said: "It means that the deployment lacked a front."[17]

King Wei said: "How can we cause the people to always listen to orders?"[18]

Sun Pin said: "Always be sincere."

King Wei said, "Good. In discussing the army's strategic power, you are inexhaustible."

T'ien Chi asked Sun Pin: "What causes trouble for the army? What causes difficulty for the enemy? How is it that walls and entrenchments are not taken? How does one lose (the advantages of) Heaven? How does one lose (the advantages of) Earth? How does one lose the people? I would like to ask if there is a Way (Tao) for these six?"

Sun Pin said: "There is. What causes trouble for the army is the terrain. What causes difficulty for the enemy is ravines.[19] Thus it is said that three *li* of wetlands will cause trouble for the army;[20] . . . . . crossing through (such wetlands) will (result in) leaving the main force behind. Thus it is said, what causes trouble for the army is terrain; what causes trouble for the enemy is ravines. If the walls and entrenchments are not taken (it is because of defensive) ditches and defiles. . . . . . ."

. . . . . "What then?"

Sun Pin said: "Drum the advance and press them; (employ) ten ways to draw them out."[21]

T'ien Chi said: "When their deployment has already been determined, how can we cause the soldiers to invariably obey?"

Sun Pin said: "Be severe and show them the profits (of their rewards)."

T'ien Chi said: "Are not rewards and punishments the most urgent matters for the military?"

Sun Pin said: "They are not. Now rewards are the means by which to give happiness to the masses and cause soldiers to forget death. Punishments are the means by which to rectify the chaotic and cause the people to fear their superiors. They can be employed to facilitate victory, but they are not urgent matters."

T'ien Chi said: "Are authority, strategic power, plans, and deception urgent matters for the military?"[22]

Sun Pin said: "They are not. Now authority is the means by which to assemble the masses. Strategic power is the means by which to cause the soldiers to invariably fight.[23] Plans are the means by which to cause the enemy to be unprepared. Deception is the means by which to put the enemy into difficulty. They can be employed to facilitate victory, but they are not urgent affairs."

T'ien Chi angrily flushed: "These six are all employed by those who excel (in military affairs), and yet you sir say they are not urgent. Then what matters are urgent?"

Sun Pin said: "Evaluating the enemy, estimating the difficulties (of terrain), invariably investigating both near and far . . . . . is the Tao of the general. Invariably attacking where they do not defend, this is the army's urgency.[24] . . . . . are the bones."

T'ien Chi asked Sun Pin: "Is there a Way (Tao) to deploy the army and not engage in battle?"[25]

Sun Pin said: "There is. Amass (troops in the) ravines[26] and increase (the height of) your fortifications, being silently alert without moving. You must not [be greedy];[27] you must not get angry."

T'ien Chi said: "If the enemy is numerous and martial but we must fight, is there a Way (Tao)?"

Sun Pin said: "There is. Augment your fortifications and expand your [soldiers'] determination.[28] Strictly order and unify the masses. Avoid (the enemy) and make him arrogant. Inveigle and tire him. 'Attack where he is not prepared; go forth where he will not expect it.'[29] You must be prepared to continue (such actions) for a long time."[30]

T'ien Chi asked Sun Pin: "What about the Awl Formation? What about the Wild Geese Formation?[31] How does one select the troops and strong officers?[32] How about the strong crossbowmen running along and firing?[33] What about the Fierce Wind Formation? What about the masses of troops?"

Sun Pin said: "The Awl Formation is the means by which to penetrate solid (formations) and destroy elite (units). The Wild Geese Formation is the means by which to abruptly assault the (enemy's) flanks and respond to [changes].[34] Selecting the troops and strong officers is the means by which to break through enemy formations and capture their generals. Strong crossbowmen running along and firing are the means by which to take pleasure in battle and sustain it. The Fierce Wind Formation is the means by which to return. . . . . . Masses of troops are used to divide the effort and achieve victory."

. . . . .
Sun Pin said: "Enlightened rulers and knowledgeable generals do not rely on masses of troops to seek success."

Sun Pin went out and his disciples asked him: "What were the questions of King Wei and T'ien Chi, minister and ruler, like?"
Sun Pin said: "King Wei asked nine questions; T'ien Chi asked seven. They are very close to knowing all about military affairs but have not yet penetrated the Tao. I have heard that those who are always sincere flourish; those who establish righteousness employ military force;[35] those without adequate preparation suffer injury; and those who exhaust their troops perish. In three generations Ch'i will be troubled."[36]

Fragment[37]

"If one excels, then the army will prepare for him."
Sun Pin said: . . . . .

"When the eight formations have already been deployed . . . . . "

" . . . . . then Confucius . . . . . "[38]
" . . . . . If you are double the enemy, halt and do not move; be full and await them. Only then . . . . . "

"One who is not prepared will suffer difficulty from the terrain; one who is not . . . . . "

" . . . . . Warriors die but . . . . . is transmitted. . . . . . "

Commentary

This chapter purportedly records Sun Pin's discussion of the crux of military affairs with King Wei and T'ien Chi, the king's famous commanding general. They pose a total of sixteen theoretical battlefield situations for which Sun Pin suggests appropriate tactical principles. Their dialogue thus participates in the growing tradition of the early military writings, which probably commenced with a few laconic sentences in *Ssu-ma Fa,* of analyzing common confrontational situations. Sun-tzu's *Art of War* offered a number of abstract general principles for conceptualizing and managing battlefield circumstances, while Wu Ch'i, who was active a generation earlier than Sun Pin, described numerous tactical situations, analyzed their inherently significant factors, and correlated them with tactical measures.

Sun-tzu, who tended to conceive of these prototypical situations largely in terms of configurations of terrain, had still been encumbered with directing forces based upon chariots as the main battle element. By Wu Ch'i's time, the infantry had become far more significant as a result of the expanding scope of battle. Sun Pin thus lived at a time when mobility and flexibility were being increasingly realized, when logistics and the imposition of new forms of organization easily allowed fielding armies of 80,000 men or more, as already seen in Chapter 1, which describes the battles between Wei and Ch'i. Thereafter the *Six Secret Teachings,* which was probably composed within a century of the *Military Methods,* advanced extensive tactical analyses targeted to commonly confronted battlefield conditions. Many of these theoretical cases are found in two or three of these books, although the responses vary. (Insofar as this chapter isolates a large number of concrete situations, comparative passages must be consigned to the Notes.)

Sun Pin's replies are all brief, no doubt simply summarized in common with other works of the period, and generally emphasize themes earlier expressed by Sun-tzu: maintaining mobility, creating opportunities by manipulating the enemy, and attacking where the enemy is unprepared, where they have not mounted an adequate defense or do not expect a strike. Some commentators detect great advances in Sun Pin's thought, with a significant stress on aggressive action, but the chapter is more balanced than at first appears, and certain defensive measures, as well as manipulation of the enemy, are advocated.

4
CH'EN CHI INQUIRES ABOUT FORTIFICATIONS

陳　忌　問　壘

T'ien Chi[1] asked Sun Pin: "If our troops, being few, unexpectedly come into mutual contact with the enemy, how should we manage it?"

Sun Pin said: "Transmit orders to have our crossbows race and to spread out our bowmen. The crossbows . . . . . "[2]

[T'ien Chi said: "Our troops in the field improve their positions and establish fortifications][3] ceaselessly. How should it be done?"

Sun Pin said: "This is the question of an enlightened general. These are things which people overlook and are not urgent about. These are the means by which we urgently [erect field defenses and raise our troops'] determination."[4]

T'ien Chi said: "May I hear about it?"

Sun Pin said: "You may. Employing these measures are the means by which one can respond to sudden distress, occupy defiles and passes, and (survive) in the midst of fatal terrain.[5] This is the way I took[6] P'ang [Chüan] and captured Imperial Prince Shen."

T'ien Chi said: "Excellent. The affair has already passed, but the form[7] is not visible."

Sun Pin said: "Caltraps are employed as ditches and moats. Chariots are employed as fortifications. . . . . . [Protective enclosures on the chariots] are employed as parapets. Shields are employed as battlements.[8] Long weapons are placed next in order to rescue any breakthrough. Short spears are placed next (inside them) in order to act as [support] for the long weapons. The short weapons follow in turn in order to make it difficult for (the enemy) to withdraw and take advantage of their weaknesses. Crossbows are placed next in order to act as trebuchets.[9] In the middle there aren't any men, so fill it with . . . . . [10]

"Once the (deployment of the) troops has been determined concretely, establish the methods (for engagement). The *Ordinances* state: 'Place the crossbows behind the caltraps; only after (the enemy has entered them),[11] shoot them according to the (predetermined) method.' The top of the fortifications should be (manned) by equal numbers of crossbows and spear-tipped halberds. The *Methods* states: 'Move only after you hear the words of the spies you dispatched.' . . . . .

"Five *li* from your defenses establish lookout posts, ordering that they be within sight of each other. If (encamped on) high ground, deploy them in a square array; if (encamped on) low ground, deploy them in a circular perimeter.[12] At night beat the drums; in the daytime raise flags."

Fragments

T'ien Chi asked Sun Pin: "You have said that with regard to (employing the) army, Generals Hsün Hsi and Sun Chen of the state of Chin did not . . . . . "[13]

[Sun Pin said: " . . . . .] Do not fail to defend on account of the army's fear."

Chi-tzu said: "Excellent."

T'ien Chi asked Sun Pin: "You said that Generals Hsün Hsi and Sun Chen of the state of Chin . . . . . "

"..... is the deployment of a strong general."
Sun Pin said: "The officers and troops ....."

T'ien Chi said: "Excellent. You are a general to exercise sole command."

"..... speak and afterward hit the mark."
T'ien Chi respectfully asked: "What is the nature of the military?"

Sun Pin said: "The squad of five ....."

"..... made it clear in Wu and Yüeh and spoke about it in Ch'i, saying that one who knows the Tao of the Sun family will invariably unite with Heaven and Earth. The Sun family ....."[14]

"..... sought its Tao and thus the state long endured." Sun Pin said .....

[T'ien Chi] asked: "How is the Tao to be known?" Sun Pin said .....

"..... To know beforehand whether one will be victorious or not victorious is termed 'knowing the Tao.' [Before] engaging in battle, knowing where their ....."

"..... The means by which to know the enemy is called wisdom. Thus the army does not ....."

Commentary

This intriguing chapter may be translated rather differently depending upon whether it is interpreted as concretely describing the tactics actually employed to achieve the famous massacre at Ma-ling or as responding generally to T'ien Chi's questions about fortifications, although with reference to such events as Ma-ling.[15] Huo Yin-chang has correctly pointed out that the contents clearly reflect the transition from the earlier, limited, and more constrained Spring and Autumn warfare, in which armies depended upon essentially fixed-piece chariot encounters and static, permanent defenses, to a period in which infantry had begun to predominate.[16] With greater troop mobility and more lengthy campaigns, a need to speedily erect temporary field fortifications arose. Instead of small numbers of troops ensconced in a few fortified cities scattered about a relatively empty countryside—rather forbidding targets even for the advancing weaponry and technology of the Spring and Autumn period—campaign armies approaching force levels of 100,000 suddenly furnished strategically important objectives. Depending upon the segmentation and deployment of the troops and the nature of the battlefield, they might be confronted by numerically superior enemies or otherwise pressed to create temporary fortifications.[17]

Whether constructed for offensive or defensive purposes, fortifications had to exploit the terrain's configuration. On open ground a closed rampart would be necessary, such as could be quickly formed by "circling the wagons," just as in early days of westward expansion in the United States. Employing the chariots as their foundation, battlements and parapets would be immediately provided by the chariots' fixed walls and additional protective coverings, as well as the various large and small shields carried by the soldiers. The different weapons, each in accord with its special characteristics, could then be deployed. For example, in this chapter Sun Pin advocates placing the long weapons (spears and halberds) right inside the chariot perimeter, no doubt because they could then thrust at any troops that survived the hail of arrows to assault the rampart itself. Shorter weapons would be arrayed farther within to contain breakthroughs, with swords and axes probably held in reserve to decimate any soldiers who succeeded in penetrating the outer defenses. The interior was apparently left completely open, either on the generally recognized principle that open spaces are necessary to encompass enemy breakthroughs, or to allow flexibility in response (such as by rapidly moving units across to reinforce other areas).[18] The chapter thus describes the organizational principles for a formidable, systematized defensive array that could quickly be effected by a mobile force operating far from its home bastion.

Naturally these organizational principles could easily be adapted to the terrain's characteristics. Temporary fortifications could be erected along the sides of a valley, utilizing the natural protection afforded by the heights to the rear. Alternatively, the mouths of ravines, especially those fronting a valley or confined road, could be occupied by constructing ramparts across the openings. Fatal terrain, a term made prominent by Sun-tzu, would present somewhat more difficult problems, especially if the troops were caught between an advancing army and a body of water.

The modern scholar Chang Chen-che interprets this chapter in a much more specific sense and is to some extent followed by Huo Yin-chang, who emphasizes that the focal theme is creating ambushes, not mounting static defenses.[19] Chang believes that it describes not just generally applicable techniques but the specific tactical details of the battle at Ma-ling, preserving vital information not otherwise recorded. Consequently, he would incorporate a fragment into the account itself and see the entire chapter recounting the historical ambush. This view requires that the "open middle" not be found within the interior of closed, defensive fortifications, but that it constitute an area at least flanked or partially surrounded by them, such as a valley floor. Historically, Sun Pin's forces would have been deployed along the sides (and probably the far end) of the valley. The soldiers bearing the short weapons would also have been concealed outside, rather than behind, the fortifications to cut off the enemy's retreat.

Although some of Chang's textual justifications are perhaps strained, the condition of the text does not permit a definitive resolution of this question, and his view merits further consideration. A significant argument against its validity would be that the chapter explicitly commences with T'ien Chi inquiring about fortifications, not ambushes, and continues with Sun Pin's response on this exact theme. Furthermore, the chapter's title, while providing a less than reliable indication, mentions "fortifications." However, even if not specifically detailing this one battle, the chapter would still retain its value as the sole summary of Sun Pin's basic tactical principles for fortifications, measures that were probably employed at Ma-ling and other engagements. Because Chang's interpretation would change the translation significantly, casting it into the past tense and incorporating another fragment, the following variant for the key paragraph is provided for the contemplation of interested readers:

Sun Pin said: "Caltraps were employed as ditches and moats. Chariots were employed as fortifications. ..... [Protective enclosures on the chariots] were employed as parapets. Shields were employed as battlements. The long weapons were placed next in order to rescue any breakthrough. Short spears were placed next (inside them) in order to act as [support] for the long weapons. The short weapons were placed next in order to make it difficult for (the enemy) to withdraw and take advantage of their weaknesses. Crossbows were placed next in order to act as trebuchets. In the middle there weren't any men, so it was filled with a tree. When we were about to engage in battle, I wrote a message on the debarked section of the tree to be the target (for our bowmen)."

5

SELECTING
THE TROOPS

Sun Pin said: "The army's victory lies in selecting the troops. Its courage lies in the regulations.[1] Its skill lies in the strategic configuration of power. Its sharpness lies in trust.[2] Its power (*te*) lies in the Tao.[3] Its wealth lies in a speedy return. Its strength lies in giving rest to the people. Its injury lies in frequent battles."

Sun Pin said: "The implementation of Virtue (*Te*) is the army's great resource.[4] Trust is the army's clear reward.[5] One who detests warfare is the army's true kingly implement.[6] Taking the masses is [the basis for][7] victory."

Sun Pin said: "There are five aspects to constantly being victorious. One who obtains the ruler's sole authority will be victorious.[8] One who knows the Tao will be victorious. One who gains the masses will be victorious. (One whose) left and right are in harmony will be victorious. One who analyzes the enemy and estimates the terrain will be victorious."[9]

Sun Pin said: "There are five aspects to constantly not being victorious. A general who is hampered (by the ruler) will not be victorious. One who does not know the Tao will not be victorious. A perverse[10] general will not be victorious. One who does not use spies will not be victorious.[11] One who does not gain the masses will not be victorious."

Sun Pin said: "Victory lies in exhausting [trust],[12] making rewards clear, selecting the troops, and taking advantage of the enemy's [weakness]. This is referred to as King Wu's treasure."[13]

Sun Pin said: "One who does not obtain the ruler's (trust) does not act as his general."

[Sun Pin, in discussing the characteristics of generals, said:] " . . . . . orders. The first is called trust, the second loyalty, the third daring. What loyalty? To the ruler. What trust? In rewards. What daring? To eliminate the bad.[14] If someone is not loyal to the ruler, you cannot risk employing him in the army.[15] One whose rewards are not trusted in the hundred surnames will not regard as Virtuous.[16] One who does not dare eliminate the bad will not be respected by the hundred surnames."

Commentary

Every sentence in this brief but pivotal chapter is capable of almost infinite expansion, being laconic compressions of fundamental concepts. Furthermore, each one is subject to widely varying interpretations depending upon how the terms are defined and what other ancient texts might be chosen to provide further illumination.

The first paragraph summarizes a number of Sun Pin's observations on the nature of military affairs and in particular the essence of the army. The individual principles, which equally draw upon Suntzu's thoughts, are self-explanatory. However, a principle underlying the entire chapter (even though it may have been cobbled together by his disciples) is the fundamental importance of the people.

Warfare in this period imposed increasingly greater burdens on the people, particularly those compelled to serve in the growing infantry, and every wise ruler therefore sought to gain their physical and psychological allegiance. This would have been accomplished by following the Tao (tantamount to imposing a benevolent dictatorship); by not interfering with the occupations and seasonal activities of the people; by implementing moral policies stressing Virtue; by reducing the hardships imposed upon the people, such as the corvée labor duties; and by eliminating whatever proved harmful. "Gaining the masses"[17] also required minimizing the length and frequency of campaigns by pursuing an enlightened military policy that allowed for adequate rest. Accordingly, Sun Pin is now praised for having opposed frequent and extensive warfare, although he was hardly alone among the military thinkers, all of whom viewed warfare as the greatest affair of state.[18]

Certain characteristics are deemed critical to forging an effective military force: selecting the men (as previously discussed), imposing regulations and developing strict discipline, fostering trust and certainty so that commands will be immediately obeyed and rewards and punishments will be effective, and creating and nurturing internal coherence, unity, and harmony. Finally, Sun Pin offers a few observations about the character and qualifications necessary for a commanding general, although not as systematically as he does in his later chapters or as several other strategists did.[19] While knowledge (of the Tao) is obviously critical, the three traits mentioned here are trust, loyalty, and daring. Virtually every military thinker emphasized courage, and some stressed knowledge, but here Sun Pin emphasizes loyalty by explicitly stating, "If someone is not loyal to the ruler, you cannot risk employing him in the army." As already witnessed in the case of T'ien Chi (in the Historical Introduction) and briefly discussed in the Notes, in this period of rising military specialization and professionalism, a powerful general posed a potentially greater threat to the ruling family than the state's external enemies. Since to be effective the commanding general had to be granted sole authority over his troops, his temptation—especially when returning triumphantly—must have posed serious problems for every political leader.

6
LUNAR WARFARE[1]

Sun Pin said: "In the region between Heaven and Earth nothing is more noble than man. Warfare [is not a matter of a][2] single factor. If the seasons of Heaven, the advantages of Earth, and the harmony of men, these three, are not realized, even though one might be victorious, there will be disaster. For this reason they must be mutually relied upon to engage in battle;[3] (thereafter) only when it is unavoidable, engage in warfare.[4] Thus if one accords with the seasons to engage in warfare, it will not be necessary to employ the masses again.[5] If one engages in battle without any basis and gains a minor victory, it is due to astrological influences."[6]

Sun Pin said: "If in ten battles someone is victorious six times, it is due to the stars.[7] If in ten battles someone is victorious seven times, it is due to the sun. If in ten battles someone is victorious eight times, it is due to the moon. If in ten battles someone is victorious nine times, the moon has . . . . .[8] If in ten battles someone is victorious ten times, the general excels, but it leads to misfortune.[9] A single . . . . . "

[Sun Pin said:] " . . . . . There are five factors which preclude victory. If any one of the five is present, you will not be victorious.[10] Thus in the Tao of warfare (there are the following common situations): Many men are killed but the company commanders are not captured. The company commanders are captured but the encampment is not taken. The encampment is taken but the commanding general is not captured. The army is destroyed and the general killed. Thus if one realizes the Tao, even though (the enemy) wants to live, they cannot."

Commentary

This chapter is marred by obliterated characters and fragmented portions and was probably assembled from disjointed paragraphs. However, as the title suggests, it preserves valuable materials that reflect a belief in the effects of astrological influences upon military affairs. At the same time, it is prefaced by Sun Pin's statement that "nothing is more noble than man," orienting him at least partially within the tradition of the military thinkers who stressed man rather than otherworldly factors.

In the second paragraph Sun Pin apparently reprises the view that, lacking the sound basis of Heaven, Earth, and man enumerated in the first paragraph, victory in a majority of engagements stems from the favorable influence of one or another of the astrological factors.[11] While this hardly seems consonant with his overall work, the thoughts of his famous ancestor, or subsequent writings that explicitly rejected such beliefs and practices, it apparently mirrors some views common to his age. One approach closely intertwined military activities with stellar and lunar influences, times, and portents, perhaps itself being derived from the ancient tradition of employing divination before all significant undertakings, including battles and campaigns. However, from Sun-tzu on, the progenitors of the *Seven Military Classics,* even when acknowledging the persistence of such customs, redirected the focus onto men, explicitly rejecting these rituals and inclinations. Confucius, roughly contemporary with Sun-tzu, when questioned about how to serve the spirits, replied, "When you are not yet able to serve men, how can you serve spirits?"[12] Sun-tzu himself, who emphasized rational preparation for warfare and human activity, stated in a different context, "Advance knowledge cannot be gained from ghosts and spirits, inferred from phenomena, or projected from the measures of Heaven, but must be gained from men, for it is the knowledge of the enemy's true situation."[13] The T'ai Kung is historically noted for having emphatically rejected the validity of divination when King Wu's campaign against the Shang dynasty in 1045 B.C. encountered heavy weather. In his well-known view, the results for a usurper mounting a bloody revolution could hardly be auspicious.[14] Finally, in the opening sequence of the *Wei Liao-tzu* (also cited by Li Ching in the *Questions and Replies*), Wei Liao-tzu points out that military success and failure are unrelated to astrological influences, concluding, "From this perspective, 'moments,' 'seasons,' and 'Heavenly Offices' are not as important as human effort."

In the light of Sun-tzu's views and those of the military thinkers just subsequent to his age, it is somewhat puzzling that Sun Pin would still strongly embrace a theory of astrological influences. Were the statements less specific in their referents, they might be understood as simply expressing the thought that not being constantly victorious is a matter of luck. However, the chapter does conclude with his assertion that one who has realized the Tao of warfare will achieve total victory, implying that his analysis applies to those who engage in warfare without the proper foundation or military acumen. This does not negate the credibility he tangentially assigns to such influences, just the domain of applicability.

In the victory hierarchy preserved in the chapter's second paragraph, the moon clearly stands out as exerting the most powerful influence, no doubt because military affairs were considered *yin* activities. (The moon is the symbol par excellence for *yin*, the sun being the paramount embodiment of *yang*.) The missing characters perhaps indicate that the moon is approaching its maximum, which would be the most auspicious time for undertaking a campaign or attack, although how that would unilaterally benefit one side remains open.[15]

The last paragraph, which may well be misplaced in this chapter, lacks an unknown number of initial sentences. Probably five factors or examples illustrating these factors have already been discussed, thereby setting the context for the summary statements that have been preserved. Even though the connective "thus" is employed to introduce the topic that follows—the Tao of warfare—only three examples of presumably incomplete victories are iterated. Therefore, despite claims by some modern commentators (who conveniently ignore the discrepancy between the five factors and the recitation of only three cases, plus what can only be regarded as a complete victory), it appears that the connective is simply employed in a general sense to bridge rather than logically conclude, much as in Sun-tzu.[16] The translation accordingly reflects our belief that Sun Pin is merely discussing some general cases, not positing five types of incomplete victory. However, some modern analysts interpret the paragraph as illustrating Sun Pin's doctrine that only an enemy's extermination can constitute a true victory—a view that would certainly mark the brutality of the last part of the Warring States period. This would contrast markedly with Sun-tzu's belief that the greatest victory is achieved when the enemy's forces are preserved intact.[17] An alternative translation based upon their interpretation can be found in the Notes.[18]

7

EIGHT FORMATIONS

Sun Pin said: "When someone whose wisdom is inadequate commands the army, it is conceit.[1] When someone whose courage is inadequate commands the army, it is bravado.[2] When someone does not know the Tao nor has engaged in a sufficient number of battles commands the army, it becomes a matter of luck.[3]

"To ensure the security of a state of ten thousand chariots,[4] to bring glory[5] to the ruler of ten thousand chariots, and to preserve the lives of the people of (a state of) ten thousand chariots, only (a general) who knows the Tao (is capable.) Above he knows the Tao of Heaven; below he knows the patterns of Earth; within the state he has gained the hearts of the people; outside it he knows the enemy's true condition; and in deploying (his forces), he knows the principles for the eight formations. If he perceives victory, he engages in battle; if he does not perceive it, he remains quiet.[6] This is the general of a true king."

Sun Pin said: "As for employing the eight formations in battle: In accord with the advantages of the terrain, use appropriate formations from among the eight.[7] Employ a deployment which segments the troops into three, each formation having an elite front, and each elite front having a rear guard. They should all await their orders before moving. Fight with one (of them); reserve (the other) two. Employ one to attack the enemy; use the other two to consolidate (the gains). If the enemy is weak and confused, use your picked troops first to exploit it. If the enemy is strong and well disciplined, use your weak troops first in order to entice them.

"The chariots and cavalry that participate in a battle should be divided into three (forces), one for the right, one for the left, and one for the rear. If (the terrain) is easy, make the chariots numerous; if difficult, make the cavalry numerous.[8] If constricted, then increase the crossbows.[9] On both difficult and easy terrain, you must know the 'tenable' and 'fatal' ground.[10] Occupy tenable ground and attack (an enemy) on fatal ground."

Commentary

This chapter—one of the few preserved intact—largely represents a continuation and explication of tactical principles first advanced by Sun-tzu in the *Art of War*. It begins by noting Sun Pin's own observations on the general's character and qualifications, although presented in terms of problems posed by three major deficiencies. Wisdom, courage, and knowledge of the Tao are commonly found in the discussions of military writers; Sun Pin's unique insight lies in tying the latter to battle experience. When a general lacks both combat experience and the requisite knowledge, the army's fate becomes simply a matter of luck. "Knowledge of the Tao," previously mentioned here and a common emphasis in most military writings, includes that of natural conditions, the desires of the people, the enemy's true situation, and the principles of military deployment. (Sun Pin apparently allowed for gaining such knowledge through means other than battlefield experience; otherwise, there would be no need to couple them.)

Perhaps the most significant characteristic that qualifies generals to actually undertake command is the ability to refrain from engaging in battle when victory is not apparent. In other words, external pressures, fear for reputation, and similar factors should not be allowed to affect the objective analysis of the enemy or force the army into action. The underlying thought of Sun-tzu's *Art of War* is that combat should be undertaken only when victory is apparent. To this end he advised that generals might have to ignore their ruler's commands: "If the Tao of warfare [indicates] certain victory, even though the ruler has instructed that combat should be avoided, if you must engage in battle it is permissible. If the Tao of warfare indicates you will not be victorious, even though the ruler instructs you to engage in battle, not fighting is permissible."[11] In summary he states, "If it is not advantageous do not move. If objectives cannot be attained, do not employ the army."[12]

The *Six Secret Teachings* states, "In military affairs nothing is more important than certain victory. One who excels at warfare will await events in the situation without making any movement. When he sees he can be victorious, he will arise; if he sees he cannot be victorious he will desist."[13] In the *Wei Liao-tzu* an even broader principle governing rulers (but with implications for commanding generals) appears: "The army cannot be mobilized out of personal anger. If victory can be foreseen then the troops can be raised. If victory cannot be foreseen, then it should be stopped."[14]

Although Sun Pin differentiates and characterizes ten formations in Chapter 16 (entitled "Ten Deployments"), whether the "eight formations" in the title and the text refer to eight different formations, such as the circular, square, and angular; the outer eight positions of a nine-space square (three horizontal, three vertical); or simply a general rubric has stimulated considerable discussion.[15] From the context it appears that eight different formations are intended, each being chosen as appropriate to the configuration of terrain. However, definitive evidence is lacking, and later discussions, such as in the *Questions and Replies,* cannot be reliably projected back into antiquity.[16]

In this chapter Sun Pin again advises segmenting the army into three forces, employing one as an active strike force, while keeping two of them in reserve. Although the character he employs also means "defense," and their role in defending the army's primary position should not be slighted, no doubt their intended function is executing ever-evolving tactics and providing a flexible response to battlefield developments, such as through the implementation of the unorthodox measures Sun Pin discusses extensively in the last chapter of the *Military Methods.* As mentioned in the Historical Introduction, Sun Pin appears to have been the first to employ the "subtracted reserve" in his campaigns; this chapter provides the first systematic articulation of the concept and its tactical realization.

The chapter also points out the importance of conforming the formation, the disposition of the force components, to the terrain's configuration. Sun-tzu had previously classified the commonly encountered configurations of terrain according to their defining features and discussed appropriate tactical measures for both engaging the enemy and temporizing. However, he did not discuss types of formations, nor are there more than a few cursory references to them in historical records from the period. Sun-tzu said, "Configuration of terrain is an aid to the army. Analyzing the enemy, taking control of victory, estimating ravines and defiles, the distant and the near, is the Tao of the superior general. One who knows these and employs them in combat will certainly be victorious. One who doesn't know these nor employ them in combat will certainly be defeated."[17] While the *Wu-tzu* also addresses the viability of certain types of terrain for various forces, the most extensive discussion is found in three chapters in the *Six Secret Teachings:* "Battle Chariots," "Cavalry in Battle," and "Infantry in Battle." Another unique chapter in that work, entitled "Equivalent Forces," analyzes the relative effectiveness of various battle elements. While Sun Pin applies Sun-tzu's concept, his insights into force and weapon specialization represent an advance that clearly places him between Sun-tzu and the *Six Secret Teachings.* However, compared with the late Spring and Autumn period, armies in Sun Pin's age were marked by a greater diversity in battlefield elements, the crossbow and cavalry having appeared since Sun-tzu's era and the role of the infantry having extensively broadened.

The final sentences expound the fundamental principle for surviving and exploiting terrain: "On both difficult and easy terrain, you must know the 'tenable' and 'fatal' ground. Occupy tenable ground and attack (an enemy) on fatal ground." In this sentence "easy" and "difficult" terrain implicitly subsume all types of terrain; therefore the principle is simply that the general must know which positions are tenable, which positions inherently fatal. Tenable terrain (which originally appeared in the *Art of War*) is terrain that will sustain life, often identified by the commentators with heights and the sunny side of mountains, although it obviously encompasses any terrain that can easily be held to advantage. Furthermore, it is generally ground that does not contain, or lie near, any of the dangers and pitfalls that can endanger and destroy an army, such as ravines, rivers, and marshes.

Fatal terrain (or ground) is explicitly described in the *Art of War*: "Where, if one fights with intensity he will survive, but if he does not fight with intensity, he will perish, it is fatal terrain. . . . On fatal terrain engage in battle."[18] "If there is no place to go it is fatal terrain. On fatal terrain I show them that we will not live."[19] "On fatal terrain you must do battle."[20]

Sun-tzu strongly believed that soldiers would attain maximum fervor when confronted by imminent, apparently inescapable death, forced into an impossible situation on fatal ground: "Cast them into hopeless situations and they will be preserved; have them penetrate fatal terrain and they will live."[21] This suggests that the final, ambiguous injunction "to attack on fatal ground" should be understood as taking the offensive and mounting an attack whenever one finds oneself on fatal ground (if it is not possible to escape intact). This differs significantly from the common—and certainly equally valid from a tactical viewpoint—assertion that the injunction refers to targeting any enemy that ventures onto fatal terrain.

8

TREASURES
OF TERRAIN[1]

Sun Pin said: "As for the Tao of terrain, *yang* constitutes the exterior; *yin* constitutes the interior.[2] The direct constitutes the warp; techniques constitute the woof.[3] When the woof and the warp have been realized, deployments will not be confused. The direct (traverses land where) vegetation thrives; techniques (take advantage of where) the (foliage is) half dead.[4]

"As for the field of battle, the sun is the essence,[5] (but) the eight winds that arise must not be forgotten.[6] Crossing rivers,[7] confronting hills,[8] going contrary to the current's flow,[9] occupying killing ground,[10] and confronting masses of trees[11]—all these that I have just mentioned, in all five one will not be victorious.[12]

"A mountain on which one deploys on the south side is a tenable mountain; a mountain on which one deploys on the eastern side is a fatal mountain.[13]

"Water that flows to the east is life-sustaining water; water that flows to the north is deadly water. Water that does not flow is death.[14]

"The conquest (relationship) of the five types of terrain is as follows: Mountains conquer high hills; high hills conquer hills; hills conquer irregular mounds; irregular mounds conquer forests and plains.[15]

"The conquest (relationship) of five types of grasses is as follows: profusion of hedges, thorny brambles, cane, reeds,[16] and sedge grass.

"The conquest relationship of the five soils is as follows: Blue conquers yellow; yellow conquers black; black conquers red; red conquers white; white conquers blue.[17]

"Five types of terrain are conducive to defeat: gorges with streams, [valleys],[18] river areas, marshes, and salt flats.[19]

"The five killing grounds[20] are Heaven's Well,[21] Heaven's Jail,[22] Heaven's Net,[23] Heaven's Fissure,[24] and Heaven's Pit.[25] These five graves are killing grounds. Do not occupy them; do not [remain on] them.

"In the spring do not descend; in fall do not ascend.[26] Neither the army nor any formation should attack to the front right.[27] Establish your perimeter to the right; do not establish your perimeter to the left."[28]

Commentary

This chapter, which discusses configurations of terrain best avoided and provides some general observations about the respective values of various physical aspects, corresponds in the "earthly sphere" to Chapter 6, "Lunar Warfare." Many of the concrete contents are identical with those raised in Sun-tzu's *Art of War* and may also be found scattered throughout the *Seven Military Classics*. Clearly every strategist and commander had to be cognizant of such dangers as well as the principles for exploiting them.

The first paragraph apparently introduces a fundamental principle of terrain by conceptualizing the topography in terms of *yin* and *yang* and advising that both—entailing the orthodox and unorthodox— must be employed for effective command.

The second paragraph emphasizes deploying one's forces to utilize the sun's rays and accord with the prevailing winds. It then describes five situations in which advancing to engage an enemy would be tactically inadvisable, all of which stem from the *Art of War* or are otherwise common to ancient military thought.

The third and fourth paragraphs, which assign survival indices apparently based upon directional values, are the subject of intense speculation. The final paragraph also assigns directional priorities, which would equally seem to lack any grounds for absoluteness, although they might have relative validity in limited situations.

The first three of the remaining five paragraphs array heights, grasses, and soils in sequences based upon relative conquest power. Heights may be understood simply in terms of greater heights being strategically superior to lesser ones, while the grasses are ranked according to their strength and ability to act as obstacles sufficient to impede an advancing force. However, the five soils are characterized in terms of one of the conquest cycles found within the theory of five-phase ("element") correlative thought. Unfortunately, while the sentence is perfectly intelligible, the underlying meaning and implications remain to be understood, prompting Marxist scholars to condemn the sentence as magical musings. Whether Sun Pin truly believed in the efficacy of such relationships, whether he merely included them for theoretical purposes, or whether they are simply later accretions remains to be studied.

Finally, Sun Pin lists two further sets of five: terrains conducive to defeat and killing grounds. Most of these are also found in other military writings, and the names of the latter all appear in the *Art of War,* having become common military knowledge thereafter. As usual Sun Pin has not provided any definitions; therefore the only recourse remains commentaries for similar terms in the other military works. However, in every case the basic principle is simply that configurations of terrain that constrict an army's movement are to be avoided; otherwise, an astute enemy will bring the awesome long-range power of bows and crossbows to bear, easily decimating a confined target.

9

PREPARATION OF STRATEGIC POWER

Sun Pin said: "Now being endowed with teeth and mounting horns, (having) claws in front and spurs in back, coming together when happy, fighting when angry, this is the Tao of Heaven; it cannot be stopped. Thus those who lack Heavenly weapons provide them themselves. This was an affair of extraordinary men. The Yellow Emperor created swords and imagized military formations upon them. Yi created bows and crossbows and imagized strategic power on them. Yü created boats and carts and imagized (tactical) changes on them. T'ang and Wu made long weapons and imagized the strategic imbalance of power on them.[1]

"Now these four (formations, power, changes, and strategic imbalance of power) are the employment of the military.[2] How do we know that swords constituted (the basis for) formations? Morning and night they are worn but not necessarily used. Thus it is said, deploying in formation but not engaging in battle, this is (how) the sword constitutes (the basis for) formations.[3] If a sword has no edge, even [someone with the courage of] Meng Pen would not dare [advance into battle with it]. If a formation has no (elite) front,[4] anyone without the courage of Meng Pen who would dare command it to advance does not know the essence of military affairs.[5] If a sword lacks a haft, even a skilled officer would be unable to advance [and engage in battle]. If a formation lacks a rear (guard), anyone who is not a skilled officer but dares command it to advance does not know the true nature of military affairs. Thus if there is an (elite) front and rear (guard), and they mutually trust each other and are unmoving, the enemy's soldiers will invariably run off. Without an (elite) front and rear (guard), [the army] will be worn out and disordered.[6]

"How do we know that bows and crossbows constituted (the basis for) strategic power? Released from between the shoulders, they kill a man beyond a hundred paces without him realizing the arrow's path. Thus it is said that bows and crossbows are strategic power.

"How do we know that [boats and carts] constituted (the basis for tactical) changes? When high . . . . .

"How do we know that long weapons constituted (the basis for) the strategic imbalance of power? In attacking, they need to strike neither from high nor from below . . . . . [but still shatter] the forehead and destroy the shoulders. Thus it is said that long weapons are (the basis for) the strategic imbalance of power.

"In general, as for these four . . . . . those who gain these four survive; those who lose these four die. . . . . . [7] [They] must be complied with in order to complete their Tao. If one knows their Tao, then the army will be successful and the ruler will be famous.[8] If someone [wants] to employ them but does not know their Tao, [the army] will lack success.[9] Now the Tao of the army is fourfold: formations, strategic power, changes, and strategic imbalance of power. Investigating these four is the means by which to destroy strong enemies and take fierce generals.[10] What is seen up close (but) strikes far off is the strategic imbalance of power. In the daytime making the flags numerous, at night making the drums many,[11] is the means by which to send[12] them off to battle. Now these four are the employment of the military. [People] all take them for their (own) use, but no one penetrates their Tao.

. . . . .

"One who has an (elite) front is extremely cautious in selecting (troops for the) formations."

Commentary

This relatively well-preserved chapter envisions combat as an inherent aspect of human and animal behavior and therefore inevitable. Although Sun Pin was not the only military thinker to believe the origin of warfare is to be found in the very roots of antiquity,[13] his attribution of four fundamental military concepts—formations, strategic power, changes, and strategic imbalance of power—to the ancient cultural heroes credited with creating the elements and artifacts of civilization is uncommon.[14] However, two subsequent philosophical works contain similar views, one of which—the *Huai-nan tzu*—incorporates an illuminating, virtually identical passage:

> When the ancients employed the military it was not to profit from broadening their lands nor coveting the acquisition of gold and jade. It was to preserve those (about to) perish, continue the severed, pacify the chaos under Heaven, and eliminate the harm affecting the myriad people.
>
> Now whatever beast has blood and *ch'i*, has teeth and bears horns, has claws in front and spurs in back—those with horns butt, those with teeth bite, those with poison sting, and those with hooves kick. When happy they play with each other; when angry they harm each other. This then is Heavenly nature.
>
> Men have a desire for food and clothes, but things are insufficient to supply them. Thus they group together in diverse places. When the division (of things) is not equitable, they fervently seek them and then conflict arises. When there is conflict the strong will coerce the weak, while the courageous will encroach upon the fearful. Since men do not have the strength of sinews and bone, the sharpness of claws and teeth, they cut leather to make armor, and smelt iron to make blades. (In antiquity) men who were greedy, obtuse, and avaricious destroyed and pillaged All under Heaven. The myriad people were disturbed and moved, none could be at peace in his place. Sages suddenly arose to punish the strong and brutal and pacify a chaotic age. They eliminated danger and got rid of the corrupt, turning the muddy into the clear, danger into peace.[15]

A section in the eclectic *Lü-shih Ch'un-ch'iu* perceives man's individual weakness in the face of natural and human threats to be the basis for social order:

> Now as for human nature, their claws and teeth are inadequate to protect themselves; their flesh and skin is inadequate to ward off the cold and heat; their sinews and bones are inadequate to pursue profit and avoid harm; their courage and daring are inadequate to repulse the fierce and stop the violent; but they still regulate the myriad things, control the birds and beasts, and overcome the wild cats, while cold and heat, dryness and dampness cannot harm them. Isn't it only because they first make preparations and group together? When groups assemble together they can profit each other. When profit derives from the group, the Tao of the ruler has been established. Thus when the Tao of the ruler has been established, profit proceeds from groups, and all human preparations can be completed.[16]

Hsün-tzu, a philosopher of the late Warring States period best remembered for his assertion that human nature is inherently evil,[17] deduced that human desire is the root cause of conflict:

> Men are born with desires. When their desires are unsatisfied, they cannot but seek (to fulfill them). When they seek without measure or bound, they cannot but be in conflict. When conflict arises there is chaos; with chaos there is poverty. The former kings hated their chaos, so they regulated the *li* (the rites and forms of social behavior) and music in order to divide them, to nourish the people's desires, supply what the people seek, and ensure that desire does not become exhausted in things, nor things bent under desire.[18]

Most of the military writings justified military activities only to defend the state against aggression and rescue the people from the suffering inflicted upon them by brutal oppressors. Accordingly, the *Ssu-ma Fa* (which probably dates from between the *Art of War* and the *Military Methods*) states:

> In antiquity taking benevolence as the foundation and employing righteousness to govern constituted uprightness. However when uprightness failed to attain the desired (moral and political) objectives (they resorted to) authority (*ch'üan*). Authority comes from warfare, not from harmony among men. For this reason if one must kill people to give peace to the people, then killing is permissible. If one must attack a state

out of love for their people, then attacking it is permissible. If one must stop war with war, although it is war it is permissible.[19]

Two somewhat contradictory depictions of society under the guidance of the ancient Sages are thus seen even in the military writings. One view holds that it was an ideal age: The realm was tranquil, the people at peace in their occupations. Accordingly, violence and perversity arose only after a precipitous decline in the ruler's Virtue, with the ensuing disorder having to be quelled through forceful military measures.[20] This interpretation of history as a devolvement from a golden age tends to characterize warfare as essentially evil, frequently echoing Lao-tzu's famous dictum, "The army is an inauspicious implement."[21] Therefore, conscientious moral rulers can undertake punitive military actions only with great reluctance, in the full recognition that Heaven abhors such violence: "The Sage King does not take any pleasure in using the army. He mobilizes it to execute the violently perverse and to rectify the rebellious. The army is an inauspicious implement, and the Tao of Heaven abhors it. However, when its use is unavoidable, it accords with the Tao of Heaven."[22] Clearly "the Tao of Heaven abhors it" dramatically contrasts with Sun Pin's view in this chapter: "This is the Tao of Heaven; it cannot be stopped."

The second view, associated here with Sun Pin, perceives the Sages and cultural heroes as having arisen in response to the world's chaos, dramatically acting to quell disorder and create security for the people. Most of the military writings, including Sun-tzu's *Art of War,* stress the deadly importance of military campaigns, while works such as the *Six Secret Teachings* emphasize the necessity for the ruler to cultivate his moral worth and initiate military actions directed toward reducing the people's suffering.[23] The *Military Methods,* however, is less concerned with such objectives than with actual theory and the science of military art.

The concept that fighting is the natural result of anger has its counterpart in the military theorists' motivational psychology of warfare. In general, most of them discussed measures designed to stimulate men's spirits, nurture their *ch'i,* and coerce them into fervently engaging the enemy. Sun-tzu had earlier said, "What (motivates men) to slay the enemy is anger."[24] Based upon this chapter, Sun Pin would certainly agree that anger is the root cause of conflict.

As for the four essential military concepts, three were previously raised by Sun-tzu, with only the topic of formations not receiving any significant discussion until such later works as the *Wei Liao-tzu*. In Sun Pin's interpretation of history, all four originated in the minds of the Sages, who derived them from concrete weapons and inventions rather than from abstract images.[25] Thus, having fashioned the first sword,[26] the Yellow Emperor modeled the concept of formations upon the concrete sword rather than creating swords in concrete imitation of some nebulous image. Similarly, bows and crossbows, which act at a distance and provide the user with a critical distance advantage, were the basis for the concept and realization of *shih*, "strategic configuration of power."[27] Boats and carts, which provide mobility and make it possible to suddenly shift a deployment or race to a position, were the basis and means to realize change.[28] And finally, long weapons (and especially missile weapons), which facilitate striking from a relatively safe distance while closing to engage an enemy, convey a temporal tactical advantage when wielded against short weapons and thus underlie the concept and provide a basis for *ch'üan*, the "strategic imbalance of power."[29] When the four are fully understood and analytically employed, the astute commander can dominate the battlefield, being active rather than passive, and effect the essential tactical principles advocated by both Sun-tzu and Sun Pin.

10

[NATURE OF THE ARMY][1]

Sun Pin said: "If you want to understand the nature of the army, the crossbow and arrows are the model. Arrows are the troops; the crossbow is the general. The one who releases them is the ruler. As for arrows, the metal is at the front; the feathers are at the rear.[2] Thus they are powerful[3] and excel in flight, for the front [is heavy and the rear is light]. Today in ordering the troops, the rear is heavy and the front light, so when deployed in formation, they are well ordered, (but) when pressed toward the enemy, they do not obey.[4] This is because in controlling the troops men do not model on the arrow.

"The crossbow is the general. When the crossbow is drawn, if the stock is not straight,[5] or if one side (of the bow) is strong and one side weak and unbalanced, then in shooting the arrow, the two arms will not be at one. (Then) even though the arrow's lightness and heaviness are correct, the front and rear are appropriate, it still will not hit [the target].

" . . . . .

" . . . . . If the general's employment of his mind is not in harmony [with the army, even though the formation's lightness and heaviness are correct, and the front and rear are appropriate], they still will not conquer the enemy.

122

"Even if the arrow's lightness and heaviness are correct, the front and [rear] are appropriate, the crossbow drawn straight, and the shooting of the arrow at one, if the archer is not correct, it still will not hit the target. If the lightness and heaviness of the troops are correct, the front [and rear appropriate, and the general in harmony with the army, but the ruler does not excel], they still will not conquer the enemy. Thus it is said for the crossbow to hit the objective, it must realize these four.[6] For the army to be successful, [there must be the ruler], the general, and the troops, these [three]. Thus it is said that an army conquering an enemy is no different from a crossbow hitting a target. This is the Tao of the military. [If the model of the arrow] is complied with, the Tao will be completed. When someone understands the Tao, the army will be successful, and the ruler will be famous."

Commentary

Although this chapter has suffered considerable damage, the extensive missing portions can be fairly well reconstructed because of the parallelism employed as Sun Pin develops his argument. "Nature of the Army" emphasizes two points: First, attack formations must be powerful to the front; second, there must be coherence and unity among all members, and at all levels, of the command hierarchy.

In previous chapters Sun Pin has already asserted the importance of employing an elite, spirited front in order to explosively penetrate the enemy's ranks. This chapter further justifies this assertion with the analogy of the arrow with its metal point and therefore weight in the front and the comparatively lighter feathers in the rear. However, this should not be viewed as an absolute principle, for earlier he has advocated withholding two-thirds of the forces as a reserve capable of executing defensive and unorthodox tactics. But within the context of battlefield situations, his concept is clear: The necessary impact in an attack will be created by concentrating mass at the front of the chosen formation rather than retaining it in the rear.

The *Ssu-ma Fa*, a work that probably attained its final form just before or about Sun Pin's time and may even have been influenced by him, resolves this apparent inconsistency regarding the main employment of forces in similar fashion. On the one hand, it advises, "When the formation is already solid do not make it heavier. When your main forces are advancing, do not commit all of them, for by doing so you will be endangered."[7] But on the other hand, "When advancing the most important thing for the ranks is to be dispersed; when engaged in battle, to be dense, and for the weapons to be of mixed types."[8]

The arrow analogy is further developed in terms of the crossbow and archer to illustrate the need for cooperation and harmony among all three, as well as for correctness to characterize them individually. If the general acts as the crossbow, then his tactics must be correct and balanced; otherwise the troops will be used in an unbalanced fashion, defeat resulting. Furthermore, his intentions must penetrate to the officers and troops, and they must be harmonized with him so that there will be no gaps or dissension. An image frequently employed by the military writers to concretely depict their relationship is that of the mind and the four limbs. For example, the *Wei Liao-tzu* states, "Now the general is the mind of the army, while all those below him are the limbs and joints. When the mind moves in complete sincerity, then the limbs and joints are invariably strong. When the mind moves in doubt, then the limbs and joints are invariably contrary. Now if the general does not govern his mind, the troops will not move as his limbs."[9]

Similarly, the ruler, who stands as the archer in ultimately determining the campaign's direction and objectives, must also be correct and, presumably, in harmony with the general. Since several of the ancient strategists, including Sun-tzu and Sun Pin, strongly condemned any interference by the ruler once the commanding general had received his mandate, the question of the ruler's correctness would have referred primarily to his role in the early councils that determined the appropriateness and feasibility of undertaking military activities. This is why the general is the crossbow rather than the archer, although the analogy cannot be pressed too far because he must become both the archer and bow in the field. As the T'ai Kung states in the *Six Secret Teachings*, "Military matters are not determined by the ruler's command; they all proceed from the commanding general."[10]

The *Wei Liao-tzu* passage just cited suggests another possible interpretation for Sun Pin's original sentence: "If the general's employment of his mind is not in harmony [with the army]." Based upon the context of the entire "Nature of the Army," "with the army" has been supplied for the missing portion. However, Sun Pin may have simply been referring to the general's own mental state, as in the *Wei Liao-tzu* quotation; accordingly the sentence would end just after "harmony." Later, in Book III of *Questions and Replies,* Li Ching asserts, "Should the commanding general have anything about which he is doubtful or fearful, (the masses') emotions will waver. When their emotions waver, the enemy will take advantage of the chink to attack." Many of the military writings, including the *Military Methods,* emphasize the need for the general to be tranquil, marked by self-control, always unperturbed by anger. For example, in the *Three Strategies* the following summary appears: "The general should be able to be pure; able to be quiet; able to be tranquil; able to be controlled; able to accept criticism; able to judge disputes; able to attract and employ men; able to select and accept advice; able to know the customs of states; able to map mountains and rivers; able to discern defiles and difficulty; and able to control military authority."[11]

The accuracy and authenticity of Sun Pin's knowledge of the crossbow also bear noting. While archers using a hand-drawn bow can compensate for subtle variations, which they immediately perceive, perhaps subliminally, after years of experience, someone using a crossbow is much more insulated from such feedback due to the mechanical nature of the equipment. Sun Pin's analogy bears witness to the crossbow's advanced development in his era. Among the many factors that can cause the arrow (or bolt, although the former is more appropriate here because of the feathers) to fly forth off line are deviations in the groove (here synonymous with the stock or, more correctly, forestock); an imbalance in the strength of the bow (the arms); an arrow not centered in the middle of the string, resulting in uneven pull to one side; and any movement by the archer while the arrow runs down the length of the groove. Assuming that composite bows were employed and all factors reasonably brought under control to achieve optimal battlefield shooting (including properly shaped, weighted, and balanced arrows), such hand-pulled early crossbows would probably have had an effective range of 250 to 300 yards and would certainly have been deadly at 60 to 80 yards. Accurate flight over this distance would have depended upon the head of the arrow being comparatively heavy and the feathers properly affixed and oriented to create true in-flight dynamics. Although the rate of fire would have been many times slower than the ordinary composite bow, the arrow's force at impact would have been greater and probably capable of fatally piercing the leather armor worn in the Warring States period. When bronze "screaming arrows" were employed, an incoming barrage would have been terrifying indeed.[12]

11
IMPLEMENTING SELECTION

Sun Pin said: "The Tao for employing the military and affecting the people is authority and the steelyard.[1] Authority and the steelyard are the means by which to select the Worthy and choose the good. *Yin* and *yang* are the means by which to assemble the masses and engage the enemy. First you must correct the balance, then the weights, and then they will have already attained the standard.[2] This is referred to as being inexhaustible. Evaluate (talent and performance) by weighing them with the standard, solely to determine what is appropriate.[3]

"Private and state wealth are one.[4] Now among the people there are those who have insufficient longevity but an excess of material goods and those who have insufficient material goods but an excess of longevity.[5] Only enlightened kings and extraordinary men know this and therefore can retain them.[6] The dead will not find it odious; those from whom it is taken will not be resentful.[7] This is the inexhaustible [Tao]. ..... The people will all exhaust their strength. Those near (the ruler) will not commit thievery; those far away will not be dilatory.[8]

"When material goods are plentiful, there will be contention;[9] when there is contention, the people will not regard their superiors as Virtuous. When goods are few, [they will incline toward (their superiors);[10] when they incline toward them] then All under Heaven will respect them. If what the people seek is the means by which I seek (their performance), this will be (the basis for) the military's endurance. In employing the army, this is the state's treasure."[11]

Commentary

This brief chapter is fraught with problems, causing the commentators to offer widely divergent interpretations for both individual sentences and the chapter's overall meaning. Apart from the obviously damaged portions, in several places it appears the original copyist may have forgotten characters or even lost paragraphs. Consequently, the reader is urged to consult the Notes for a detailed discussion of the various possibilities, although the major differences are outlined here.

The first paragraph is reasonably clear: In order to effectively select and employ men, standards must be employed. Appropriate standards make it possible to judge not only an individual's abilities and moral qualifications but also his character and personal inclinations. Although various criteria and impromptu tests for evaluating men are found throughout the military classics,[12] only Sun Pin, in common with the contemporary Legalist Lord Shang, made the concept of employing standards explicit.[13] According to the fragmentary work attributed to Lord Shang, he said, "The Former Kings suspended the weights and beam for the steelyard and established the (measures of) feet and inches which, up to the present time, have been taken as the model because their divisions are clear. Now if one abandons the steelyard when deciding lightness and heaviness, or discards the foot and inch when measuring length and shortness, even the most perspicacious merchant would not employ such (a step) because it would be uncertain."[14]

The reference to *yin* and *yang* in the first section of the chapter commands attention. Unfortunately, Sun Pin has not indicated what sort of concrete measures they would entail; however, between them *yin* and *yang* encompass all possible government policies, tactics of battle, and methods for employing men. *Yin* may well refer to coercive measures, the dark virtue of government, and *yang* to rewards and incentives, the positive side. As translated, the basis—authority and the steelyard—would subsume rewards and punishments, commonly recognized as the twin handles of power in antiquity, under the ruler's authority. Without the power to punish and reward, his authority would lack an effective foundation, his commands would be unenforceable, and his appointees would be unaccountable.[15] However, as indicated in the Notes, the commentators all believe that *ch'üan*, the character translated as "authority," appears here in its original meaning of "steelyard" (or more accurately, "weights" for a steelyard), and thus the two characters translated as "authority/steelyard" refer either to steelyards in general or to two types of steelyards. However, without authority—the power to appoint men to office and impose rewards and punishments—the steelyard would be useless for selecting men. Again, the *Book of Lord Shang* provides an illuminating passage:

> The means by which a state is governed are three: The first is called law; the second is called sincerity; and the third is called authority (*ch'üan*). The law is what the ruler and ministers grasp together. Sincerity is what the ruler and his ministers establish together. Authority is what the ruler controls alone.[16] If the master of men fails to preserve (authority), then he will be endangered. If the ruler and ministers discard the law and rely on personal (opinion), there will inevitably be chaos. Thus by establishing the law, making distinctions clear, and not harming the law with the personal, (the state) will be governed. When the governance of authority is solely decided by the ruler, then he will be awesome. When the people trust in the rewards then affairs will be successfully completed; when they believe in the punishments then licentiousness will not have a beginning. Only the enlightened ruler loves authority and emphasizes sincerity, and does not harm the law with the personal.[17]

The second paragraph of "Implementing Selection" contains some statements of possibly great historical import depending upon how the sentences are understood. The translation is fairly literal, suggesting the enigmatic character of the original. The first sentence, "Private and state wealth are one," expresses Sun Pin's view that there should not be any distinction between state and personal material wealth. Rather, utilizing all sources of wealth for government ends, including military activities, is critical. Policies that exploit the imbalance between goods and longevity, between possession and desire, that marks human existence will successfully provide the state with inexhaustible resources. While most commentators would understand the sentence "Now among the people there are those who have insufficient longevity but an excess of material goods and those who have insufficient material goods but an excess of longevity" as describing a felt dissatisfaction with either life or material goods, Sun Pin appears to be simply contrasting their paired nature.[18] Obviously people with excess wealth have to be motivated in other ways, while those lacking the essentials for life can be coerced to disregard their lives in pursuit of government rewards. Assuming the introduction of an appropriate system of rewards and punishments, men who die under arms will not resent their deaths, and those from whom goods are confiscated will not complain. This will create a tranquil, orderly society in which the people all exhaust their energies, while not presuming upon their positions to commit thievery. (However, two modern commentators believe that "excessive wealth" refers to the riches remaining after a person's death and "the taking" of such riches, to a virtual inheritance tax. If so, this would be a remarkable historical record.)[19]

The last paragraph is interesting because it suggests, at least as translated, a doctrine starkly in contrast to the theories noted in the Commentary to the previous chapter that assert scarcity inevitably leads to conflict. Rather, Sun Pin appears to be saying that a plethora of material goods stimulates the people to disregard the incentives offered by government and thus their superiors, slighting the "Virtue" of the ruler.[20] Conversely, when goods are scarce and mainly available through government incentives, people will be compelled to focus upon them and will accordingly value them. Moreover, they will be forced to compete for them by behaving in a structured way, whereas when goods are widely available, people can freely accumulate and contend for them. A similar view is preserved in some portions of the *Book of Lord Shang:* "When punishments are made heavy and rewards made few the ruler loves the people, and the people will die for the rewards.[21] When rewards are made generous and punishments light, the ruler does not love the people, and the people will not die for the rewards."[22] Since only the state has the resources to satisfy the people, and if the implementation of incentive programs continually satisfies such basic conditions as credibility and timeliness, the state can manipulate the populace by exploiting the power of desire. Accordingly, it can attain the ultimate objective of military thinkers— being able to mount an extensive campaign effort and endure prolonged exposure to combat conditions.

12
KILLING
OFFICERS[1]

Sun Pin said: "Make rewards and emoluments clear and then . . . . .

"If you kill the officers, then the officers will certainly . . . . .

" . . . . . know it. Knowledgeable officers can be trusted, so do not allow the people to depart from them. Only when victory is certain does one engage in battle, but do not let the soldiers know it. When engaged in battle do not forget the flanks. . . . . .[2]

"If one invariably investigates and implements them, the officers will die. . . . . .

"If you treat them deferentially, then the officers will die (for you). . . . . . The officers will die, but [their names] will be transmitted. . . . . . If you encourage them with fundamental pleasures, they will die for their native places. . . . . . [If you importune them with] family relationships, they will die for the ancestral graves. . . . . . [If you honor them] with feasts, they will die for food and drink. . . . . . If you have them dwell in tranquility, they will die in the urgency (of defense). If you inquire about their febrile diseases, they will die [for your solicitude]."[3]

Commentary

The remnants of this chapter, identified by a notation on the back of the strips as "Killing Officers," are too fragmented to permit more than speculative analysis. However, additions and deletions made by the 1985 edition dramatically alter the contents, requiring a brief overview of the divergent interpretations as well as extensive annotation of the various contextual possibilities.

The very title of the chapter is "*Sha Shih*," "Kill *Shih*." By Sun Pin's era the scope of the term *shih* had changed dramatically; having originally designated the sons of the nobility and the minor nobility in the early Chou, it now encompassed "officers," "warriors," and "soldiers." As the infantry grew in importance and the status of the old nobility declined, talent and professional specialization became important to the state's survival, with the *shih* frequently becoming professional military men, serving as officers and eventually noncommissioned officers. Thus in such military writings as the *Art of War* and the *Six Secret Teachings*, the term is often contrasted with the character for "troops," the latter being composed of common (although free) men. With the dawn of the Warring States period, it frequently designated men noted for their courage and martial abilities and eventually came to simply refer to "soldiers" in general. The history of the term is neither linear nor simple, and it continued to appear in virtually all these meanings even late in the Warring States period.[4] However, in this chapter, even though infantry had already developed, Sun Pin seems to be speaking about the officers as distinguished from the ordinary troops. Based upon the original contents of the chapter (as will be discussed later), the title would seem to be best translated as "Killing Officers." However, because of the emphasis upon motivating warriors in the revised contents, the possibility arises that the chapter may well have instead discussed officers of great military prowess, capable of easily killing others—in other words, "killer (or elite) officers" or simply courageous, highly skilled warriors.[5]

The first sentence asserts one of the two or three fundamental principles found in most of the military writings for implementing an effective system of rewards and incentives: "Make rewards and emoluments clear." Only when the actual rewards are extensively promulgated and thoroughly understood will they motivate the soldiers to fervently advance into battle. The T'ai Kung said, "In general, in employing rewards one values credibility; in employing punishments one values certainty. When rewards are trusted and punishments inevitable wherever the eye sees and the ear hears, then even where they do not see or hear there is no one who will not be transformed in their secrecy."[6] The *Wei Liao-tzu* adds, "People do not take pleasure in dying, nor do they hate life, but if the commands and orders are clear, and the laws and regulations carefully detailed, you can make them advance. When, before combat, rewards are made clear, and afterward punishments are made decisive, then when the troops issue forth they will be able to realize an advantage, when they move they will be successful."[7]

The second sentence, "If you kill the officers," may refer to subjecting any transgressors among the officers to capital punishment. Some of the commentators draw attention to statements in Sun-tzu's chapter "Planning Offensives" and the *Wei Liao-tzu*'s "Army Orders." The former deplores the high casualty rate likely to ensue when a general recklessly attacks a fortified city: "If the general cannot overcome his impatience, but instead launches an assault wherein his men swarm over the walls like ants, he will kill a third of his officers and troops, and the city still will not be taken." The latter has caused considerable controversy over its interpretation and historical importance: "I have heard that in antiquity those who excelled in employing the army could (bear to) kill half their officers and soldiers. The next could kill thirty percent, and the lowest ten percent." As is evident, the terminology is similar, but the conclusion that follows shows that the passage refers to losses sustained on the battlefield rather than through the excessive and brutal application of capital punishment, as many Confucians and other detractors of the military writings have averred is the meaning of the passage.[8] It concludes, "The awesomeness of one who could sacrifice half of his troops affected all within the Four Seas. The strength of one who could sacrifice thirty percent could be applied to the feudal lords. The orders of one who could sacrifice ten percent would be implemented among his officers and troops." In contrast, from the original chapter makeup Sun Pin appears to be discussing the execution of disobedient officers as an example to the remainder.[9]

When punishment, especially capital punishment, is visibly inflicted upon an officer, the ordinary troops will become fearful. In fact, the military writings generally stressed that in the administering of punishment, the great and noble should never be spared, thereby striking awe into the troops and causing them to obey commands and fight aggressively in battle. Historically, Lord Shang emphasized universal and impartial application of the law and concomitant punishments to the noble, powerful, and famous. Wu Ch'i perished for his aggressive posture toward the entrenched nobility, while the early military classics advocated the need not only for impartiality but also for a dramatic striking down of the powerful. For example, the *Wei Liao-tzu* paraphrases a passage from the *Six Secret Teachings*: "In general, executions provide the means to illuminate the martial. If by executing one man the entire army will quake, kill him. If by rewarding one man ten thousand men will rejoice, reward him. In executing, value the great; in rewarding, value the small. If someone should be killed, then even though he is honored and powerful, he must be executed, for this will be punishment that reaches the pinnacle."[10]

The third and only substantial paragraph—eliminated in the 1985 edition—explicitly advances the rarely voiced but certainly assumed principle that "knowledgeable officers can be trusted." Whether this knowledge refers simply to their personal knowledge of rewards and punishments[11] (and thus to their certain fidelity) or to their command expertise is not clear. However, the thought that subordinates can be relied upon, and that the people should cling to them, is unusual, even though ancient Chinese forces all had strong chains of command and clear hierarchical organization.

The principle, also removed from the 1985 edition, that "only when victory is certain does one engage in battle" no doubt summarizes the Sun family approach to initiating combat. In his fourth chapter, which explicates the nature of victory and defeat, Sun-tzu stated, "The victorious army first realizes the conditions for victory, and then seeks to engage in battle. The vanquished army fights first, and then seeks victory." Sun-tzu further advocated making oneself unconquerable before attempting to conquer the enemy. Consequently, whenever prospects for victory prove uncertain, temporizing and defensive measures should be employed. Furthermore, the ruler should not mobilize the army except under appropriate conditions: "If it is not advantageous do not move. If objectives cannot be attained, do not employ the army. Unless endangered do not engage in warfare. The ruler cannot mobilize the army out of personal anger. The general cannot engage in battle because of personal frustration. When it is advantageous move, when not advantageous stop. Anger can revert to happiness, annoyance can revert to joy, but a vanquished state cannot be revived, the dead cannot be brought back to life."[12] Similar sentiments are also found in several other important texts from the period.[13]

"Do not let the soldiers know it," a concept that starkly contrasts with having knowledgeable officers, also reflects one of Sun-tzu's thoughts: "In accord with the enemy's disposition we impose measures on the masses that produce victory, but the masses are unable to fathom them."[14] Sun-tzu also said, "At the moment the general has designated with them, it will be as if they ascended a height and abandoned their ladders. The general advances with them deep into the territory of the feudal lords and then releases the trigger. He commands them as if racing a herd of sheep—they are driven away, driven back, but no one knows where they are going."[15]

The next fragment returns to the theme of rewards and punishments, pointing out that properly investigating affairs will result in appropriately implemented rewards and punishments and thus (it may be assumed) certainty among the officers. Therefore, just as mentioned in the last chapter, they will be willing to die in their positions without regret.

The concluding paragraph—whose translation represents our best attempt to puzzle out the overall meaning of a series of disjointed, fragmentary strips—dramatically changes the chapter's complexity, supporting the argument that it was originally about motivating warriors to die for the state. (In which case "kill" really stands for "death," referring to "death warriors," a possibility that existed in any case.) Basically this paragraph entails a discussion of motivation: what will stimulate men to act, what they will kill to protect. Apart from anger (which has already been discussed), the most prominent factors raised by the military writings encompass shame, rewards and punishments, family, and native place. Wu Ch'i, in Chapter 6 of the *Wu-tzu*, advised repeatedly feasting those distinguished for their courage and achievements in order to motivate other men to strive for similar honor for themselves and their families. However, the *Wei Liao-tzu* provides a succinct contextual passage isolating a number of these motivators:

> In order to stimulate the soldiers, the people's material welfare cannot but be ample. The ranks of nobility, the degree of relationship in death and mourning, the activities by which the people live cannot but be made evident. One must govern the people in accord with their means to life, and make distinctions clear in accord with the people's activities. The fruits of the field and their salaries, the feasting of relatives through the rites of eating and drinking, the mutual encouragement in the village ceremonies, mutual assistance in death and the rites of mourning, sending off and greeting the troops—these are what stimulate the people.[16]

The various military writings also emphasized that the commanding general should evince an ongoing solicitude for the welfare and physical condition of his men, being certain to set a personal example and share their hardships.[17] Treating them with appropriate courtesy, as Sun Pin indicates, will also accord them the requisite respect.

13
EXPANDING "CH'I"

Sun Pin said: "When you form the army and assemble the masses, [concentrate upon stimulating their *ch'i*].[1] When you again decamp and reassemble the army, concentrate upon ordering the soldiers[2] and sharpening their *ch'i*. When you approach the border and draw near the enemy, concentrate upon honing their *ch'i*.[3] When the day for battle has been set, concentrate upon making their *ch'i* decisive. When the day for battle is at hand, concentrate upon expanding their *ch'i*. . . . . [4]

" . . . . . in order to overawe the warriors of the Three Armies, the means by which one stimulates their *ch'i*. The commanding general orders . . . . . [5]

" . . . . . his orders, the means by which to sharpen their *ch'i*.[6] The commanding general then . . . . .

" . . . . . The short coat and coarse clothes, which encourage the warriors' determination, are the means by which to hone their *ch'i*.[7] The commanding general issues orders to have every single man prepare three days' rations.[8] As for the state's soldiers, their families . . . . .

" . . . . . hope. Emissaries do not come from the state; emissaries from the army do not go forth[9] in order to make their *ch'i* decisive.

The commanding general summons the commander of the camp security forces and informs him: 'As for food and drink, do not . . . . . in order to expand their *ch'i*.'"

Displaced Strips

" . . . . . the encampment. When (encamping) upon easy terrain,[10] you must be numerous and esteem the martial, for then the enemy will certainly be defeated. If their *ch'i* is not sharp, they will be plodding.[11] When they are plodding, they will not reach (their objective). When they do not reach (their objective), they will lose the advantage. . . . . . "

" . . . . . When their *ch'i* is not honed, they will be. frightened. When they are frightened, then they will mass together. When they mass together[12] . . . . . When their *ch'i* is not decisive, then they will be slack. When they are slack, they will not be focused[13] and will easily disperse. If they easily disperse, when they encounter difficulty they will be defeated."

" . . . . . If their *ch'i* is not . . . . . then they will be lazy. If they are lazy, it will be difficult to employ them. If it is difficult to employ them, they will not be able to converge on their objective. . . . . . "[14]

" . . . . . Then they will not know to constrain themselves. When they do not know to constrain themselves, affairs . . . . . "[15]

" . . . . . And do not rescue him; they themselves will die, and their families will be exterminated.[16] The commanding general summons his subordinates and exhorts them and then attacks. . . . . . "

Commentary

This intriguing chapter, which merits a study in itself, preserves the outlines of a psychology of battlefield motivation conceptualized in terms of *ch'i*, the essential pneuma (spirit) of life. The ancient military writers were acutely aware that an army's performance in battle—irrespective of its equipment, training, and general condition—would depend mainly upon the motivation and commitment of its soldiers. Numerous concrete measures were therefore systematically implemented to direct their preparation for combat from the earliest training stages through the final drumming of the advance and the actual engagement. Remnants of these are recorded in the *Seven Military Classics,* together with comments upon their effectiveness and manipulation. In "Expanding *Ch'i,*" Sun Pin has described the normative sequence of *ch'i* states that must be realized, although the actual techniques for attaining them unfortunately remain unknown because of the damaged condition of the strips.

Even though the six early Military Classics were composed over a span of two centuries or more, during which concepts and tactics evolved significantly, the extant texts generally recognize and agree upon the basic assumptions and underlying role of *ch'i.* The early *Ssu-ma Fa* contains an oft-quoted sentence: "In general, in battle one endures with strength, and gains victory through spirit."[17] Later the *Wei Liao-tzu* explicitly identified *ch'i* as the decisive component: "Now the means by which the general fights is the people; the means by which the people fight is their *ch'i.* When their *ch'i* is substantial they will fight; when their *ch'i* has been snatched away they will run off."[18]

Realizing that a loss of *ch'i* renders an army susceptible to defeat, the astute general focuses upon formulating strategies and tactical principles to manipulate the enemy, causing his forces to suffer just such a loss. This was one of the main thrusts of the *Art of War,* for Sun-tzu emphasized being active rather than passive, controlling the development of events rather than being compelled into movement by others:[19] "The *ch'i* of the Three Armies can be snatched away, the commanding general's mind can be seized. For this reason in the morning their *ch'i* is ardent; during the day their *ch'i* becomes indolent; at dusk their *ch'i* is exhausted. Thus one who excels at employing the army avoids their ardent *ch'i,* and strikes when it is indolent or exhausted. This is the way to manipulate *ch'i.*"[20]

Wu-tzu, for whom *ch'i* was one of the four vital points of warfare, believed it "ebbed and flourished," and therefore he advised a policy directed at "snatching away" the enemy's *ch'i.* A concrete example is seen in his strategy for defeating the state of Ch'u: "Ch'u's character is weak; its lands are broad; its government troubling (to the people) and its people weary. Thus while they are well ordered they do not long maintain their positions. The Tao for attacking them is to suddenly strike and cause chaos in the encampments. First snatch away their *ch'i*, lightly advancing and quickly retreating, tiring and laboring them, never actually joining battle with them. Then their army can be defeated."[21]

Troops far from home, which are physically tired and mentally exhausted, furnish easy opportunities that should be exploited.[22] However, if the enemy remains vigorous and spirited, two other possibilities can be either awaited or created: doubt and fear. About the former the strategists constantly warned: "Of disasters that can befall an army none surpasses doubt."[23] Therefore, launching an attack when the enemy is beset by doubt, when the commanding general is puzzled, was suggested by virtually all of them.[24]

Fear causes even greater paralysis in an enemy, rendering it easy prey. The T'ai Kung cited fear among the "perceived opportunities" for mounting an attack,[25] advising that "taking advantage of their fright and fear is the means by which one can attack ten."[26] The *Ssu-ma Fa* similarly states, "Attack when they are truly afraid, avoid them when they (display) only minor fears."[27] Wu-tzu frequently suggested measures designed to harry and frighten enemy troops, advancing to strike them when they grew fearful.[28]

Within this context the commanding general had, of necessity, to wrestle with the difficult question of how to instill spirit and develop courage. While much motivational theory, as discussed elsewhere, was founded upon the draconian implementation of rewards and punishments, a policy that essentially caused the soldiers to fear officers and death at their hands more deeply than the enemy,[29] there were several other measures designed to stimulate their *ch'i*, their spirit, at the appropriate stages. "Expanding *Ch'i*" provides the most systematic overview found in the extant ancient writings, even though the techniques themselves are lacking.

As background it should be noted that the *Ssu-ma Fa* expressly contrasted the basic, underlying attitude distinguishing the civil and martial spheres: The former is the realm of propriety and deference, whereas the latter is the realm of action and straightforwardness.[30] Thus, the true warrior's demeanor radically differs from ordinary demeanor, and his *ch'i* is appropriately constrained.[31] However, *ch'i* is explicitly understood to be a subject for manipulation, and there are appropriate techniques for raising it,[32] for "when the heart's foundation is solid, a new surge of *ch'i* will bring victory."[33]

The critical solution to the age-old question of what motivates individuals to fight in military engagements and measures to ensure that each man's effort is maximized were founded upon the basic perception that courage, which is a manifestation and function of *ch'i*, is the key. Assuming that life under arms has instilled the basic discipline and attitude, thereafter fostering an intense commitment that will not admit any possibility but fighting to the death is required. Two analytical illustrations found in the Military Classics vividly depict the nature of this commitment. In the earlier one, Wu Ch'i speaks about a "murderous villain": "Now if there is a murderous villain hidden in the woods, even though a thousand men pursue him they all look around like owls, and glance about like wolves. Why? They are afraid that violence will erupt and harm them personally. Thus one man oblivious to life and death can frighten a thousand."[34] Nearly two centuries later the *Wei Liao-tzu* echoed this sentiment: "If a warrior wields a sword to strike people in the marketplace, among ten thousand people there will not be anyone who doesn't avoid him. If I say that it's not that only one man is courageous, but that the ten thousand are unlike him, what is the reason? Being committed to dying and being committed to seeking life are not comparable."[35]

Although the *Ssu-ma Fa* discusses numerous measures for stimulating and nurturing spirit and resolve, and the other Military Classics also offer scattered suggestions, only the *Wei Liao-tzu* briefly characterizes the (idealized) progression of the soldiers' mental states as they advance into battle:

> Soldiers have five defining commitments: for their general they forget their families; when they cross the border they forget their relatives; when they confront the enemy they forget themselves; when they are committed to die they will live; while urgently seeking victory is the lowest. A hundred men willing to suffer the pain of a blade can penetrate a

line and cause chaos in a formation. A thousand men willing to suffer the pain of a blade can seize the enemy and kill its general. Ten thousand men willing to suffer the pain of a blade can traverse under Heaven at will.[36]

Against this background, which, although spanning nearly two centuries, outlines the common conceptions of motivation and *ch'i* in the ancient period, Sun Pin's chapter may be clearly interpreted. In the initial period when a campaign army is being formed and the troops assembled, an attitude of seriousness and constraint is required. The soldiers' willing commitment, their initiative and voluntary participation, must be stimulated. As the other writings note, baleful omens, fears, and doubts must be prevented. However, unbridled courage is equally disruptive, leading to excessive displays of bravado and a tendency to unruliness.[37]

As the army advances into the field, "order," understood as military discipline and the strict governance of the army's hierarchical organization of responsive units, must be maintained. Simultaneously, the soldiers' anticipation and commitment should be made sharper. This sharpening might best be understood as nurturing their general enthusiasm and commitment to the military enterprise because it would be detrimental for their *ch'i* to become too "sharp." As the Taoists point out, what is too sharp will easily become blunted and broken, especially if such sharpness is not wielded in action.

Once the day for battle has been set, a commitment to dying, the decisiveness just described, has to be created. All vestiges of fear must be eliminated, with the soldiers manifesting the desperate resolve that Sun-tzu elicits by thrusting them into hopeless situations. As the *Ssu-ma Fa* states, "When men have their minds set on victory, all they see is the enemy. When men have minds filled with fear, all they see is their fear."[38]

Finally, when they enter battle focused on the grim task of dealing death and dying themselves—rather than simply being swept along by high-spirited enthusiasm—the general must "expand their *ch'i*." Sun Pin's analysis emphasizes this distinction between ebullient, unrealistic enthusiasm and the decisive, committed to "death as if returning home" attitude required as the foundation for violent combat. He apparently recognized the danger of bringing men to this ultimate point, which, while absolutely necessary, may prove too brittle. Such fervency might be quickly dissipated by the initial moments of battle, perhaps following the first concerted thrust at the enemy, or even result in the sort of impulsive action that Wu Ch'i brutally condemned.[39] Therefore, the general's final task is to expand the soldiers' *ch'i* so that their courage will be sustained throughout the day's conflict rather than broken precipitously. It is perhaps a fundamental aspect of the "new surge of *ch'i*" that will bring the victory already noted.

Unfortunately, the details of Sun Pin's methods have been lost, with only glimpses remaining. Rewards and punishments, made the foundation of the ruler's and commander's awesomeness in the Military Classics, were the basic tools for stimulating the troops. Their clothes, which clearly offered little protection against the cold and doubtlessly even less comfort, coupled with the general hardship of military service, clearly "hone" their *ch'i* just like a whetstone sharpening a blade through slow grinding. Furthermore, issuing minimal rations would ensure that they would soon be compelled to aggressively wrest their supplies from the enemy and could not defer taking the initiative.

The dangers posed by any lack of courage have also been extensively noted. Sun Pin points out that an absence of "sharp *ch'i*"' results in defeat. When men are dispirited, they perform poorly and without vigor, presenting the enemy with a compelling opportunity. Similarly, when their *ch'i* has not been honed, they are susceptible to becoming frightened. Although Sun Pin's conclusion is missing, it is obvious that even an unperceptive enemy will rush to take advantage of the situation to frighten and scare them and thereafter attack. The *Wei Liao-tzu* captures these two aspects: "Those from whom the initiative has been taken have no *ch'i*; those who are afraid are unable to mount a defense."[40]

The last fragment gives evidence of Sun Pin's ascription of overarching stimulative power to the implementation of punishments. Clearly the members of each unit, whatever the level—although likely the squad of five was the basis—were bound by being mutually responsible for one another. Other works preserve regulations for a squad suffering the death of a member in battle; this chapter is unusual in probably referring to the capture of a unit member or possibly the unit leader. Men were thus coerced into fervently fighting by what has probably been the prime motivation over the millennia: their relationship with their fellow soldiers. Although the military writings generally discuss coercion in terms of the threat of capital punishment, sometimes supplemented by great rewards, many of the ancient thinkers also understood the power of shame in motivating men to fight.[41] Reinforcing the duress of personal punishment was the constant threat that an individual's family, equally at risk, could suffer for his failure. Punishment to be meted out for battlefield transgressions could be remitted only through outstanding individual and squad performance in subsequent combat.

The three fragments augmented from the 1985 revised bamboo strip edition (marked by footnotes 13–16) explicate the effects obtained when the requisite *ch'i* states are not appropriately realized. Whether in sum these fragments constitute a second paragraph or should be integrated with the topics that appear in the main body remains unclear; however, they obviously focus on the dimensions and effects of these *ch'i* states. As Sun Pin notes, when their *ch'i* is not decisive, soldiers will easily be scattered by enemy pressure, whereas when they are lazy (probably because their *ch'i* is not stimulated), they will not respond to their orders. Being unresponsive, it will be impossible to command them to initiate attacks or converge upon a designated target (in accord with Sun Pin's basic doctrine of segmenting forces and Sun-tzu's principle of segmenting and reuniting to concentrate mass upon weak points). Finally the sentence that speaks about constraints raises a critical issue for the military, one particularly emphasized in the earlier texts such as the *Ssu-ma Fa*. Without constraint and proper measure, the army will not only lack order and discipline but also wastefully dissipate its energy. Although unstated, the resulting inability to wage sustained battles will ultimately result in defeat.

14
OFFICES, I

Sun Pin said: "In general, to command troops,[1] make formations advantageous, and unify[2] the mailed soldiers, establish offices as appropriate to the body.[3] Implement orders with colored insignia;[4] (have the chariots) carry pennants to distinguish the relationships of things;[5] arrange the rows by [...]; organize the troops by hamlets and neighborhoods;[6] confer leadership in accord with the towns and villages; settle doubts with flags and pennons; disseminate orders with gongs and drums; unify the soldiers with tight marching;[7] and form them into close order, shoulder to shoulder.[8]

"To hunt down the (enemy's) army, use an elongated formation;[9] labor and exhaust them by constraining and contravening them.[10]

"To deploy the regiments, use an endangering [...] formation.[11]

"Engage in arrow warfare with the Cloud Formation.

"Defend against and surround (the enemy) with an entangled, flowing formation.[12]

"Seize the (enemy's) fierce beak with a closing envelopment.[13]

"Attack the already defeated by wrapping and [seizing them].

"When racing to rescue an army, employ a close formation.[14]

"In fierce combat, use alternated rows.[15]

"Employ [heavy troops] in order to attack [light troops].[16]

"Employ light (troops) in order to attack the dispersed.[17]

"When attacking mountain cliffs, employ the 'Arrayed Walls.'[18]

"On [...] terrain employ a square (formation).

"When you confront heights and deploy (your forces), employ a piercing formation.[19]

"For ravines . . . . . employ a circular formation.

"When engaged in combat on easy terrain, to effect a martial retreat, employ your soldiers (in a rear-guard action).[20]

"When your strategic power exceeds (the enemy's), when deploying to approach[21] them, employ a flanking attack on the wings.

"In ordinary[22] warfare when the (short) weapons clash, employ a sharp, piercing front.

"When the enemy is bottled up in a ravine, release the mouth in order [to entice] them farther away.[23]

"Amidst grasses and heavy vegetation, use *yang* (visible) pennants.[24]

"After being victorious in battle, deploy in formation in order to rouse the state.[25]

"To create awesomeness, deploy with mountains as the right wing.[26]

"When the road is thorny and heavily overgrown, use a zig-zag advance.

"To facilitate exhausting the enemy, use the Awl Formation.[27]

"In ravines and gullies, use intermixed elements.[28]

"When turning about and withdrawing, use measures to entangle (the enemy).[29]

"When circumventing mountains and forests, use segmented units in succession.

"Attacking state capitals and towns with water will prove effective.[30]

"For night withdrawals, use clearly (written) bamboo strips.[31]

"To maintain alertness at night, use passes with counterauthorizations.

"(To counter) raiding forces that forcefully penetrate the interior, use 'Death Warriors.'[32]

"To go against short weapons, use long weapons and chariots.[33]

"Use chariots to mount incendiary attacks on supplies under transport.

"To realize a sharp-edged deployment, use the Awl Formation.

"To deploy a small number of troops, use united, intermixed (forces); combining mixed (forces) is the means by which to resist being surrounded.[34]

"Rectifying the ranks and systematizing the pennants are the means by which to bind the formations together.

"Breaking apart and intermixing like clouds are the means by which to create a tactical imbalance of power and explosive movement.[35]

"Turbulent winds and shaking formations are the means by which to exploit doubts.[36]

"Hidden plans and concealed deceptions are the means by which to inveigle (the enemy) into combat.[37]

"Descending dragons (hidden power) and deployed ambushes are the means by which to fight in the mountains.[38]

"[Unusual movements] and perverse actions are the means by which to crush the enemy at fords.[39]

" . . . . . the troops, the means by which to . . . . .

"Being unexpected and relying on suddenness are the means by which (to conduct) unfathomable warfare.[40]

"Preventative ditches[41] and [. . .] formations are the means by which to engage (a superior enemy in battle) with a few troops.

"Spreading out the pennants and making the flags conspicuous are the means by which to cause doubt in the enemy.[42]

"The Whirlwind Formation and swift chariots are the means by which to pursue a fleeing (enemy).[43]

"When under duress,[44] shifting the army is the means by which to prepare for a strong (enemy).

"The Floating Marsh Formation and flank attacks are the means by which to fight an enemy on a confined road.[45]

"Slow movements and frequent avoidance are the means by which to entice an enemy to (try to) trample you.[46]

"Zealous training and whirlwind alacrity are the means by which to counter piercing thrusts.

"Solid formations and massed [battalions] are the means to attack an enemy's fiery strength.[47]

"Analytically positioning fences and screens is the means by which to bedazzle and make the enemy doubtful.[48]

"Deliberate tactical errors and minor losses are the means by which to bait the enemy.[49]

"Creating heavily disadvantageous circumstances is the means by which to trouble and exhaust (the enemy).[50]

"Patrolling in detail and verbal challenges are the means by which to (maintain security for) the army at night.[51]

"Numerous supply sources and dispersed provisions are the means by which to facilitate victory.[52]

"The resolute are the means by which to defend against invasion.

"(The various units) moving in turn are the means by which to pass over [bridges].

"[. . .] are the means by which to defend against . . . . .

" . . . . . are the means by which to repress (enemy) formations.

"Reckless withdrawals and (roundabout) entries are the means by which to release (the army) from difficulty."[53]

Commentary

Among the thirty-one chapters of the *Military Methods*, "Offices, I" remains the most difficult to fathom and translate because—apart from the opening paragraph—it consists of a series of disparate pronouncements upon concrete tactical principles. While the strips themselves have not suffered extensive damage, much of the language is reasonably obscure and requires imaginative reconstruction. Over the past two decades analysts and modern commentators have expended hundreds, perhaps thousands, of hours upon the text, seeking out collateral passages and similarly illuminating phrases. Many of their suggestions have been incorporated in our translation; others, however, have been foregone in favor of a more direct interpretation of the sentences.[54]

The chapter's title, "Offices, I," is subject to varying interpretations ranging from "Unifying the Army's Offices" to indicating that it is one of two versions of the text, the second now represented only by numerous fragments of identical sentences.[55] If the former view has merit, then the first paragraph may have made up the original kernel; Sun Pin either deliberately added the tactical principles to expand the fundamental thought, or they are simply later accretions. How-

ever, a third view with some validity asserts that *kuan,* the term translated as "offices," refers to the "sensory organs" of the body and that
the organs of government, including the structure and positions (or
officers) of the military, should be founded upon them. Citing the
first-paragraph clause "establish offices as appropriate to the body" as
a basis, this interpretation divides the chapter into five sections, with
each one postulated as reflecting an active aspect of the military.[56]
However, while all three possibilities enjoy a certain degree of textual
support, none is overwhelmingly convincing. The "organic" assumption appears particularly flawed, remaining a futile attempt to
artificially integrate concrete tactical principles around nonexistent
common themes.

The tactical principles enumerated in "Offices, I" easily fall within
the overall framework of Sun Pin's *Military Methods* and are consistent with the integrated discussion provided in the Historical Introduction. Many of them express concerns and solutions found in the
other military writings of the Warring States period; accordingly, numerous illustrative examples are cited in the Notes. Perhaps the most
significant question stems from the unusually large number of formations appearing in the chapter. Several commentators identify the various two character terms as in fact naming particularized formations
rather than designating general characteristics. However, while some
of them (such as the Cloud Formation) clearly belong to the former
category, in general it must be doubted that armies could have mas-
tered so many individual, complex formations. More likely, apart
from the basic ones (such as square, circular, and so forth) and some
particularized deployments designed to incorporate unique
strengths, they refer to temporal ordering arrangements that a general might effect using his basic building blocks. However, because
no other source or even combination of writings provides such an extensive list, the degree of tactical flexibility characterizing these ancient armies needs to be studied.

As the scope of warfare evolved and the numbers of men thrown
onto the battlefield increased dramatically, exerting effective control—a problem that also plagued Greek and Roman warfare for
much of its history and saw many famous generals still leading their
troops rather than being able to exercise command once a battle

commenced—became a critical issue. The answer in ancient China was organization, articulation, and an emphasis upon communication, which made the flexible execution of tactics possible. The foundation was grouping the men into squads, companies, and armies according to a rigid hierarchical system. Thereafter, the key became identifying them and developing their sense of larger unit identity, for then they could be commanded to act as integral parts of greater organizations and effect suitable tactical deployments. Insignia, mentioned in the first paragraph, identified men and made the flexible execution of commands possible.

Insignia were associated with every level, from the lowest to the highest, including dynastic houses. For example, the *Ssu-ma Fa* records that "for flags, the Hsia had a black one at the head representing control of men. The Shang's was white for the righteousness of Heaven. The Chou used yellow for the Tao of Earth. For insignia the Hsia used the sun and moon, valuing brightness. The Shang used the tiger, esteeming awesomeness. The Chou used the dragon, esteeming culture."[57] The *Wei Liao-tzu* preserves some indication of the extensive use of emblems: "Generals have different flags; companies have different emblems. The Army of the Left wears their emblems on the left shoulder; the Army of the Right wears their emblems on the right shoulder; the Central Army wears their emblems on the front of the chest. Record their emblems as 'a certain armored soldier' and 'a certain officer.' From front to rear, for each platoon of five lines the most honored emblems are placed on the head, the others accordingly lower and lower."[58]

Related to the emblems and insignia were personal flags: "From the commandant down, every officer has a flag. When the battle has been won, in each case look at the ranks of the flags that have been captured in order to stimulate their hearts with clear rewards."[59] Emblems, flags, pennants, and insignia not only reflect order; they also ensure it. They were numbered among the twelve essential matters leading to the Tao of certain victory: "The seventh, 'five emblems,' refers to distinguishing the rows with emblems so that the troops will not be disordered. The twelfth, 'strong troops,' refers to regulating the flags and preserving the units. Without the flags signaling an order, they do not move."[60]

Sun-tzu cited an earlier text on the origin of flags and drums—the two always being used in conjunction, although drums had to suffice at night—and discussed the immediate effect they have upon individual soldiers:

> The *Military Administration* states: "Because they could not hear each other, they made gongs and drums; because they could not see each other, they made pennants and flags." Gongs, drums, pennants, and flags are the means to unify the men's ears and eyes. When the men have been unified, the courageous will not be able to advance alone; the fearful will not be able to retreat alone. This is the method for employing large numbers. Thus in night battles make the fires and drums numerous, and in daylight battles make the flags and pennants numerous in order to change the men's ears and eyes.[61]

Wu Ch'i depicted the effects even more explicitly:

> Now the different drums, flags, gongs, and bells are the means to awe the ear; flags and banners, pennants and standards the means to awe the eye; and prohibitions, orders, punishments, and fines the means to awe the mind. When the ear has been awestruck by sound it cannot but be clear. When the eye has been awestruck by color it cannot but be discriminating. When the mind has been awestruck by penalties it cannot but be strict. If these three are not established, even though you have the support of the state you will invariably be defeated by the enemy. Thus it is said that wherever the general's banners are, everyone will go, and wherever the general points, everyone will move forward—even unto death.[62]

When all the elements are appropriately synthesized and implemented, the individual soldier's actions will be solely and explicitly directed by the flags and drums, creating the articulated maneuverability and essential unity needed to vanquish the enemy in combat. As the *Ssu-ma Fa* states, "When the Three Armies are united as one man they will conquer."[63] Wu Ch'i, who dramatically ordered a great warrior's execution despite his bravery because he acted impulsively, not on orders, appropriately concluded:

> In general it is a rule of battle that during daylight hours the flags, banners, pennants, and standards provide the measure, while at night the gongs, drums, pipes, and whistles provide the constraints. When left is

signaled they should go left; when right, then right. When the drum is beaten, they should advance; when the gongs sound, they should halt. At the first blowing they should form ranks; at the second assemble together. Execute anyone who does not follow the orders. When the Three Armies submit to your awesomeness and the officers and soldiers obey commands, then in combat no enemy will be stronger than you, nor will any defenses remain impenetrable to your attack.[64]

The importance of flags, pennants, and drums cannot be overestimated. Accordingly, the commanding general personally directed them: "Wu Ch'i said: 'The general takes sole control of the flags and drums, and that is all. Approaching hardship he decides what is doubtful, controls the troops, and directs their blades. Such is the work of a general. Bearing a single sword, that is not a general's affair.'"[65] Moreover, contrary to normal expectation that the strongest men would be placed in the forefront of the fighting, the flags—whose manipulation could decide the fate of hundreds, if not thousands—were given priority: "The basic rule of warfare that should be taught is that men short in stature should carry spears and spear-tipped halberds, while the tall should carry bows and crossbows. The strong should carry the flags and banners; the courageous should carry the bells and drums. The weak should serve in supply work, while the wise should supervise the planning."[66] Naturally in every situation and at all times—whether on difficult terrain, in tall grass, or along mountains—the flags had to be positioned to be clearly visible.[67]

Since an army's organization, discipline, and command would all be reflected by flags and drums, they immediately provided clues for evaluating enemy forces and estimating the possibilities for victory. For example, the *Six Secret Teachings* preserves an example of an assessment based upon the flags and drums combined with more conventional factors:

When the Three Armies are well ordered; the deployment's strategic configuration of power solid—with deep moats and high ramparts—and moreover they enjoy the advantage of high winds and heavy rain; the army is untroubled; the signal flags and pennants point to the front; the sound of the gongs and bells rises up and is clear; and the sound of the small and large drums clearly rises—these are indications of having obtained spiritual, enlightened assistance, foretelling a great victory.

When their formations are not solid; their flags and pennants confused and entangled with each other; they go contrary to the advantages of high wind and heavy rain; their officers and troops are terrified; their war horses have been frightened and run off, their military chariots have broken axles; the sound of their gongs and bells sinks down and is murky; the sound of their drums is wet and damp—these are indications foretelling a great defeat.[68]

In a famous dictum Sun-tzu himself said, "Do not intercept well-ordered flags; do not attack well-regulated formations."[69]

However, in contrast to feigned disorder, true confusion and chaos are an invitation to victory and a sign that even numerically superior forces can be successfully attacked, as Wu Ch'i observed in two well-known passages:

If their troops approach yelling and screaming, their flags and pennants in confusion, while some of their units move of their own accord and others stop, some weapons held vertically, others horizontally—if they pursue our retreating troops as if they are afraid they will not reach us, or seeing advantage are afraid of not gaining it, this marks a stupid general. Even if his troops are numerous they can be taken.[70]

If the enemy approaches in reckless disarray, unthinking; if their flags and banners are confused and in disorder; and if their men and horses frequently look about, then one unit can attack ten of theirs, invariably causing them to be helpless.[71]

In an attack on the enemy, in order to befuddle his targeting and confuse any estimations based upon the flags and pennants, several writers advised multiplying the number of flags and drums and otherwise manipulating them, just as Sun Pin does here by increasing their spacing. Typical admonitions include:

Change our flags and pennants several times; also change our uniforms. Then their army can be conquered.[72]

When at dusk the enemy is turning back while his soldiers are extremely numerous, his lines and deployment will certainly become disordered. We should have our cavalry form platoons of ten and regiments of one hundred, group the chariots into squads of five and companies of ten, and set out a great many flags and pennants intermixed with strong crossbowmen. Some should strike their two flanks, others cut off the front and rear, and then the enemy's general can be taken prisoner.[73]

Multiply the number of flags and pennants, and increase the number of gongs and drums.[74]

In the daytime set up five colored pennants and flags. At night set out ten thousand fire-cloud torches, beat the thunder drums, strike the war drums and bells, and blow the sharp-sounding whistles.[75]

Sun Pin himself stated, "In the daytime making the flags numerous, at night making the drums many are the means by which to send them off to battle."[76] Moreover, in the body of this chapter he added, "Rectifying the ranks and systematizing the pennants are the means by which to bind the formations together."

15
[STRENGTHENING THE ARMY]

King Wei asked Sun Pin: "In instructing me how to strengthen the army, none of the *shih* in Ch'i (espouse) the same Tao. . . . . . Some instruct me about government; some instruct me about [restraint] in making impositions;[1] and some instruct me to dispense provisions (to the people). Some instruct me about tranquility. . . . . . [Among what they] teach, what should [I] put into practice?"

[Sun Pin said]: " . . . . . None of them are urgent for strengthening the army."[2]

[King] Wei said: . . . . .
Sun Pin said: "Enrich the state."

King Wei said: "[How should I go about] enriching the state?"
" . . . . .

Fragments

" . . . . . generously. (This was the way) King Wei and King Hsüan conquered the feudal lords and even . . . . . "[3]

" . . . . . severely defeated Chao."[4]

" . . . . . will be victorious over them. This was the means by which Ch'i severely defeated Yen. . . . . . "[5]

156

" . . . . . The masses will then know it. This was the basis by which Ch'i severely defeated Ch'u's soldiers and turned around to . . . . . "[6]

" . . . . . [defeated Sung's] soldiers at Nieh-sang and captured Fan Kao."[7]

" . . . . . captured T'ang [Wei]."[8]

" . . . . . captured [. . .] Huan."

Commentary

The topic of this severely fragmented chapter remains just visible in King Wei's question: How should Ch'i go about strengthening its army? His complaint about the divisive approaches being suggested attests to the panoply of views espoused at the court and the Chi-hsia Academy, the result of Ch'i's avowed openness to exploring different disciplines in the quest to discover strategic paradigms and essential programs critical to the state's survival. Of particular importance here is the focus of the king's interest—not simply administrative measures or general approaches to government, but concrete steps that can be implemented to strengthen the army. Thus framed, every response is inevitably coerced into focusing on this crucial problem rather than enjoying the latitude to advise deemphasizing the army and instead strengthening the state through the cultivation of Virtue or similar practices. This topic fully accords with Sun Pin's discussion in "Audience with King Wei," where he notes that the great moral paragons suffered the encroachment of evildoers and were compelled to employ military force to impose order in the world, thus emphasizing the innate nature of conflict in human affairs. The fragments at the end of "Strengthening the Army," no doubt added by Sun Pin's disciples early in the next century,[9] strongly suggest that Sun Pin's advice, and that of the Legalists and military thinkers in general, was seriously heeded and generally implemented, becoming fundamental in reforming the state and nurturing a powerful military arm.

In King Wei's era, the mid-fourth century B.C., proponents of Taoism, early Confucianism, Mohism, early Legalism, the Logicians, the agriculturalists, *yin-yang* and other naturalists, and the military thinkers were eagerly in evidence. Among the several thousand mendicant advisers in Ch'i, virtually every philosophical perspective and political doctrine could certainly be found. Moreover, at this stage even the fundamental principles of what came to be identified in traditional Chinese thought as schools, such as Legalism, were still in the formative stage. Accordingly, the proponents of government affairs mentioned by King Wei might equally well have been individuals advocating proto-Legalist principles, such as the two figures subsequently regarded as the founders of Legalism, Shen Pu-hai in Han and Lord Shang in Ch'in, or perhaps early Confucians. In the fourth century B.C., Confucianism's chief representative would have been the youthful Mencius (371–289 B.C.), who is known to have visited Ch'i late in his career after failing to properly sway King Hui of Liang.[10] Naturally numerous minor figures—emphasizing one aspect or another of government practice—would have propounded their views in an attempt to influence the king and gain ministerial office, much as Tsou Chi early in King Wei's reign. The Confucians, in addition to emphasizing a rule of Virtue and the practice of righteousness, would have advocated minimizing the impositions on the people (identified by the king of Wei as the "proponents of frugality" in government) and the welfare of the people.[11]

Although "dispensing provisions" is generally identified as a Mohist doctrine,[12] it may well have been a temporary Confucian-oriented policy or one raised by the military thinkers since both emphasized the welfare of the people and the latter stressed the need to attract immigrants from other states, thereby increasing the population and size of the army.[13] Mo-tzu himself (fl. 479–438 B.C.) is traditionally equated with the doctrine of "universal love," but his philosophy was neither as simplistic nor as naive as often portrayed. Moreover, early in the Warring States period the Mohists, a highly organized and dedicated band of disciples, practiced their doctrine of trying to deter warfare and ameliorate suffering by venturing from place to place to offer their services to the besieged.[14] No doubt they were suitably rewarded if successful, but certainly their motivation cannot have been simply pecuniary, for they clearly risked their lives in mounting such defensive efforts. Furthermore, siege warfare in the Warring States period gradually shifted to favor the massive, well-equipped attacker that was compelled to capture the enemy's key cities because they had become military and economic centers and therefore tactically significant obstacles, thereby increasing the difficulty of the Mohist's self-assigned task.[15]

Finally, "tranquility" probably refers to the doctrines of the early Taoists just entering the formative period characterized by Chuang-tzu's appearance (ca. 399–295 B.C.).[16] Whether the proposals emphasized quietistic doctrines of simple inaction (such as identified with recluses and Yang Chu) or a more complex formulation that integrated military thought (as in the *Tao Te Ching*) can only be speculated upon.[17] Sun Pin's own views of course can be extrapolated from the principles found throughout the *Military Methods*, although (as noted in the Historical Introduction) apart from this chapter, which emphasizes "enriching the state," he never discusses such basic topics as the people's welfare.

The views advanced by these various schools are well documented in the secondary literature and easily accessible. However, insofar as the general principles and perspectives of the military thinkers in the Warring States period have received virtually no coverage, abstracting the core ideas from the *Seven Military Classics* will provide a context for Sun Pin's succinct reply: "Enrich the state." The beliefs synthesized in the *Six Secret Teachings*, *Wu-tzu*, *Ssu-ma Fa*, and *Wei Liao-tzu* clearly preserve late Warring States military thought and may be summarized as follows: Fundamentally, the state needs to garner the willing allegiance of the populace in order to convert it into enthusiastic, well-trained soldiers capable of defending the state and defeating the enemy. Only a reasonably prosperous, satisfied, and well-ordered populace—one free of onerous labor services and excessive taxes—will be physically and emotionally capable of undertaking the hardships of military duty.[18] Moreover, the state that focuses upon developing an adequate degree of material prosperity will be able to afford the vast expenditures and great waste of military campaigns, as well as the luxury of removing able-bodied males from the active farming population to undergo training and serve for prolonged periods.[19] In general, only benevolent governments that espouse such fundamental virtues as righteousness and sincerity, whose rulers cultivate Virtue and thereby develop personal charisma and power, will significantly affect the people, motivating those within the state to fight for it and those without to emigrate into it (particularly if uncultivated lands are offered as incentives).[20] The degree to which punishments rather than rewards might be emphasized varied among the theorists; however, the military thinkers believed in the strict implementation of rewards and punishments as well as the severe but impartial practice of law (in contrast to the Confucian emphasis upon personal morality). Organizational concepts such as the mutual guarantee system and the inescapable grouping by hamlets and villages held prominent places in their thought and were the basis for the draconian implementation of punishments both in civil and military life.[21]

16
TEN DEPLOYMENTS

In general, there are ten deployments: square, circular, diffuse, concentrated, Awl, Wild Geese, hooked, Dark Rising, incendiary, and aquatic. Each of them has its advantages.

The square deployment is for cutting.[1]

The circular deployment is for unifying.[2]

The diffuse deployment is for rapid (flexible) response.[3]

The concentrated deployment is to prevent being cut off (and taken).[4]

Deployment into the Awl Formation is for decisively severing the enemy.[5]

Deployment into the Wild Geese Formation is for exchanging archery fire.[6]

Deployment into the hooked formation is the means by which to change targets and alter plans.[7]

The Dark Rising deployment is for causing doubts in the (enemy's) masses and difficulty for his plans.[8]

The incendiary deployment is the means to seize (enemy encampments).

The aquatic deployment is the means to inundate the solid.[9]

The tactics for square deployment: You must thin out the troops in the middle and make those on the sides thicker. The reserve (ready response) formations are at the rear.[10] By thinning out the middle, the general can effect a rapid response.[11] By [expanding] and making [the sides] heavy, the general can cut (the enemy).[12] Retaining the (ready response) reserves in the rear is the means by which to . . . . . [13]

[The tactics for the circular deployment:] . . . . .

[In the tactics for the diffuse deployment] armor is scarce and men are few. For this reason make it firm. Martial prowess lies in the flags and pennants; showing (large numbers of) men lies in your weapons.[14] Thus (the soldiers) must disperse and maintain their internal separation. Make the flags, banners, and feathered pennants numerous; sharpen your blades to act as your flanks.[15] For them not to be compressed (by the enemy) when diffuse or surrounded when concentrated lies in exercising great care. The chariots do not race; the infantry does not run. The tactics for diffuse deployment lie in creating numerous small (operational) units. Some advance; others retreat. Some attack; others hold and defend.[16] Some launch frontal assaults; others press their developing weaknesses.[17] Thus the diffuse (deployment) is able to seize (the enemy's) elite forces.

The tactics for concentrated deployment: Do not augment the spacing (between the men). When they are compressed, gather your blades at the head of the formation and then extend it forward while the front and rear mutually preserve each other.[18] Amidst the changes [(of battle) do not alter it].[19] If the mailed soldiers are afraid, have them sit.[20] Use sound to direct them to sit [and arise].[21] Do not dispatch (any forces) after (enemy troops) that go off; do not stop those who come forth.[22] Some (of our troops) should attack their circuitous routes (of approach);[23] others should "insult" their elite troops.[24] Make them as dense as feather down without any gaps; when they turn about and retreat, they should be like a mountain.[25] Then the concentrated deployment cannot be taken.

Deployment into the Awl Formation should be like a sword.[26] If the tip is not sharp, it will not penetrate; if the blade is not thin, it will not cut; if the foundation is not thick, you cannot deploy the formation.[27] For this reason the tip must be sharp, the blade must be thin, and the foundation must be substantial. Only then can a deployment into the Awl Formation decisively sever (the enemy).

[Deployment into the Wild Geese Formation] . . . . . middle. This refers to the function of the Wild Geese deployment. The front ranks should be like a baboon; the rear ranks should be like a wildcat. [At-

tack from] three [sides, not letting the enemy] escape your net to preserve themselves.[28] This is referred to as the function of the Wild
Geese deployment.

When deployed into the hooked formation the front ranks must be
square, while those conjoined on the left and right must be hooked.[29]
When the three sounds (of the drums, gongs, and pipes) are already
complete, (flags in) the five colors must be prepared. When the
sounds of our commands are clearly discriminated and (the troops
all) know the five flags, and there is no front or rear, no . . . . . [30]

In the Dark Rising deployment you must make the flags, pennants,
and feathered banners numerous; the drums should be integrated
and resounding. If the mailed troops are confused, have them sit; if
the chariots are disordered, array them in rows. When they have been
ordered [. . .], with a great pounding and tumult,[31] as if descending
from Heaven, as if coming out from Earth, the infantry should come
forth and be unwavering.[32] Throughout the day they will not be
taken. This is referred to as the Dark Rising deployment.

The tactics for incendiary warfare: When your ditches and ramparts
have already been completed, construct another outer ring of ditches
and moats. Every five paces pile up firewood,[33] being certain to equal-
ize the quantities (in each pile).[34] A designated number of attendants
should be assigned to them. Order men to make linked *chevaux-de-
frise*;[35] they must be light and sharp.[36] If it is windy, avoid . . . . . and
if the vapors from the fire overspread you,[37] while if you engage in
battle you will not conquer them, stand down and retreat.[38]

The tactics for incendiary warfare: If (the enemy) is downwind in
an area abundant with dry grass where the soldiers of their Three Armies would not have anywhere to escape, then you can mount an incendiary attack.[39] When there is a frigid fierce wind, abundant vegetation and undergrowth, and firewood and grass (for fuel) already
piled up while their earthworks have not yet been prepared, in such
circumstances you can mount an incendiary attack. Use the flames to
confuse them; loose arrows like rain. Beat the drums and set up a
clamor to motivate your soldiers. Assist (the attack) with strategic
power.[40] These are the tactics for incendiary warfare.

The tactics for (defensive) aquatic warfare: You must make the infantry numerous and the chariots few. Command them to fully pre-

pare all the necessary equipment, such as hooks, repelling poles, cypress wood, pestles, light boats, oars, baskets, and sails.[41] When advancing, you must follow close on; when withdrawing, do not press together.[42] When mounting a flank attack, follow the current's flow, taking their men as the target.[43]

The tactics for (aggressive) aquatic warfare: Nimble boats should be used as flags; swift boats should be used as messengers. When the enemy goes off, pursue him;[44] when the enemy comes forth, press him. Resist or yield[45] . . . . . in accord with the situation organize against them. When they shift (their forces), make them change (their plans); when they are deploying, [strike] them; when they are properly assembled, separate them.[46] Accordingly, the weapons include spades,[47] and the chariots have defensive infantry.[48] You must investigate their numerical strength as many or few, strike their boats, seize the fords, and show the people that the infantry is coming.[49] These are the tactics for aquatic warfare.

Commentary

This chapter considers the critical question of deployments, the essence of any successful military engagement. Among the pre-Ch'in *Seven Military Classics* the square and round formations are frequently mentioned, almost always in tandem, but none of the others appears until the late *Questions and Replies,* where the "square, round, curved, straight, and angular dispositions" are briefly discussed for training the troops.[50] In addition, the first book of the *Questions and Replies* analyzes the nature of formations and debunks the origin of the names appended to the well-known "eight formations"—Heaven, Earth, wind, cloud, dragon, tiger, bird, and snake.[51] This set of formations is generally, although without critical basis, considered to be the oldest series, for it is found in the *T'ai-pai-yin ching,* where its creation is attributed to the Yellow Emperor. Insofar as popular tradition commonly acclaims the Yellow Emperor as the progenitor of Chinese military history because of his epoch-making battles with the Red Emperor and Ch'ih Yu, this would project the set of formations well back into the mists of antiquity.[52] Furthermore, the great achievements of outstanding historical generals such as Wu Ch'i and Sun-tzu are sometimes said to have been made possible through their mastery of these formations. However, another series of eight is also well known: square, round, female, male, striking, Wheel, Floating Obstacle, and Wild Geese arrays.[53] Over the centuries various diagrams have been created to characterize them; however, most seem incongruous and merely the product of late imagination.

The Awl and Wild Geese formations have already been encountered in Chapter 3, "The Questions of King Wei," while Chapter 7, "Eight Formations," discussed in general terms the principles governing the constitution and deployment of all formations, such as creating a sharp front and maintaining adequate reserves. Unfortunately, the present chapter offers only a few salient characteristics for each of the suggestive names rather than itemizing the constituent elements and clearly characterizing the deployment's shape. No doubt these elements and shapes were well known to the ancients, allowing Sun Pin to merely emphasize certain critical aspects. Only the Awl Formation is particularly clear, while the sword analogy was previously employed in Chapter 9. The contrast between the diffuse and the dense deployments merits noting, with the attendant problems of preventing the former from becoming compressed and the later from being too dense and therefore disorganized. Certainly the diffuse deployment's main advantage should be its mobility and quick responsiveness because small units can move freely—especially across their own internal terrain—whereas heavy ones, having great inertia, require both time and extensive logistical support. Unmentioned are the dangers of being penetrated by a concentrated enemy force—perhaps in the Awl Formation—or the small units encountering superior forces that may simply overwhelm them. However, some tactics for responding to such situations, as well as for attacking them, are found in the next chapter and are scattered throughout the book.

In Chapter 12, "Incendiary Attacks," Sun-tzu discussed employing fire to realize tactical objectives. Apparently he felt incendiary attacks could facilitate the capture of fortified positions, whereas water could not: "Using fire to aid an attack is enlightened, using water to assist an attack is powerful. Water can be used to sever, but cannot be employed to seize." Sun-tzu never specified the formations appropriate to mounting such attacks, nor are any indicated in the *Seven Military Classics*. Only a few measures essential to defending against incendiary attacks are noted in the *Six Secret Teachings*—mainly setting backfires and then occupying the scorched but viable ground with a strong, defensive array. One example with parameters common to this chapter is the following impossible situation:

> King Wu asked the T'ai Kung: "Suppose we have led our troops deep into the territory of the feudal lords where we encounter deep grass and heavy growth which surround our army on all sides. The Three Armies have travelled several hundred *li*; men and horses are exhausted and have halted to rest. Taking advantage of the extremely dry weather and a strong wind, the enemy ignites fires upwind from us. Their chariots, cavalry, and elite forces are firmly concealed in ambush to our rear. The Three Armies become terrifed, scatter in confusion, and run off. What can be done?"
>
> The T'ai Kung said: "Under such circumstances use the cloud ladders and flying towers to look far out to the left and right, to carefully investigate front and rear. When you see the fires arise, then set fires in front of our own forces, spreading them out over the area. Also set fires to the rear. If the enemy comes, withdraw the army and take up entrenched positions on the blackened earth to await their assault. In the same way, if you see flames arise to the rear, you must move far away. If we occupy the blackened ground with our strong crossbowmen and skilled soldiers protecting the left and right flanks, we can also set fires to the front and rear. In this way the enemy will not be able to harm us.[54]

Just as with incendiary tactics, those for defensive and aggressive aquatic engagements are analyzed separately. Moreover, rather than being directed just toward naval engagements between boats, these tactics focus upon defending against amphibious attacks and even inundation (which accounts for the odd admixture of tools and materials). The assault tactics encompass not only engaging the enemy on the water but also simultaneously thwarting his attack. Remarkably, despite an extensive history of combat in the Spring and Autumn period among states with numerous rivers, lakes, and marshes, tactics for "water warfare" receive scant attention in the extant early military writings.[55] This perhaps stemmed from the chariot-based nature of early tactics and the consequent predilection to view bodies of water as entanglements and obstacles to be vigorously avoided.

17
TEN QUESTIONS

Inquiring About the Military

"Suppose our army encounters the enemy and both establish encampments. The provisions and foodstuffs (for both sides) are equal and ample, our men and arms are balanced with the enemy's, while both 'guest' (the invader) and 'host' (the defender) are afraid.[1] If the enemy has deployed in a circular formation in order to await us and relies upon it for his solidity, how should we strike them?"

"To strike them, the masses of our Three Armies should be divided to comprise four or five (operational groups). Some of them should assault them and then feign retreat, displaying fear to them. When they see we are afraid, they will divide up their forces and pursue us with abandon, thereby confusing and destroying their solidity. The four drums should rise up in unison;[2] our five operational forces should all attack together. When all five arrive simultaneously, the Three Armies will be united in their sharpness. This is the Tao for striking a circular formation."

"Suppose our army encounters the enemy and both establish encampments. The enemy is rich, while we are poor; the enemy is numerous, while we are few; the enemy is strong, while we are weak. If they approach in a square formation, how should we strike them?"

"To strike them, deploy in the [diffuse][3] formation and [fragment][4] them; if they are properly assembled, separate them;[5] engage

167

them (in battle) and then feign retreat; and kill the general for their rear (guard) without letting them become aware of it.[6] This is the Tao for striking a square formation."

"Suppose our army encounters the enemy and both establish encampments. If the enemy's troops are already numerous and strong, muscular, agile, and resolute and have deployed into a sharp formation in order to await us, how should we strike them?"[7]

"To strike them, you must segment into three (operational groups) to separate them. One should be stretched out horizontally,[8] two should [go off to strike their flanks. Their upper ranks][9] will be afraid and their lower ranks confused. When the lower and upper ranks are already in chaos, their Three Armies will (then) be severely defeated. This is the Tao for striking a sharp deployment."

"Suppose our army encounters the enemy and both establish encampments. The enemy is already numerous and strong and have assumed an extended horizontal deployment. We have deployed and await them, but our men are few and incapable (of withstanding them). How should we strike them?"

"To strike them, you must segment our soldiers into three (operational groups) and select the 'death warriors.'[10] Two groups should be deployed in an extended array with long flanks; one should consist of talented officers and selected troops. They should assemble (to strike) at the enemy's critical point. This is the Tao for killing their general and striking horizontal deployments."

"Suppose our army encounters the enemy and both establish encampments. Our men and weapons are numerous, (but) our chariots and cavalry are few. If the enemy's men are ten times ours,[11] how should we attack them?"

"To attack them, you should conceal yourselves in the ravines and take the defiles as your base,[12] being careful to avoid broad, easy terrain.[13] This is because easy terrain is advantageous for chariots, while ravines are advantageous to infantry. This is the Tao for striking chariots (in such circumstances)."

"Suppose our army encounters the enemy and both establish encampments. Our chariots and cavalry are numerous, but our men and weapons are few. If the enemy's men are ten times ours, how should we attack them?"

"To attack them, carefully avoid ravines and narrows; break open (a route) and lead them, coercing them toward easy terrain. Even though the enemy is ten times (more numerous), (easy terrain) will be conducive to our chariots and cavalry, and our Three Armies can attack. This is the Tao for striking infantry."

"Suppose our army encounters the enemy and both establish encampments. Our provisions and food supplies have been disrupted. Our infantry and weapons are inadequate to be relied upon. If we abandon our base and attack, the enemy's men are ten times ours. How should we strike them?"

"To strike them when the enemy's men have already [deployed into] and are defending the narrows,[14] we should . . . . . turn about and inflict damage upon their vacuities.[15] This is the Tao for striking (an enemy) on contentious [terrain]."[16]

"Suppose our army encounters the enemy and both establish encampments. The enemy's generals are courageous and difficult to frighten. Their weapons are strong, their men numerous and self-reliant. All the warriors of their Three Armies are courageous and untroubled. Their generals are awesome, their soldiers are martial, their officers strong, and their provisions well supplied.[17] None of the feudal lords dares contend with them.[18] How should we strike them?"

"To strike them, announce that you do not dare (fight). Show them that you are incapable; sit about[19] submissively and await them in order to make their thoughts arrogant and (apparently) accord with their ambitions.[20] Do not let them recognize (your ploy). Thereupon strike where [unexpected], attack where they do not defend, apply pressure where they are indolent, and attack their doubts.[21] Being both haughty and martial,[22] when their Three Armies break camp the front and rear will not look at each other. Therefore strike their middle just as if you had the infantry (strength) to do it. This is the Tao for striking a strong, numerous foe."

"Suppose our army encounters the enemy and both establish encampments. The enemy's men have concealed themselves in the mountains and taken the passes as their base.[23] Our distant forces cannot engage them (in battle), but nearby we have no foothold. How should we strike them?"

"To strike them, you must force them to move from some of the passes they have taken . . . . . and then they will be endangered. Attack positions that they must rescue. Force them to leave their strongholds in order to analyze their tactical thinking, (and then) set up ambushes and establish support forces. Strike their masses when they are in movement. This is the Tao for striking those concealed in strongholds."

"Suppose our army encounters the enemy and both 'guest' and 'host' have deployed. The disposition of the enemy's men is like a (woven, flat) basket.[24] If I estimate the enemy's intentions, he seems to want us to penetrate (his lines) and be overwhelmed. How should we strike them?"

"To strike them, the thirsty should not drink; the hungry should not eat. Segment into three (operational groups), and employ two to assemble (and strike) their critical point. When (the enemy) has already [initiated a response toward the middle], our talented officers and selected soldiers should then strike their two flanks. . . . . . Their Three Armies will be severely defeated. This is the Tao for striking basketlike deployments."

Commentary

In "Ten Questions" an unknown interlocutor, perhaps King Wei or even Sun Pin himself, describes a series of ten possible battlefield situations for which an unidentified source—presumably Sun Pin—provides appropriate tactical solutions. Consequently, this chapter falls into the analytical tradition initiated in the *Art of War* and found extensively embodied in the *Wu-tzu* and the *Six Secret Teachings*. Since many of these hypothetical problems are similar or in some aspects even identical to those resolved in these other writings, further comparative study would be merited to determine directions of influence or identify borrowing from a common tradition of tactical thought. A couple of illustrative situations are examined here, and additional references and interesting examples can be found in the Notes.

Certain principles germane to Sun Pin's thought (as summarily discussed in the Historical Introduction) dominate the ten theoretical responses. Most of the situations raised pose difficulties because the "enemy" force is invariably stronger, better equipped, or well entrenched, while the interlocutor's solution force may suffer from additional, pronounced handicaps, such as disproportionately fewer chariots, cavalry, or infantry. Sun Pin's suggested resolutions inherently depend upon two principles: Divide forces into operational units, and manipulate the enemy. Examples of the former have already been seen in the earlier chapters and are particularly characteristic of his thought.

In general, Sun Pin apparently believed it would be extremely difficult for an opponent to observe, anticipate, counter, and then successfully neutralize several simultaneous threats. Naturally segmented forces suddenly appearing in several locations or descending from several directions would rend the enemy's hard-forged unity, compound his logistical problems, and disrupt his command structure. A well-planned effort could then take advantage of the confusion and resulting fragmentation of forces to wrest victory in whole or part, particularly by rapidly focusing the combined power of the operationally segmented units on one point. Methodologically, this approach adheres to Sun-tzu's basic dicta: "If they are united cause them to be separated."[25] "If we are concentrated into a single force while he is fragmented into ten, then we attack him with ten times his strength."[26] "The army is established by deceit, moves for advantage, and changes through segmenting and reuniting."[27]

The second principle underlying Sun Pin's tactics is simply "manipulate the enemy." This was also one of Sun-tzu's fundamental doctrines, thoroughly implemented by the tactics found throughout the *Art of War*, but especially in the chapter entitled "Vacuity and Substance." In summary Sun-tzu said, "One who excels at warfare compels men and is not compelled by other men."[28] Even the methods suggested in this chapter for manipulating the enemy are found in the *Art of War* as well as the other works mentioned previously. In particular, a strong enemy's alertness should be undermined by feigning weakness, retreats, and general submissiveness or stupefied inaction.[29] Measures designed to disorder his forces or command structure before swiftly attacking them should also be employed, while headlong attacks and battles of attrition should always be avoided. Whenever the enemy is ensconced in strongpoints or has otherwise exploited the natural advantages of terrain, the enemy must be moved onto more favorable ground—whether through temptation or a threat to a critical position.[30] Outnumbered forces should avail themselves of ravines and confined spaces, as Sun-tzu and virtually every other thinker advocated,[31] while those possessing great superiority in men and especially chariots should engage the enemy on "easy" terrain, that is, open spaces where mobility will allow superior numbers to concentrate mass appropriately. Only then will the army have a reasonable chance of victory.

The *Art of War* preserves the earliest analysis of terrain types, categorizing them by their main features and associated tactical possibilities. This analysis is more succinct than the theoretical analyses found in the *Wu-tzu* and the *Six Secret Teachings* and are well known. A typical example relevant to this chapter is "constricted configurations of terrain," for which Sun-tzu's fundamental tactics run as follows: "If we occupy them first we must fully deploy throughout them in order to await the enemy. If the enemy occupies them first and fully deploys in them, do not follow them in. If they do not fully deploy in them, then follow them in."[32]

Conversely, more detailed descriptions of potential battlefield situations and their resolutions are found in the *Wu-tzu* and the *Six Secret Teachings*. Two passages, among many, analyze situations closely related to those found in this chapter:

> Marquis Wu asked: "If the enemy is numerous while we are few, what can I do?"
>
> Wu Ch'i replied: "Avoid them on easy terrain, attack them in narrow quarters. Thus it is said, for one to attack ten nothing is better than a

narrow defile. For ten to attack one hundred, nothing is better than a deep ravine. For one thousand to attack ten thousand, nothing is better than a dangerous pass. Now if you have a small number of troops, should they suddenly arise—striking the gongs and beating the drums—to attack the enemy on a confined road, then even though his numbers are very great, they will all be startled and move about. Thus it is said, when employing large numbers concentrate on easy terrain; when using small numbers concentrate on naturally confined terrain."[33]

King Wu asked the T'ai Kung: "Both the enemy and our army have reached the border where we are in a standoff. They can approach and we can also advance. Both deployments are solid and stable; neither side dares to move first. We want to go forth and attack them, but they can also come forward. What should we do?"

The T'ai Kung said: "Divide the army into three sections. Have our advance troops deepen the moats and increase the heights of the ramparts, but none of the soldiers should go forth. Array the flags and pennants, beat the leather war drums, and complete all the defensive measures. Order our rear army to stockpile supplies and foodstuffs without causing the enemy to know our intentions. Then send forth our elite troops to secretly launch a sudden attack against their center, striking where they do not expect it, attacking where they are not prepared. Since the enemy does not know our real situation, they will stop and not advance."[34]

This chapter also has considerable historical importance insofar as it discusses the problems of disproportionate and imbalanced component forces. Given that the cavalry only became an integral part of Warring States armies early in the third century B.C., and probably tactically significant only late in the Warring States period (near the end of the third century), "Ten Questions" either must have been penned by a late disciple or the role of cavalry in Sun Pin's time has to be considerably revised. The last few chapters in the *Six Secret Teachings* also discuss the relative value and appropriate tactics for the three components of infantry, chariots, and cavalry and include an examination of problematic and advantageous situations. The criteria encompass force composition, conditions, and disposition, while the chapters include analyses of operational characteristics for terrain.

Among the salient principles, the following have particular relevance for this chapter:

> Chariots are the feathers and wings of the army, the means to penetrate solid formations, to press strong enemies, and to cut off their flight. Cavalry are the army's fleet observers, the means to pursue a defeated army, to sever supply lines, to strike roving forces.
>
> Now chariots and cavalry are the army's martial weapons. Ten chariots can defeat one thousand men; one hundred chariots can defeat ten thousand men. Ten cavalrymen can drive off one hundred men, and one hundred cavalrymen can run off one thousand men.[35]

The infantry values knowing changes and movement; the chariots value knowing the terrain's configuration; the cavalry values knowing the side roads and unorthodox Tao.

In general, in chariot battles there are ten types of terrain on which death is likely and eight on which victory can be achieved. (For example):

If after advancing there is no way to withdraw, this is fatal terrain for the chariots.

Penetrating into narrow and obstructed areas from which escape will be difficult, this is terrain on which the chariots may be cut off.

When the chariots are few in number, the land easy, and one is not confronted by enemy infantry, this is terrain on which the chariots may be defeated.[36]

For the cavalry there are ten situations that can produce victory and nine that will result in defeat. (For example):

When the enemy's lines and deployment are well-ordered and solid while their officers and troops want to fight, our cavalry should outflank them but not go far off. Some should race away, some race forward. Their speed should be like the wind, their explosiveness like thunder, so that the daylight becomes as murky as dusk. Change our flags and pennants several times; also change our uniforms. Then their army can be conquered.

When the way by which we enter is constricted but the way out is distant; their weak forces can attack our strong ones; and their few can attack our many—this is terrain on which the cavalry will be exterminated.[37]

When infantry engage in battle with chariots and cavalry, they must rely on hills and mounds, ravines and defiles. The long weapons and strong crossbows should occupy the fore; the short weapons and weak cross-bows should occupy the rear, firing and resting in turn. Even if large numbers of the enemy's chariots and cavalry should arrive, they must maintain a solid formation and fight intensely while skilled soldiers and strong crossbowmen prepare agai:.st attacks from the rear.[38]

18
REGULATING MAILED TROOPS

Fragments

As for the method for regulating mailed troops,[1] when the enemy's men have deployed in a square formation . . . . .

. . . . . If you want to strike them but their strategic power does not allow it, in cases such as these force them to submit by . . . . .

. . . . . with the seals of state and if you want to engage in battle, act as if deranged.[2] In cases such as these, deploying a few . . . . .

. . . . . reverse. In cases such as these, follow them with masses of troops. Select your troops in accord with (the situation), being certain to . . . . .

. . . . . Select your troops in accord with (the situation). . . . . .

. . . . . When the left and right flanks attack by rapidly converging, this is termed a "sharp hooking strike."

. . . . . The *ch'i* of [. . .] is not stored in the heart, while the masses of the Three Armies [. . .] follow them, knowing not . . . . .

. . . . . about to divide the [Three] Armies in order to reform their [ranks. If the officers and troops] are few but the people . . . . .

. . . . . awesomeness . . . . . will find it difficult to . . . . . Divide up their masses; bring about chaos in . . . . .

. . . . . Deployment is not sharp; thus their ranks[3] will not . . . . .

. . . . . If you entice them far off, the enemy will roll up (their armor) in order to (race) far off. . . . . . [4]

Control and isolate their general, rattle his mind, strike . . . . .
. . . . . Their generals are courageous, their troops numerous.
. . . . .
. . . . . Their great masses are about to . . . . .
. . . . . The Tao of the troops . . . . .

Commentary

This chapter is so badly fragmented that some editions do not even include it, while others fail to provide any notes.[5] One or two of the fragments are of interest, particularly "If you want to engage in battle, act as if deranged," but otherwise little can be abstracted. Without additional context for each of the sentences, the translations are uncertain at best but are provided for the sake of completeness and for giving a sense of what might have been incorporated.

19

DISTINCTION BETWEEN GUEST AND HOST

Armies are distinguished as being a "guest" or a "host." The guest's forces are (comparatively) numerous, the host's forces (comparatively) few.[1] Only if the guest is double and the host half can they contend as enemies. To defeat ..... [The host is the one who] establishes his position [first]; the guest is the one who establishes his position afterward.[2] The host ensconces himself on the terrain and relies on his strategic power to await the guest who contravenes mountain passes and traverses ravines to arrive. Now if they contravene mountain passes [and traverse ravines] only to retreat and thereby dare to cut their own throats[3] rather than advancing and daring to resist the enemy, what is the reason? It is because their strategic configuration of power is not conducive (to attacking) and the terrain is not advantageous.[4] If their strategic power is conducive and the terrain advantageous, then the people by themselves [will advance. If their strategic power is not conducive and the terrain not advantageous, the people][5] will retreat by themselves. Those who are referred to as excelling in warfare make their strategic power conducive and the terrain advantageous.

If the mailed troops are counted by the hundreds of thousands,[6] while the people have a surplus of grain that they are unable to eat, they have an excess ..... [7] If the number of troops dwelling (in a state) is numerous, but the number employed is few, then the standing forces are excessive and those employed (in combat) insufficient.[8] If there are several hundred thousand mailed soldiers going forth a thousand by a thousand, continuing thousands after thousands[9] ..... ten thousand after ten thousand are thereby dispatched toward us.[10] Those who are referred to as excelling at warfare excel at cutting and severing them, just as if [a hand] happened to wipe them away.[11] One who can divide up the enemy's soldiers, who can repress the enemy's soldiers, will have enough (men) even with the smallest amounts.[12] One who cannot divide up the enemy's soldiers, who cannot repress the enemy's soldiers, will be insufficient even if several times more numerous.

Is it that the more numerous will be victorious? Then calculate the numbers and engage in battle. Is it that the richest will win? Then measure the grain supplies and engage in battle. Is it that the sharpest weapons and stoutest armor will win? Then it will be easy to foretell victory. Since (this is not the case), the rich still do not dwell in security; the poor do not yet dwell in danger; the numerous have not yet attained victory; the few [are not yet defeated. Now] what determines victory or defeat, security or danger, is the Tao.

If the enemy's men are more numerous but you can cause them to become divided and unable to rescue each other; those that suffer attacks unable to [know about each other; moats that are deep and fortifications that are high unable][13] to be taken as secure; stout armor and sharp weapons unable to be taken as strength;[14] and courageous, strong warriors unable to protect their general, then your victory will have realized the Tao. Thus enlightened rulers and generals who know the Tao will certainly first [calculate] whether they can attain success before the battle, so they will not lose any opportunity for achievement after engaging in battle.[15] Thus if when the army goes forth it achieves success, while when it returns it is unharmed, (the commander) is enlightened about military affairs.[16]

..... make them tired. If the Three Armies' warriors can be forced to completely lose their determination, victory can be attained and maintained.[17] For this reason repress the left while you hit the right; then when the right is being defeated, [the left] will not be able to rescue them. Repress the right while you hit the left; then when the left is being defeated, the right will not be able to rescue them.[18] For this reason if the army sits about and does not get up,[19] if they avoid battle and are not employed, those close by being few and inadequate for employment, while the distant are dispersed and incapable . . . . .

Fragments

When you are the guest of others, you should [. . .] first . . . . .

The *Tactics* states: The host counters the guest at the border.

. . . . . When the guest loves to engage in combat, [he will certainly be defeated].[20]

Commentary

This chapter introduces the interesting distinction of "guest" and "host," the former generally referring to an invader, the latter to a defender normally fighting on his home territory or on territory he already occupies. These two terms are not otherwise found in the early military classics but figure prominently in later thought and general writings and are also briefly discussed in the *Questions and Replies,* where Li Ching indicates that the distinctions are not invariably set.

> The T'ai-tsung said: "The army values being the 'host'; it does not value being a 'guest.' It values speed, not duration. Why?"
>
> Li Ching said: "The army is employed only when there is no alternative, so what advantage is there in being a 'guest' or fighting long? Sun-tzu says: 'When provisions are transported far off, the common people are impoverished.' This is the exhaustion of a 'guest.' He also said: 'The people should not be conscripted twice, provisions should not be transported thrice.' This comes from the experience of not being able to long endure. When I compare and weigh the strategic power of host and guest, then there are tactics for changing the guest to host, changing the host to guest."
>
> The T'ai-tsung said: "What do you mean?"
>
> Li Ching said: "By foraging and capturing provisions from the enemy, you change a guest into a host. 'If you cause the sated to be famished and the rested to be tired,' it will change a host into a guest. Thus

the army is not confined to being a host or guest, slow or fast, but only focuses upon its movements invariably attaining the constraints and thereby being appropriate."[21]

The concept of guest should not be delimited solely to invaders moving into foreign territory but rather should be considered another tactical designation useful for comparative purposes because, in essence, it refers to a force in movement striking one already emplaced. Throughout history defensive forces have generally realized often insurmountable advantages through choosing the battlefield, exploiting the terrain, and establishing fortifications, forcing their guests to possess overwhelming superiority in firepower or numbers, as Sun Pin indicates, to create even the possibility of victory. As the *Ssu-ma Fa* earlier noted, comparative strength is required: "In general, in warfare: If you advance somewhat into the enemy's territory with a light force it is dangerous. If you advance with a heavy force deep into the enemy's territory you will accomplish nothing. If you advance with a light force deep into enemy territory you will be defeated. If you advance with a heavy force somewhat into the enemy's territory you can fight successfully. Thus in warfare the light and heavy are mutually related."[22]

Naturally the number of soldiers brought to bear in an encounter and the sequence of their arrival on a battlefield would determine the nature of the conflict. Even within a state, especially at the time of Sun Pin's writing, although cities were developing economically and proliferating, the countryside was still fairly open, and armies could range for some distance before encountering a stronghold. Therefore, it would be possible to choose, even if less freely than in the Spring and Autumn period, sparsely populated routes in accord with Sun-tzu's dicta.[23] Thus, despite the growth of extensive walls in this period, even within its own state the defender would be compelled to mobilize and dispatch troops to an appropriate location in order to oppose an invader rather than being able to rely upon heavy static defenses already in place. Accordingly, the *Six Secret Teachings* devotes most of two divisions—the Tiger and Leopard Secret Teachings—to the tactics for invading forces and for countering responses.

Chapter 69 of the *Tao Te Ching* opens with a remarkable but puzzling passage that turns on the distinction of guest and host: "Among those who employ the military there is a saying which runs: 'I do not dare to act as the host but act as the guest; I do not dare to advance an inch but withdraw a foot.'" Within the context of Sun Pin's chapter the meaning of guest and host is apparent, but the two parts of the saying are mutually contradictory. Clearly the first half advises assuming an aggressive posture rather than maintaining a defensive stance, a view very much in accord with the overall tenor of Sun Pin's *Military Methods*. However, for twenty centuries numerous commentators have explained host as meaning "first" and have understood the clause as referring to the one that initiates military activities. Accordingly, the state or army that responds to an offensive acts as the guest, exactly opposite to its meaning in Sun Pin's chapter. Furthermore, this interpretation is consistent with the second clause's advocacy of yielding rather than advancing. However, it should be remembered that the *Tao Te Ching* also characterizes the military as inherently unorthodox, while the chapter itself continues on to discuss the deployment of the formless formation. The questions raised by these distinctly divergent views obviously merit pondering.

"Distinction Between Guest and Host" also reiterates the fundamental principle that victory can be wrought through fragmenting the enemy's forces. As this has been well discussed in previous chapters, little need be added here. Similarly, the importance of ravines for defense is tangentially evident in this chapter, for Sun Pin cites them to emphasize the difficulty of the invader's advance and the invader's courage in the face of such obstacles.

Two other concepts are of fundamental importance: the Tao and strategic power, both of which are discussed in the Historical Introduction and in our other works.[24] In this chapter Tao primarily refers to the Tao of warfare, although the ruler must also be concerned with the Tao of government. Sun Pin has previously said that the army's "power (*te*) lies in the Tao"[25] and that "if one has realized the Tao, even though (the enemy) wants to live, they cannot."[26] Furthermore, in concord with other thinkers, he believed that "only (a general) who knows the Tao (is capable)."[27] Wu Ch'i constantly spoke of the Tao of warfare, while Sun-tzu placed the Tao first among his critical factors in the opening chapter, "Initial Estimations":

Warfare is the greatest affair of state, the basis of life and death, the Tao
to survival or extinction. It must be thoroughly pondered and analyzed.

   Therefore, structure it according to the following five factors, evaluate it comparatively through estimations, and seek out its true nature.
The first is termed the Tao, the second Heaven, the third Earth, the
fourth generals, and the fifth the laws for military organization and discipline.

   The Tao causes the people to be fully in accord with the ruler. Thus
they will die with him; they will live with him and not fear danger.

Sun-tzu has long been mistakenly identified with espousing a simplistic position because of another important statement about the Tao of
warfare: "Warfare is the Tao of deception."[28]

The role of strategic power, a concept prominently identified with
Sun-tzu, has also been previously mentioned in this work by Sun Pin
who said, "Strategic power is the means by which to cause the soldiers to invariably fight."[29] This echoes Sun-tzu's earlier statement:
"One who excels at warfare seeks victory through the strategic configuration of power, not from reliance upon men. Thus he is able to
select men and employ strategic power."[30] When there is an imbalance in strategic power, it is easy to prevail; moreover the soldiers, realizing their advantages, are both motivated and willing to risk the
dangers required to wrest victory. However, by Sun Pin's time if
these strategic advantages were lacking, the soldiers would apparently
rather face disgrace, punishment, and even death than uselessly advance directly upon the enemy.

   Finally, this chapter elaborates a principle only glimpsed in Sun-
tzu's *Art of War*: the ability of the outnumbered and comparatively
impoverished to sustain the battle and even gain victory (through
such techniques as dividing and segmenting). While the *Wu-tzu* contained some suggestions for harassing a larger foe and thereby prevailing,[31] Sun-tzu generally advocated avoiding such battles in favor
of assuming a defensive posture, except in one critical paragraph:
"The army does not esteem the number of troops being more numerous for it only means you cannot aggressively advance. It is sufficient for you to muster your own strength, analyze the enemy, and
take them. Only someone who lacks strategic planning and slights an

enemy will inevitably be captured by others."[32] Sun-tzu's method for engaging a superior foe consisted of fragmenting the enemy's forces, keeping the enemy ignorant, and forcing him to defend across an extremely broad front, thereby turning many into a few and the outnumbered few into many.[33]

20
THOSE WHO EXCEL

Those who excel (in warfare, even) when the enemy's forces are strong[1] and numerous, can force them to divide and separate, unable to rescue each other, and suffer enemy attacks without mutually knowing about it. Thus ditches that are deep and ramparts that are high will be unable to provide security; chariots that are sturdy and weapons that are sharp unable to create awesomeness; and warriors of courage and strength unable to make one strong. Those who excel (in warfare) control the ravines and evaluate the narrows,[2] incite the Three Armies, and take advantage of contracting and expanding.[3] Enemy troops that are numerous they can make few. Armies fully supplied and well provisioned they can make hungry.[4] Those securely emplaced, unmoving, they can cause to become tired. Those who have gained All under Heaven they can cause to become estranged.[5] When the Three Armies are united, they can cause them to become rancorous.[6]

Thus the army has four roads and five movements. Advancing is a road, withdrawing is a road, left is a road, right is a road. Advancing is a movement, withdrawing is a movement, left is a movement, right is a movement. Silently emplaced is also a movement. For someone to excel these four roads must be penetrating; these five movements must be skillful. Thus when advancing he cannot be contravened to the fore; when withdrawing he cannot be cut off to the rear. To the left and right he cannot be forced into ravines.[7] Silently [remaining in position he cannot be troubled][8] by the enemy's men. Accordingly, he causes the enemy's four roads to be impoverished, his five move-ments to be invariably troubled. If (the enemy) advances, he will be pressed to the fore; if he withdraws, he will be cut off to the rear. To the left and right he will be forced into ravines, while if he remains quietly encamped, his army will not avoid misfortune.

Those who excel (in warfare) can cause the enemy to roll up his ar-mor and race far off; to travel two days' normal distance at a time; to be exhausted and sick but unable to rest; to be hungry and thirsty but unable to eat.[9] An enemy emaciated[10] in this way certainly will not be victorious! Sated, we await his hunger; resting in our emplacement, we await his fatigue; in true tranquility, we await his movement.[11] Thus our people know about advancing but not about withdrawing. They will trample on naked blades and not turn their heels.

Commentary

This is perhaps the most lucid chapter in the entire work, focusing en-tirely upon the commander's ability to manipulate the enemy and thereby ensure the realization of the conditions for victory. The tacti-cal principles essentially derive from Sun-tzu's concepts of maneuver warfare, requiring mastery of command and control while stressing the implementation of measures to debilitate the enemy both physi-cally and morally. In "Nine Terrains" Sun-tzu states, "In antiquity those who were referred to as excelling in the employment of the army were able to keep the enemy's forward and rear forces from connecting; the many and few from relying on each other; the noble and lowly from coming to each other's rescue; the upper and lower ranks from trusting each other; the troops to be separated, unable to reassemble, or when assembled, not to be well-ordered. They moved when it was advantageous, halted when it was not advantageous."[12]

The skills required are symmetrical, but concrete tactics designed to compel the enemy must not be successfully applied by the enemy to one's own army. Every effort must be made to thwart the enemy's movements, coercing them into ineffectual and enervating activities, while preserving the army's strength and ensuring it remains free from maneuver constraints. Only then will the commander excel and his army prove consistently victorious.

21

FIVE NAMES,
FIVE RESPECTS

五名五恭

Armies have five names: The first is "Awesomely Strong," the second is "Loftily Arrogant," the third is "Firmly Unbending," the fourth is "Fearfully Suspicious," and the fifth is "Doubly Soft."[1]

In the case of the Awesomely Strong army, be pliant and soft and await them.

In the case of the Loftily Arrogant army, be respectful and outlast them.

In the case of the Firmly Unbending army, entice and then seize them.

In the case of the Fearfully Suspicious army, press them to the fore, set up a clamor on the flanks, deepen your moats and increase the height of your fortifications, and cause difficulty for their supplies.

In the case of the Doubly Soft army, set up a clamor to terrorize them; shake and disrupt[2] them. If they go forth, then strike them. If they do not go forth, surround them.

Five names.[3]

Armies have five (manifestations of) "respect" and five of "brutality." What is meant by the five (manifestations of) respect?

When it crosses the (enemy's) border and is respectful, the army loses its normality.[4]

If it acts respectfully twice, the army will not have anywhere to forage.[5]

If it acts respectfully three times, the army will lose its (appropriate) affairs.[6]

If it acts respectfully four times, the army will not have any food.

If it acts respectfully five times, the army will not attain its objective.[7]

When it crosses the border and acts brutally, (the army) is referred to as a guest.

If it acts brutally twice, it is termed glorious.[8]

If it acts brutally three times, the host's men are afraid.[9]

If it acts brutally four times, the troops and officers have been deceived.[10]

If it acts brutally five times, the soldiers invariably have been greatly wasted.[11]

The five respects and five brutalities must be mutually implemented.

Five respects.

Commentary

The early military writings occasionally appended succinct, poignant names to various types of armies, often coupled with suggested means for confronting and defeating them. This practice mirrored similar evaluations found in the painfully real world of statecraft and political diplomacy, for enemies of both types had to be accurately analyzed and characterized before tactics for manipulating them could be evolved. A passage in the *Wu-tzu* defining five types of armies typifies such descriptions:

> In general the reason troops are raised are five: to contend for fame; to contend for profit; from accumulated hatreds; from internal disorder; and from famine. The names of the armies are also five: "righteous army," "strong army," "hard army," "fierce army," and "contrary army." Suppressing the violently perverse and rescuing the people from chaos is termed "righteousness." Relying on the strength of the masses to attack is termed "strong." Mobilizing the army out of anger is termed "hard." Abandoning the forms of propriety and greedily seek-

ing profit is termed "fierce." While the country is in turmoil and the people are exhausted, embarking on military campaigns and mobilizing the masses is termed "contrary."

These five each have an appropriate Tao. In the case of the righteous you must use propriety to subdue them. Toward the strong you must be deferential to subjugate them. Against the hard you must use persuasive language to subjugate them. Against the fierce you must use deceit to subjugate them. Against the contrary you must use the tactical balance of power to subjugate them.[12]

Orthodox theory, both as found in the early military classics and the Confucian thinkers, stressed that warfare was a righteous activity, not to be undertaken lightly, and then only when compelled by self-preservation or the need to extirpate the evil.[13] Naturally, determining whether military action might be necessary was fundamentally affected by the analyst's mind-set and point of view. Consequently, just as with twentieth-century history, perceived threats frequently justified preemptive action. However, apart from those Legalists who viewed warfare as the necessary means for enlarging the state and augmenting its wealth and power, most thinkers interpreted military activities from within a reasonably benign framework. Sun-tzu himself was a strong exponent of avoiding frequent, prolonged military actions because they would debilitate the state,[14] and Sun Pin's era, although far more precarious, generally accorded nominal respect to proper motives. However, perhaps because he lived in an age of ever-escalating warfare, Sun Pin's tactics for defeating the five different armies are more engagement oriented than Wu Ch'i's more abstract, largely verbal approach.

The second part of the chapter, which may have originally been separate, characterizes an invading army's behavior in terms of two fundamental modes of action: respectful (or constrained) versus cruel and brutal. Naturally the latter was normally associated with invading forces and was inescapably witnessed in the massively destructive Warring States period. However, the perspective prevailing in Sun Pin's era (or somewhat thereafter when this book may have been compiled) still retained vestiges of earlier conceptualizations and values, such as are found in the *Ssu-ma Fa*.[15] To conquer, an invading army had to manifest a severe, fearful image and act decisively, with strength. The frequent warnings found in the *Six Secret Teachings* and other Warring States works against rampaging, plundering, and wantonly destroying the countryside no doubt reflect reactions against excesses that, while effectively striking terror and possibly cowering foes into rapid submission, increased hatreds and hardened resistance.[16] In contrast, Sun Pin's analysis emphasizes that in general being respectful provokes disaster (and is identified with a weak, incapable army), whereas brutality, if not carried to extremes, proves effective as well as expected.

Concretely, when the army crosses the border it has to act strongly and is accordingly termed a *guest*, the normal designation. Manifesting brutal behavior a second time makes the army "glorious." A third time and everyone in the defender's territory is terrified. However, exceeding three such expressions of violence creates difficulties. The soldiers, by then no doubt exhausted (although perhaps personally profiting from plundering the countryside),[17] feel deceived, probably having expected a much simpler mission. (This would be particularly true when, contrary to propaganda, the campaign meets with determined resistance from the populace rather than merely being directed against the ruler, as in the famous case of King Wu vanquishing the "depraved" Shang.) A fifth time and their strength is finished, their efforts wasted. This echoes a passage in the *Wu-tzu:* "Now being victorious in battle is easy, but preserving the results of victory is difficult. Thus it is said that among the states under Heaven that engage in warfare, those that garner five victories will meet with disaster; those with four victories will be exhausted; those with three victories will become hegemons; those with two victories will be kings; and those with one victory will become emperors. For this reason those who have conquered the world through numerous victories are extremely rare, while those who thereby perished are many."[18]

Of course victory in battle and brutality are not necessarily synonymous, nor is the range of behavior encompassed by Sun Pin's employment of the term otherwise indicated.[19] However, the closing remark, which asserts that respect and brutality need to be balanced and implemented in tandem, is somewhat problematic because this chapter essentially condemns all the manifestations of respect, as well as too frequently repeated acts of brutality. Therefore, the conclusion is suspect; perhaps the words *five* respects appear incorrectly.

22
THE ARMY'S LOSSES

If you want to employ that in which the enemy's people are not secure, you should rectify the customs which ..... [1]

[If you want to strengthen and augment the shortcomings in your state's army in order to] cause difficulty for the enemy's army in what he is strong, it will be a wasted army.[2]

If you want to strengthen and multiply that in which your state has a paucity in order to respond to that in which the enemy is numerous, it will be a rapidly subjugated army.[3]

If your preparations and strongholds are unable to cause difficulty for the enemy's (assault) equipment, it will be an "insulted" army.[4]

If your (assault) equipment is not effective against the enemy's preparations and strongholds, it will be a frustrated army.[5]

When the army .....

If someone excels at deployments, knows the (appropriate) orientations for forward and rear, and knows the configuration of the terrain, but yet the army frequently suffers difficulty, he is not enlightened about (the distinction between) states conquering and armies conquering.[6]

The people .....

[If .....] an army cannot flourish great achievements, it is because it does not know about assembling.[7]

If an army loses the people, it does not know about excess.[8]

If an army employs great force but the achievements are small, it does not understand time.[9]

An army that is unable to overcome great adversity is unable to unite the people's minds.

An army that frequently suffers from regret trusts the doubtful.

An army that is unable to discern good fortune and misfortune in the as-yet-unformed does not understand preparations.[10]

When the army sees the good but is dilatory, when the time comes but it is doubtful, when it expels perversity but is unable to dwell (in the results), this is the Tao of stopping.[11]

To be lustful, yet scrupulous; to be a dragon, yet respectful; to be weak, yet strong; to be pliant, yet [firm], this is the Tao of arising.[12]

If you implement the Tao of stopping, then (even) Heaven and Earth will not be able to make you flourish. If you implement the Tao of arising, then (even) Heaven and Earth [will not be able to obstruct you].[13]

Fragments

. . . . . the army. If you want to use the state to . . . . .

. . . . . an internally exhausted army. Numerous expenditures of energy do not result in solidity. . . . . .

. . . . . see the enemy is difficult to subdue. If the army still acts wantonly between Heaven and Earth . . . . .

. . . . . and the army strong, the state . . . . .

. . . . . The army is unable to . . . . .

Commentary

This chapter is the first of four consisting of collected observations on the strengths and errors of the military and generals such as are commonly found in the early military writings. As the individual lines are generally clear, further commentary on the individual propositions is consigned to the Notes. However, two topics—the vital importance of the critical moment and the coupled concepts of the hard and soft, firm and pliable—together with some similar passages from other books merit exploring.

Several sentences in this chapter are virtually identical with Chapter 65 of the *I-Chou-shu* (*Lost Books of the Chou*), purporting to be works from the Early Chou period, but probably dating from the late fourth century—Sun Pin's era—through the Former Han. In fact, two previously problematic characters in one sentence of this book have now been clarified by the Sun Pin version.[14] For comparison purposes, the entire chapter may be translated as follows:

> What the king wears (at his waist) lies in Virtue.[15] Virtue lies in profiting the people. The people lie in according with the ruler. Unification lies in according with the time. By responding to affairs, they will be easily completed. The completion of plans lies in constant sincerity. Realizing success lies in strength being numerous. Flourishing greatness lies in conquering oneself. Not going to excess lies in frequent correction. Not being put in difficulty lies in anticipating and assiduously discerning disaster in the as-not-yet-formed. Eliminating harm lies in being able to be decisive. Giving security to the people lies in knowing excess. Employing the army lies in knowing the time. Overcoming great adversity lies in uniting the people's minds. Frequent regret[16] lies in trusting the doubtful. (Problems from) the sons of concubines stem from listening to talk in the inner apartments. Transforming behavior lies in knowing harmony. Bespreading beneficence lies in tranquilizing the heart. The unfortunate lies in not hearing about one's errors. Good fortune lies in accepting criticism. The foundation lies in loving the people. Solidity lies in bringing Worthies near. Misfortune and good fortune lie in what is secret. Advantage and disadvantage (profit and harm) lie in what is near. Survival and perishing lie in what one employs. Separating and uniting (the people) lie in the edicts one issues. Honor lies in being careful about majesty. Security lies in making oneself respected. Being endangered and perishing lie in not knowing the time. If one sees good but is dilatory, if the time arrives and one is doubtful, if one loses uprightness and dwells in the perverse, it is not something that can be sustained. This is the crux of gaining and losing the time; it cannot but be investigated.

Several of these pithy sentences also appear in the *Six Secret Teachings,* suggesting that they all derive either from Sun Pin's work or, more likely, from a common text, such as a proto-text for the *Six Secret Teachings.*[17] The phrases illuminate each other, and the contexts further understanding. The following passage from the *Six Secret Teachings* also integrates the critical concept of timeliness with those for the hard and soft:

The T'ai Kung said: "If one sees good but is dilatory, if the time for action arrives and one is doubtful, if you know something is wrong but you sanction it—it is in these three that the Tao stops. If one is soft and quiet, dignified and respectful, strong yet genial, tolerant yet hard—it is in these four that the Tao begins. Accordingly, when righteousness overcomes desire one will flourish; when desire overcomes righteousness one will perish. When respect overcomes dilatoriness it is auspicious; when dilatoriness overcomes respect one is destroyed."[18]

The military writings, including Sun Pin's work, emphasize the concept of timeliness, stressing the need to recognize and exploit the fleeting moment for action. The *Six Secret Teachings* offers an analogy phrased in terms of the common activities of material life: "When the sun is at midday you must dry things. If you grasp a knife you must cut. If you hold an ax you must attack. If, at the height of day, you do not dry things in the sun, this is termed losing the time. If you grasp a knife but do not cut anything, you will lose the moment for profits. If you hold an ax but do not attack, then bandits will come."[19] Wu Ch'i said, "If the general is not quick-witted and acute, the Three Armies will lose the moment,"[20] while the T'ai Kung concluded, "One who excels in warfare will not lose an advantage when he perceives it or be doubtful when he meets the moment. One who loses an advantage or lags behind the time for action will, on the contrary, suffer from disaster. Thus the wise follow the time and do not lose an advantage; the skillful are decisive and have no doubts."[21] Initiating action at the precise moment is so critical that any enemy force that fails to recognize and exploit such opportunities immediately becomes just as vulnerable as if it suffered from other fundamentally disabling conditions:

In employing the army you must ascertain the enemy's voids and strengths and then race to his endangered points. When the enemy has just arrived from afar and their battle formations are not yet properly deployed, they can be attacked. If they have eaten but not yet established their encampment, they can be attacked. If they are running about wildly, they can be attacked. If they have labored hard, they can be attacked. If they have not yet taken advantage of the terrain, they can be attacked. When they have lost the critical moment and not followed up on opportunities, they can be attacked.[22]

The other concept prominently found in this chapter and many military writings is the inherently dynamic concept of the hard and the soft, the firm and the pliable (or flexible). In "The Army's Losses" they are coupled with two other paired conditions, but it is these two that command attention. Sun Pin's era witnessed the growth of Taoist thought and its evolution into different perspectives, including eventually the so-called Huang-Lao school.[23] The core text underlying much of Taoist philosophy is the famous *Tao Te Ching*, in which the importance of the soft and pliable is advanced, subtly juxtaposed with normal worldviews that expect—and predicate actions accordingly—the hard to dominate the soft, the strong to brutalize the weak. Portions of two sections are particularly illuminating:

> Alive man is pliable and weak,
> Dead he is firm and strong.
> Alive the myriad things, grasses, and trees are pliable and fragile;
> Dead they are dry and withered.
> Thus the firm and strong are the disciples of death,
> The pliant and weak are the disciples of life.
> For this reason armies that are strong will not be victorious;
> Trees that are strong will break.
> The strong and great dwell below,
> The pliant and weak dwell above.[24]

> Under Heaven there is nothing more pliant and weak than water, but for attacking the firm and strong nothing surpasses it, nothing can be exchanged for it. The weak being victorious over the strong, the pliant being victorious over the firm—there isn't anyone under Heaven who does not know this. (Yet) no one is able to implement it.[25]

These insights are clearly embraced by portions of the *Wei Liao-tzu:* "The army that would be victorious is like water. Now water is the softest and weakest of things, but whatever it collides with—such as hills and mounds—will be collapsed by it for no other reason than its nature is concentrated and its attack is totally committed."[26] Thus formulated, the concept is more complex than originally expressed in the *Tao Te Ching*, for it recognizes that it is not just "softness" that works the change, but rather the water's focus and endurance—its unremitting pressure over time—cannot be withstood.

In general, the military writers perceived a need to employ each one of the four—the soft, hard, pliant, and firm—appropriately: "The *Military Pronouncements* states: 'The soft can control the hard, the weak can control the strong.' The soft is Virtue. The hard is a brigand. The weak is what the people will help, the strong is what resentment will attack. The soft has situations in which it is established; the hard has situations in which it is applied; the weak has situations in which it is employed; and the strong has situations in which it is augmented. Combine these four and implement them appropriately."[27] As the concluding sentence from this same chapter ("Superior Strategy") in the *Huang Shih-kung* notes, the four must be integrated and combined. Perversely adopting a single one will doom the state: "The *Military Pronouncements* states: 'If one can be soft and hard, his state will be increasingly glorious! If one can be weak and strong, his state will be increasingly glorious! If purely soft and purely weak, his state will inevitably decline. If purely hard and purely strong, his state will inevitably be destroyed.'"[28] Even Sun-tzu, whose *Art of War* betrays Taoist influence but was written somewhat earlier, noted, "Realize the appropriate employment of the hard and soft through patterns of terrain."[29] Finally, Wu Ch'i, who was highly concerned about the problems of command and control, indicated the necessity for any qualified general to embrace such capabilities: "Now the commanding general of the Three Armies should combine both military and civilian abilities. The employment of soldiers requires uniting both hardness and softness."[30]

23

THE GENERAL'S RIGHTEOUSNESS

The general cannot but be righteous.[1] If he is not righteous, then he will not be severe. If he is not severe, then he will not be awesome.[2] If he is not awesome, then the troops will not die (for him). Thus righteousness is the head of the army.

The general cannot but be benevolent. If he is not benevolent, then the army will not conquer. If the army does not conquer, it will lack achievement.[3] Thus benevolence is the belly of the army.

The general cannot be without Virtue.[4] If he lacks Virtue, then he will not have any strength. If he lacks strength, the advantages of the Three Armies will not be realized. Thus Virtue is the hands of the army.

The general cannot be without credibility. If he is not trusted, then his orders will not be implemented. If his orders are not implemented, then the army will not be unified. If the army is not unified, then it will not attain fame. Thus credibility is the feet of the army.

The general cannot but know victory. If he does not know victory
. . . . . [5]

. . . . . the army will not be [decisive]. Thus decisiveness is the tail of the army.

Commentary

This chapter and the succeeding ones concretely discuss the essential qualities characterizing effective commanders and detail the commonly found faults and flaws that result in significant errors. Most of them also appear in the other early military writings because, with the exception of the *Ssu-ma Fa,* the *Seven Military Classics* all discuss the critical question of the commander's qualifications. Commentators and contemporary historians tend to focus upon the differences between Sun-tzu's and Sun Pin's five essential characteristics, stressing that Sun-tzu emphasized courage; Sun Pin, righteousness and Virtue. However, a careful reading of each thinker's entire work will reveal that they essentially agreed and that they considered knowledge foremost.[6] This is to be expected since the battlefield environment invariably elicits certain qualities that, if absent, would result in obvious failures in command and control.

Many of the extant military books emphasize the necessity for personal leadership, admonishing the general to set an obvious example and visibly lead by sharing every hardship with his men rather than assuming a regal position to the rear in comfort and glory.[7] However, by Sun Pin's era, when campaign armies clearly began to exceed 100,000 soldiers, generals had long since abandoned any personal participation in the actual fighting.[8] The late Spring and Autumn period had earlier seen the slow evolution of professional commanders, while Wu Ch'i (who was active at the start of the Warring States period) was the last significant figure to combine both civil and martial abilities. Sun-tzu's time obviously marked a turning point in the status and authority of "expert" generals, who obviously sought (of necessity) to become visibly independent of the ruler once having been commissioned and assigned troops to command. Thus, Sun-tzu emphasized the commander's necessary independence, while the general's character and virtues, including loyalty, became increasingly important to the political rulers of the day.[9]

Even though this chapter is remarkably clear, parallel passages from other works may prove illuminating. Sun-tzu, who stressed that a general's talents should be "all-encompassing,"[10] also succinctly stated, "The general encompasses wisdom, credibility, benevolence, courage, and strictness."[11] Wu Ch'i noted similar traits: "In general, warfare has four vital points: *ch'i*, terrain, affairs, and strength. One who knows these four is qualified to be a general. However, his awesomeness, Virtue, benevolence, and courage must be sufficient to lead his subordinates and settle the masses. Furthermore, he must frighten the enemy and resolve doubts. When he issues orders, no one will dare disobey them. Wherever he may be, rebels will not dare oppose him."[12]

The *Six Secret Teachings* contains a chapter entitled "A Discussion of Generals," which enumerates both virtues and vices, correlating them with their effects. Deferring the weaknesses until the relevant Sun Pin chapters, the virtues can be arrayed here: "What we refer to as the five critical talents are courage, wisdom, benevolence, trustworthiness (credibility), and loyalty. If he is courageous he cannot be overwhelmed. If he is wise he cannot be forced into turmoil. If he is benevolent he will love his men. If he is trustworthy he will not be deceitful. If he is loyal he will not be of two minds."

The *Wei Liao-tzu* notes that the commanding general "should be composed so that he cannot be stimulated to anger. He should be pure so that he cannot be inveigled by wealth."[13] And the late *Three Strategies of Huang Shih-kung* cites a purportedly ancient book entitled the *Military Pronouncements* in its discussion: "Now the general is the fate of the state. If he is able to manage the army and attain victory, the state will be secure and settled. The *Military Pronouncements* states: 'The general should be able to be pure; able to be quiet; able to be tranquil; able to be controlled; able to accept criticism; able to judge disputes; able to attract and employ men; able to select and accept advice; able to know the customs of states; able to map mountains and rivers; able to discern defiles and difficulty; and able to control military authority.'"[14] Obviously by the time of the *Three Strategies*—perhaps the end of the Former Han—the general's specific abilities and his knowledge of military measures and essential techniques were prominently emphasized due to the complexity of advanced warfare.

24

[THE GENERAL'S VIRTUE]

[If he regards the troops] like an infant, loves them like a handsome boy, respects them like a severe teacher, and employs them like clumps of earth,[1] the general . . . . .

. . . . . not lost, it is the general's wisdom. If he does not slight the few or suffer incursions from the enemy, if he is as cautious about the end as about the beginning, the general . . . . .

. . . . . is not interfered with; and if the ruler's commands do not enter the army's gate,[2] these are the general's constants.[3] When he enters the army . . . . .

(In combat) the two (commanding) generals will not (both) live; the two armies will not (both) survive. The general of the army's . . . . .

. . . . . the general of the army's beneficence. When the bestowing of rewards does not extend past the day, the imposition of punishments is as quick as turning the face, and they are not affected by the man or subject to external threats, this is the general of the army's Virtue.[4]

Commentary

This chapter is badly fragmented, and the reconstruction is questionable because there are virtually no clues to strip sequence or justifications for even including them all. However, the content of each strip taken in isolation remains reasonably clear and can be contextually illuminated by similar passages in the other early military writings. The first passage focuses upon questions of discipline, understood as the overall treatment of troops and the means to bind them to the commander. Several measures, such as sharing hardship, have previously been discussed in the commentaries and Notes. The crux of the matter is balancing the soldiers' fear of the general's power and awesomeness with a devotion forged through emotional allegiance, thereby precluding the danger of simple desertion. Sun-tzu's chapter entitled "Configurations of Terrain" contains a passage that may well underlie this one: "When the general regards his troops as young children, they will advance into the deepest valleys with him. When he regards the troops as his beloved children, they will be willing to die with him. If they are well treated but cannot be employed, if they are loved but cannot be commanded, or when in chaos they cannot be governed, they may be compared to arrogant children and cannot be used."[5] The final sentence in the fragment indicates the new, realistic attitude that pervaded Sun Pin's era: The troops, however much loved, are to be used like "clumps of earth." Any other approach, while temporarily saving lives, would result in greater losses and possibly the state's own demise.

Commentators generally cite a passage from the *Tao Te Ching* in conjunction with the second fragment: "In their management of affairs people constantly defeat them just as they are about to be completed. If one is as cautious about the end as the beginning, then there will not be any defeated states."[6] Wu Ch'i included focused caution as one of the five important affairs for generals: "Now the affairs to which the general must pay careful attention are five: first, regulation; second, preparation; third, commitment; fourth, caution; and fifth, simplification. Regulation is governing the masses just as one controls the few. Preparation is going out the city gate as if seeing the enemy. Commitment means entering combat without any concern for life. Caution means that even after conquering, one maintains the same control and attitude as if entering a battle. Simplification means the laws and orders are kept to a minimum and are not abrasive."[7]

The rise of the professional commander and the assertion of his necessary independence from the ruler, previously noted, are reflected in the third fragment. Sun-tzu said, "One whose general is capable and not interfered with by the ruler will be victorious."[8] The *Three Strategies of Huang Shih-kung,* a later text, states, "When the army is mobilized and advances into the field, the sole exercise of power lies with the general. If in advancing or withdrawing the court interferes, it will be difficult to attain success."[9] The *Six Secret Teachings* also contains a chapter, "Appointing the General," devoted solely to formalizing the commissioning of the commanding general. The chapter emphasizes these themes and would even have the general-designate, upon accepting his mandate, stress his necessary independence, saying, "I have heard that a country cannot follow the commands of another state's government, while an army cannot follow central government control. Someone of two minds cannot properly serve his ruler; someone in doubt cannot respond to the enemy." The T'ai Kung then concludes, "Military matters are not determined by the ruler's commands; they all proceed from the commanding general. When he approaches an enemy and decides to engage in battle, he is not of two minds. In this way there is no Heaven above, no Earth below, no enemy in front, and no ruler to the rear. For this reason the wise make plans for him, the courageous fight for him. Their spirits soar to the blue clouds; they are swift like galloping steeds. Even before the blades clash, the enemy surrenders submissively."[10]

This trend away from civilian or political control starkly contrasts with present thinking and practices, at least as seen in the early 1990s in the United States. (Naturally the communications revolution has radically affected the dimensions and possibilities of real-time battlefield control. Whenever new potentials appear for authorities to exercise power, inappropriately or not, such potentials tend to be exploited. In contrast, in antiquity the fastest message or prearranged signaling system could require hours or days—far too slow to match the pace of battle.) Furthermore, there are two issues here. One is interfering with the commander's exercise of authority by directing him to implement externally generated orders. The second is undermining command authority by issuing orders to the army itself or allowing senior officers to insubordinately presume upon their relationship with the ruler.

Combat in Sun Pin's era increasingly developed into battles of annihilation, and actual clashes generally entailed the defeat of one side or the other and the subsequent death of the losing commander—whether on the battlefield or as punishment for losing.[11] "Thus it is said that two armies will not be victorious, nor will both be defeated. When the army ventures out beyond the borders, before they have been out ten days—even if a state has not perished—one army will certainly have been destroyed and the general killed."[12] Of course, Sun-tzu, Wu Ch'i, and many other strategists felt it necessary to warn against foolishly engaging an enemy force out of blind courage or the fear of being accused of cowardice.

Rewards and punishments were understood by all the military writers as providing the foundation for troop control. Generally the Legalists, such as Lord Shang and Han Fei-tzu, provide the most extensive analyses of their systematization and psychology, but the military theorists also embraced the Legalists' fundamental insights and psychological principles. Two of the most basic are that the implementation of rewards and punishments must be immediate and that they should be effected without regard to rank or position. (In fact, many military writers emphasized that punishments should be visibly imposed especially on the noble and powerful in order to cause the entire army to quake, to show that no one may presume upon position to escape punishment or should fear being ignored because of low rank.)[13] The *Ssu-ma Fa* is cited for its parallels with this chapter: "Rewards should not be delayed beyond the appropriate time for you want the people to quickly profit from doing good. When you punish someone do not change his position for you want the people to quickly see the harm of doing what is not good."[14]

25

THE GENERAL'S DEFEATS

As for the general's defeats (defects):
First, he is incapable but believes himself to be capable.
Second, arrogance.
Third, greedy for position.
Fourth, greedy for wealth.
. . . . .
Sixth, light.[1]
Seventh, obtuse.
Eighth, has little courage.
Ninth, courageous but weak.
Tenth, has little credibility.
. . . . .
Fourteenth, rarely decisive.
Fifteenth, slow.
Sixteenth, indolent.
Seventeenth, [oppressive].[2]
Eighteenth, brutal.
Nineteenth, selfish.
Twentieth, induces confusion.[3]
When the defeats (defects) are numerous, the losses will be many.

Commentary

This chapter obviously enumerates what Sun Pin believed to be the twenty most common or most important defects in commanders. The chapter's title, "The General's Defeats," has been translated to preserve the fundamental meaning of the character *pai* ("defeat") and thereby cohere with both the military context and its appearance in the previous chapter, but in this usage *pai* is synonymous with "defects."

As already mentioned, most of the military writings identified character defects that would adversely affect command and impact the fate of battles and campaigns. A few early books also offered criteria for evaluating officers, sometimes unique but generally falling within the general tradition of "knowing men."[4] Chapter 20, "Selecting Generals," of the *Six Secret Teachings* is the most sophisticated and systematic essay on evaluating potential commanders and resolving the problem posed by external appearance being seriously misleading. For example, "He appears adept at planning but is indecisive; he appears to be decisive and daring but is incapable." Many of them are not just simple discrepancies, such as "He appears courageous but is a coward," but are instead more complex and subtle instances, with behavioral manifestations and underlying character being radically askew. (The chapter goes on to suggest techniques for testing and evaluating men and thereby piercing misleading appearances.)[5] Chapter 19, "A.D.scussion of Generals," of the *Six Secret Teachings* enumerates both the talents required (as already noted in the previous commentary) and the serious defects, many of which are common to this chapter:

> What are referred to as the ten errors are as follows:
>   being courageous and treating death lightly:[6]
>   being hasty and impatient;
>   being greedy and loving profit;
>   being benevolent but unable to inflict suffering;
>   being wise but afraid;
>   being trustworthy and liking to trust others;
>   being scrupulous and incorruptible but not loving men;

being wise but indecisive;
being resolute and self-reliant;
being fearful while liking to entrust responsibility to other men.

As enumerated, these complex "defects" frequently combine a strength with a weakness, the strength being the factor that allows an undesirable tendency to materialize.

The *Six Secret Teachings* chapter also explicates the reasons that these combined defects may prove behaviorally inappropriate on the battlefield and open to exploitation by an astute enemy:

One who is courageous and treats death lightly can be destroyed by violence.

One who is hasty and impatient can be destroyed by persistence.

One who is greedy and loves profit can be bribed.

One who is benevolent but unable to inflict suffering can be worn down.

One who is wise but fearful can be distressed.

One who is trustworthy and likes to trust others can be deceived.

One who is scrupulous but incorruptible but does not love men can be insulted.

One who is wise but indecisive can be suddenly attacked.

One who is resolute and self-reliant can be confounded by events.

One who is fearful and likes to entrust responsibility to others can be tricked.[7]

In the *Art of War*, Sun-tzu also identified various inappropriate traits and listed five excesses found in generals together with their specific dangers:

One committed to dying can be slain.

One committed to living can be captured.

One [easily] angered and hasty can be insulted.

One obsessed with being scrupulous and untainted can be shamed.

One who loves the people can be troubled.[8]

Wu Ch'i emphasized the importance of evaluating an enemy's commanding general and exploiting any recognizable defects:

In general the essentials of battle are as follows: You must first attempt to divine the enemy's general and evaluate his talent. In accord with the situation exploit the strategic imbalance of power; then you will not labor but still achieve results.

A commanding general who is stupid and trusting can be deceived and entrapped.

One who is greedy and unconcerned about reputation can be given gifts and bribed.

One who easily changes his mind and lacks real plans can be labored and distressed.[9]

Finally, even the *Three Strategies of Huang Shih-kung* reprises some flaws and their effects:

If the general stifles advice, the valiant will depart. If plans are not followed, the strategists will rebel.

If the good and evil are treated alike, the meritorious officers will grow weary.

If the general relies solely on himself, his subordinates will shirk all responsibility.

If he brags, his assistants will have few attainments.

If he believes slander, he will lose the trust of the people.

If he is greedy, treachery will be unchecked.

If he is preoccupied with women, then the officers and troops will become licentious.[10]

The paragraph then concludes as if it were expanding Sun Pin's statement "When the defeats (defects) are numerous, the losses will be many":

If the general has a single one of these faults, the masses will not submit; if he is marked by two of them, the army will lack order; if by three of them, his subordinates will abandon him; if by four, the disaster will extend to the entire state.

26

[THE GENERAL'S LOSSES]

The general's losses:

First, if he has lost the means for going and coming, he can be defeated.[1]

Second, if he gathers together turbulent people and immediately employs them, if he stops retreating troops and immediately engages in battle with them, or if he lacks resources but acts as if he has resources, then he can be defeated.[2]

Third, if he constantly wrangles over right and wrong, and in planning affairs is argumentative and disputatious, he can be defeated.[3]

Fourth, if his commands are not implemented, the masses not unified, he can be defeated.[4]

Fifth, if his subordinates are not submissive and the masses not employable, he can be defeated.

Sixth, if the people regard the army with bitterness, he can be defeated.

Seventh, if the army is "old," he can be defeated.[5]

Eighth, if the army is thinking (about home), he can be defeated.[6]

Ninth, if the soldiers are deserting, he can be defeated.[7]

Tenth, if the soldiers . . . . . he can be defeated.

Eleventh, if the army has been frightened several times, he can be defeated.[8]

Twelfth, if the soldiers' route requires difficult marching and the masses suffer, he can be defeated.

Thirteenth, if the army is focusing upon ravines and strongpoints and the masses are fatigued, he can be defeated.[9]

Fourteenth, [if he engages in battle but is] unprepared, he can be defeated.

Fifteenth, if the sun is setting and the road is far while the masses are dispirited, he can be defeated.[10]

Sixteenth . . . . . he can be defeated.

Seventeenth . . . . . the masses are afraid, he can be defeated.

Eighteenth, if commands are frequently changed and the masses are furtive, he can be defeated.[11]

Nineteenth, if the army is disintegrating while the masses do not regard their generals and officials as capable, he can be defeated.[12]

Twentieth, if they have been lucky[13] several times and the masses are indolent, he can be defeated.

Twenty-first, if he has numerous doubts (so) the masses are doubtful, he can be defeated.[14]

Twenty-second, if he hates to hear about his excesses, he can be defeated.

Twenty-third, if he appoints the incapable, he can be defeated.

Twenty-fourth, if their *ch'i* (spirit) has been injured from being long exposed on campaign, he can be defeated.[15]

Twenty-fifth, if their minds are divided at the appointed time for battle, he can be defeated.

Twenty-sixth, if he relies upon the enemy becoming dispirited, he can be defeated.[16]

Twenty-seventh, if he focuses upon harming others and relies upon ambushes and deceit, he can be defeated.[17]

Twenty-eighth, if the army's chariots lack . . . . . , [he can be defeated].

[Twenty-ninth, if he] deprecates the troops and the minds of the masses are hateful, he can be defeated.

Thirtieth, if he is unable to successfully deploy (his forces) while the route out is constricted, he can be defeated.

Thirty-first, if in the army's forward ranks are soldiers from the rear ranks and they are not coordinated and unified with the forward deployment, he can be defeated.[18]

Thirty-second, if in engaging in battle he is concerned about the front and the rear is (therefore) empty; or concerned about the rear, the front is empty; or concerned about the left, the right is empty; or concerned about the right, the left is empty—his engaging in battle being filled with worry, he can be defeated.[19]

Commentary

This chapter appears to be an admixture of material from several sources rather than simply a series of observations from Sun Pin's own school. While many of them are found scattered about other writings, the most telling evidence is the use of three different terms for "army," including the original character *shih*.[20] However, "The General's Losses" does preserve nearly thirty criteria for evaluating potential courses of action against an enemy, some attributable to command failures, others simply conditions of disorder or weakness. Specific comments will be found in the Notes; here, a few of the more famous and condensed enumerations incorporated in roughly contemporary works will be cited for comparative purposes.

Several chapters in the *Six Secret Teachings* describe basic exploitable situations, many similar to the shortcomings loosely identified with the commanding general in Sun Pin's chapter. "Military Vanguard" succinctly enumerates vulnerable moments:

When the enemy has begun to assemble they can be attacked.
When the men and horses have not yet been fed they can be attacked.
When the seasonal or weather conditions are not advantageous to them they can be attacked.
When they have not secured good terrain they can be attacked.
When they are fleeing they can be attacked.
When they are not vigilant they can be attacked.
When they are tired and exhausted they can be attacked.
When the general is absent from the officers and troops they can be attacked.
When they are traversing long roads they can be attacked.
When they are fording rivers they can be attacked.
When the troops have not had any leisure time they can be attacked.
When they encounter the difficulty of precipitous ravines or are on narrow roads they can be attacked.

When their battle array is in disorder they can be attacked.
When they are afraid they can be attacked.

A second series, found in "Battle Chariots," is significant enough to merit reproducing here:

When the enemy's ranks—front and rear—are not yet settled, strike into them.

When their flags and pennants are in chaos, their men and horses frequently shifting about, then strike into them.

When their battle array is not yet solid, while their officers and troops are looking around at each other, then strike into them.

When in advancing they appear full of doubt, and in withdrawing they are fearful, strike into them.

When the enemy's Three Armies are suddenly frightened, all of them rising up in great confusion, strike into them.

When you are fighting on easy terrain and twilight has come without being able to disengage from the battle, then strike into them.

When, after traveling far, at dusk they are encamping and their Three Armies are terrified, strike into them.

The *Ssu-ma Fa* also summarizes various principles of warfare and describes the campaign of the weak and chaotic in the following two passages:

In general, in warfare: Attack the weak and quiet, avoid the strong and quiet. Attack the tired, avoid the well trained and alert. Attack the truly afraid, avoid those that display only minor fears. From antiquity these have been the rules for governing the army.[21]

When in attacking, waging battle, defending, advancing, retreating, and stopping, the front and rear are ordered and the chariots and infantry move in concord, this is termed a well-planned campaign. If they do not follow orders; do not trust their officers; are not harmonious; are lax, doubtful, weary, afraid; avoid responsibility; cower; are troubled, unrestrained, deflated, or dilatory, it is termed a "disastrous campaign." When they suffer from extreme arrogance, abject terror, moaning and grumbling, constant fear, or frequent regrets over actions being taken, they are termed "destroyed and broken."[22]

Similarly, the *Wu-tzu* preserves a number of tactical evaluations with principles for response. Two particularly related to those collected in Sun Pin's chapter are as follows:

If the upper ranks are wealthy and arrogant while the lower ranks are poor and resentful, they can be separated and divided. If their advancing and withdrawing are often marked by doubt and the troops have no one to rely on, they can be shocked into running off. If the officers despise the commanding general and are intent on returning home, by blocking off the easy roads and leaving the treacherous ones open, they can be attacked and captured. If the terrain over which they advance is easy but the road for withdrawal difficult, they can be forced to come forward. If the way to advance is difficult but the road for retreating easy, they can be pressed and attacked. If they encamp on low wetlands where there is no way for the water to drain off, if heavy rain should fall several times, they can be flooded and drowned. If they make camp in a wild marsh or fields dense with a heavy tangle of grass and stalks, should violent winds frequently arise you can burn the fields and destroy them. If they remain encamped for a long time—the generals and officers growing lax and lazy, the army becoming unprepared—you can sneak up and spring a surprise attack.[23]

In general, when evaluating the enemy there are eight conditions under which one engages in battle without performing divination:

First, in violent winds and extreme cold, they arise early and are on the march, while barely awake, breaking ice to cross streams, unfearing of any hardship.

Second, in the burning heat of midsummer, they arise late and without delay press forward in haste, through hunger and thirst, concentrating on attaining far-off objectives.

Third, the army has been out in the field for an extended period; their food supplies are exhausted; the hundred surnames are resentful and angry; and numerous baleful portents have arisen, with the superior officers being unable to squash their effects.

Fourth, the army's resources have already been exhausted; firewood and hay are scarce; the weather frequently cloudy and rainy; and even if they wanted to plunder for supplies, there is nowhere to go.

Fifth, the number mobilized is not large; the terrain and water not advantageous; the men and horses both sick and worn out; and no assistance comes from their allies.

Sixth, the road is far and the sun setting; the officers and men have labored long and are fearful. They are tired and have not eaten; having cast aside their armor, they are resting.

Seventh, the generals are weak; the officials irresponsible; the officers and troops not solid; the Three Armies are frequently frightened; and the forces lack any assistance.

Eighth, their formations are not yet settled; their encampment is not yet finished; or they are traveling dangerous territory and narrow defiles, half concealed and half exposed.

In these eight conditions attack them without any doubts.[24]

27

[MALE AND FEMALE CITIES]¹

If a city lies amidst small marshes, lacks high mountains and notable valleys, but has moderate-sized mounds² about its four quarters, it is a "male city" and cannot be attacked.

If an army drinks from flowing water, [it is water that will sustain life, and they cannot be attacked].³

If before a city there is a notable valley while it has a high mountain behind it, it is a male city and cannot be attacked.

If (the terrain) within a city is high while it falls away outside it, it is a male city and cannot be attacked.

If within a city there are moderate-sized mounds, it is a male city and cannot be attacked.

An army that is encamping after being on the march without avoiding notable rivers, whose *ch'i* has been harmed and determination weakened, can be attacked.⁴

A city with a notable valley behind it that lacks high mountains to its left and right is a vacuous city and can be attacked.⁵

Thoroughly incinerated terrain is deadly ground; (an army occupying it) can be attacked.⁶

If an army drinks stagnant water,⁷ it is water that will result in death, and they can be attacked.

If a city lies amidst vast marshes and lacks notable valleys and moderate-sized mounds, it is a "female city" and can be attacked.

If a city lies between high mountains and lacks notable valleys and moderate-sized mounds, it is a female city and can be attacked.

If there is a high mountain in front of a city and a notable valley behind it, while before it (the ground) ascends and to the rear it descends, it is a female city and can be attacked.

Commentary

This chapter is extensively cited—including by those who fail to translate it in their modern Chinese editions—as evidence that the scope and concepts of warfare significantly evolved over the century between Sun-tzu and Sun Pin.[8] As is generally known, Sun-tzu had strongly cautioned against precipitously and wastefully attacking cities in this famous passage:

> This tactic of attacking fortified cities is adopted only when unavoidable. Preparing large movable protective shields, armored assault wagons, and other equipment and devices will require three months. Building earthworks will require another three months to complete. If the general cannot overcome his impatience, but instead launches an assault wherein his men swarm over the walls like ants, he will kill one-third of his officers and troops, and the city will still not be taken. This is the disaster that results from attacking fortified cities.[9]

For centuries thereafter his view was frequently but incorrectly simplified to merely "Do not attack cities." However, the passage's initial section indicates that rather than simply condemning such attacks outright, he advocated the implementation of more effective tactics: "The highest realization of warfare is to attack the enemy's plans; next is to attack their alliances; and the lowest is to attack their fortified cities."[10] Furthermore, tomb fragments recovered with Sun Pin's *Military Methods* that appear to be an integral part of Sun-tzu's *Art of War* expand his view further: "As for fortified cities that are not assaulted: We estimate that our strength is sufficient to seize it. If we seize it, it will not be of any advantage to the fore; if we gain it we will not be able to protect it to the rear. If our strength (equals) theirs, the city certainly will not be taken. If, when we gain the advantage of a forward position the city will then surrender by itself, while if we do not gain such advantages (the city) will not cause harm to the rear—in such cases, even though the city can be assaulted, do not assault it."[11] Sun-tzu thus emphasized calculating the potential advantages and employing methods other than frontal assaults, such as drawing the enemy out so that it will be forced to fight on more advantageous and open terrain.

Sun Pin's dual categorization of "male" and "female" cities is generally contrasted with Sun-tzu's reluctance to assault fortified cities and is interpreted as reflecting the growth of cities as economic and strategic centers. Unlike in the Spring and Autumn era, when campaign armies could move relatively unhindered through the sparsely populated open countryside, in the Warring States period they could be thwarted by the fortified strongholds that had concurrently assumed much greater military and economic value. Among these cities the strategically weaker ones, classified as female, could—and by implication—should be attacked, while the stronger, or male, ones should be avoided. (However, note that Sun Pin never explicitly stated that the female cities should invariably be attacked or designated them as more than preferred targets.) His classificatory principles appear to be simply topographical; however, other situations similarly categorized for their attack potential have also been intermixed. The latter are similar to those raised in previous chapters, and to Wu-tzu's tactical analyses, and will be appropriately discussed in the Notes.

The *Wei Liao-tzu*, a military classic probably composed in the century following Sun Pin's death, specifically discusses the importance of cities and notes the economic importance of markets (centered in cities) for sustaining the armed forces.[12] The text further states that "land is the means for nourishing the populace; fortified cities the means for defending the land; combat the means for defending the cities."[13] Consequently, it identifies cities as primary targets, particularly if "the cities are large and the land narrow":[14]

Thus, in general, when the troops have assembled and the general has arrived, the army should penetrate deeply into (the enemy's) territory, sever their roads, and occupy their large cities and large towns. Have the troops ascend the walls and press the enemy into endangered positions. Have the several units of men and women each press the enemy in accord with the configuration of the terrain and attack any strategic barriers. If you occupy the terrain around a city or town and sever the various roads about it, follow up by attacking the city itself.[15]

Thus, a clear historical progression is witnessed from Sun-tzu through Sun Pin to Wei Liao-tzu, from viewing assaults on cities as the lowest tactical option to emphasizing the need to defend and attack them as the highest.[16]

Sun Pin's principles for categorizing the cities in this chapter are less apparent. In antiquity, both in the West and China, it was axiomatic that one should "value high terrain and disdain low ground."[17] Thus, the T'ai Kung said, "Occupying high ground is the means by which to be alert and assume a defensive posture."[18] Sun-tzu himself said, "Do not approach high mountains, do not confront those who have hills behind them."[19] In addition, as a general principle he stated, "To cross mountains follow the valleys, search out tenable ground, and occupy the heights. If the enemy holds the heights, do not climb up to engage them in battle. This is the way to deploy an army in the mountains."[20] Clearly, since it is decidedly disadvantageous to mount an uphill assault, to rush against an enemy directing its fire downward or taking advantage of gravity in wielding its shock weapons while benefiting from the attacker's rapid exhaustion, as brought about by the greater exertion necessary to ascend heights, it would be even more foolhardy to cross a deep valley and then attempt to storm the walls. Not only would the troops be tired, but also the terrain would constrict the number that might be focused upon the walls, making it difficult to achieve the historically attested ratio of about four to one for the assault to prevail.

Whenever a city incorporates higher terrain (such as moderate-sized mounds) that will allow missile weapons to be directed downward while forcing the attackers to direct their fire upward, the city should be considered strong and therefore not easily approached and overwhelmed. Even if the walls are ·penetrated, the interior terrain will provide natural vantage and strongpoints for mounting a concerted defense.

Less clear is the relationship of mountains located to the sides of a city. Mountains to the rear would seem to prevent the employment of large numbers of men, just as marshes would, but at the same time, if undefended, would allow an enemy the possibility of height advantage.[21] Merely being able to employ incendiary arrows would endanger the city from above; this sort of tactical thought perhaps motivated Sun Pin to classify "a city lying between high mountains" as a female city. However, the mirrored relationship of the third and last classifications raises questions: "If before a city there is a notable valley while it has a high mountain behind it, it is a male city and cannot be attacked. . . . If there is a high mountain in front of a city and a notable valley behind it, while before it (the ground) ascends and to the rear it descends, it is a female city and can be attacked." In the second case, if all the defenses were directed toward the front—an incredible and therefore unthinkable blunder—an effective downward assault on their fixed arrays would be possible. However, front and rear, while reflecting the city's orientation, simply represent a rotation in perspective for an attacker. A city with a high mountain behind it would seem to be equally easy prey for assault forces raining missiles downward, particularly if the defenders had not established preemptive defensive positions upon it. While Sun-tzu warned against having valleys, gorges, and similar depressions behind one's forces, and stressed maneuvering the enemy so that he would find himself constricted by them to the rear, as a tactical principle this applies to field forces, not fixed citadels. Unfortunately, the reasoning here remains unclear, perhaps awaiting the discovery of further texts on topography and configuration.

The presence of mounds appears to be the distinguishing factor in the first male city (situated amidst small marshes), for later Sun Pin explicitly states that a city with moderate-sized mounds is a male city. Accordingly, a city lacking them and other natural defenses, even though protected by marshes (which all the military writers warned would impede progress and enmire vehicles, as previously discussed), falls under the female classification.

The remaining situations, which may have been included by later compilers, overlaying a core discussion about "male and female cities," indicate the importance of potable water to the army's survival and suggest the problems posed by thoroughly savaged ground that is unable to sustain any life at all. Finally, there is one passage noting the plight of those exhausted from working their way along or across large rivers, who are exhausted and suffering from a loss of morale and determination, making them easy prey.

28

[FIVE CRITERIA, NINE SEIZINGS]

五度九奪

. . . . . And when the rescuers arrive, they are also severely defeated. Thus the essential principle for the army[1] is that those fifty *li* apart do not rescue each other. How much more so is this the case when the nearest are [a hundred *li* apart, the farthest] several hundred *li*. These are the extremes for weighing the army's (possibilities). Thus the *Tactics* states: "If your provisions are unlike theirs, do not engage them in protracted (battles). If your masses are unlike theirs, do not engage them in battle. [If your . . . . . is unlike theirs, do not . . . . . If your . . . . . is unlike theirs], do not contend with them across a broad front.[2] If your training is unlike theirs, do not oppose them in their strength.[3] When these five criteria are clear, the army will be able to forcefully advance unhindered. Thus the army . . . . .

. . . . . As for the techniques for forcing the enemy to rush about: The first is called seizing provisions. The second is called seizing water. The third is called seizing fords. The fourth is called seizing roads. The fifth is called seizing ravines. The sixth is called seizing easy terrain. The seventh is called [. . . . . The eighth is called . . . . . The ninth] is called seizing what he solely values.[4] In general these nine "graspings"[5] are the means by which to force the enemy to hasten about.

Commentary

Only slightly more than one-fourth of this chapter remains, with no indication of what the other topics beyond the two clearly expressed in these sections may have been. The beginning reflects an emphasis found in Sun-tzu's first chapter, "Initial Estimations," upon analyzing the enemy and calculating the possibilities for victory and defeat. While the terms differ somewhat, the concept of "measuring" is found throughout Sun-tzu and most other military writers, particularly with regard to determining the number of men appropriate to configurations of terrain and upcoming campaigns.[6] For example, "As for military methods: the first is termed measurement; the second, estimation of forces; the third, calculation of numbers of men; the fourth, weighing relative strength; and the fifth, victory. Terrain gives birth to measurement; measurement produces the estimation of forces. Estimation of forces gives rise to calculating the numbers of men. Calculating the numbers of men gives rise to weighing strength. Weighing strength gives birth to victory."[7]

The missing portion of Sun Pin's initial discussion probably cited historical examples of forces being defeated after hasty but ill-conceived rescue efforts, leading to the conclusion that forces only 50 *li* apart already exceed the possible range for mutual aid. Assuming that one force rushes to assist a separate, embattled unit (rather than both of them racing toward each other, thereby cutting their respective distances in half), the full 50 *li* would require approximately two days based upon a normal march speed of 30 *li* per day. Therefore, depending upon the terrain's traversability, the engagement site could theoretically be reached in a single day at double pace, just as at the battle of Ma-ling. However, the famous general Wu Ch'i (Wu-tzu) stressed the need for a measured advance in order not to wear out the men and horses,[8] while among Sun Pin's measures for exhausting an enemy is a forced double march: "Those who excel (in warfare) can cause the enemy to roll up his armor and race far off, to travel two days' normal distance at a time, to be exhausted and sick but unable to rest, to be hungry and thirsty but unable to eat. An enemy emaciated in this way certainly will not be victorious."[9] Sun-tzu stated even more explicitly:

If you abandon your armor and heavy equipment to race forward day and night without encamping, covering two days normal distance at a time, marching forward a hundred *li* to contend for gain, the Three Armies' generals will be captured. The strong will be first to arrive, while the exhausted will follow. With such tactics only one in ten will reach the battle site. If one contends for gain fifty *li* away, it will cause the general of the Upper Army to stumble, and by following such tactics half the men will reach the objective. If you contend for gain at thirty *li*, then two-thirds of their army will reach the objective.[10]

The fundamental conclusion, expressed in concrete terms by the *Methods,* is simple: Do not engage forces for which one is not a match. This embodies Sun-tzu's basic principle that one must first evaluate the enemy and then implement appropriate tactics and Sun Pin's warning not to attack strength with weakness.[11] Directly engaging a potent enemy would also dramatically contravene Sun-tzu's dictum, "If it is not advantageous, do not move. If objectives cannot be attained, do not employ the army."[12] However, Sun Pin addressed the necessity for manipulating superior forces so that they might be successfully engaged in two chapters, "Distinction Between Guest and Host" and "Those Who Excel." Sun-tzu himself previously provided some basic parameters: "In general, the strategy for employing the military is this: If your strength is ten times theirs, surround them; if five, then attack them; if double, then divide your forces. If you are equal in strength to the enemy, you can engage him. If fewer, you can circumvent him. If outmatched, you can avoid him. Thus a small enemy that acts inflexibly will become the captives of a larger enemy."[13]

The last part of this chapter discusses some concrete means for manipulating the enemy—forcing him to hasten about—that are virtually identical to Sun-tzu's own measures. The thrust of these "seizures" is to compel the enemy to precipitously act by seizing what he values, what is essential to him, thereby taking the initiative against the enemy and ensuring that any engagements will be with a weakened, confused, and tired opponent. In "Ten Questions" Sun Pin stated, "Attack positions that they must rescue." The list of critical targets, several previously identified in "Those Who Excel," encompasses such fundamentals as food, water, and strategic points. By striking and capturing them, the army not only prevents the enemy from benefiting from them but also compels him to attempt a defense, mounting a desperate effort aimed at simple self-preservation.

29
[THE DENSE
AND DIFFUSE]

[The dense] conquer the diffuse;[1] the full conquer the vacuous;[2] by-ways (shortcuts) conquer roads;[3] the urgent conquer the slow; the numerous conquer the few; the rested conquer the weary.

If they are dense, make them denser; if they are diffuse, disperse them; if they are full, make them fuller; if they are vacuous, [make them more vacuous; if they are taking shortcuts, make them shorter]; if they are on the road, make the road (longer); if they are urgent, make them more urgent; [if they are slow, make them slower; if they are numerous, make them more numerous]; if they are few, make them fewer; if they are rested, make them more rested;[4] if they are tired, make them more tired.

The dense and diffuse mutually change into each other; the full and vacuous [mutually change into each other]; the urgent and slow mutually change into each other; the numerous and few mutually [change into each other; the rested and tired mutually] change into each other.

Do not oppose the dense with the dense;[5] do not oppose the dispersed with the dispersed; do not oppose the full with the full; do not oppose the vacuous with the vacuous; do not oppose the urgent with the urgent; do not oppose the slow with the slow; do not oppose the numerous with the numerous; do not oppose the few with the few; do not oppose the rested with the rested; do not oppose the weary with the weary.

The dense and diffuse mutually oppose each other; the full and vacuous mutually oppose each other; [shortcuts and roads mutually oppose each other; the urgent and slow mutually oppose each other; the numerous and few] mutually oppose each other; the rested and weary mutually oppose each other.

A dense enemy can be dispersed; the full can be made vacuous; one taking shortcuts can (be forced onto) main roads; the urgent [can be slowed; the numerous can be made few; the rested can be fatigued.]

Commentary

Although the characters and phrases in this chapter are all simple and clear, the meaning remains somewhat enigmatic, turning upon how the referents are defined. Furthermore, because of Sun Pin's penchant for employing laconic parallels in series, it appears that some terms may have been inappropriately included, such as shortcuts (the direct?) and main roads (the indirect?). Depending upon perspective and inference, radically different interpretations are possible.

The entire chapter revolves around six correlated pairs, all of which are primarily found in Sun-tzu's *Art of War.*[6] The first paragraph introduces them by positing conquest relationships between the first and second items: dense/diffuse, full/vacuous, shortcut/road, urgent/slow, numerous (masses)/few, rested/tired. Except perhaps for the problematic shortcut/road pair, they all are obvious and reflect commonly held views derived from empirical battlefield experience and codified into fundamental, orthodox theory. For example, Sun-tzu said, "In order await the disordered; in tranquility await the clamorous. This is the way to control the mind. With the near await the distant; with the rested await the fatigued; with the sated await the hungry. This is the way to control strength."[7] Sun Pin's sentences follow the basic pattern "Density conquers dispersion," but the implication, as expressed in the translation, is simply their concrete embodiment: "The dense conquer the diffuse."

The second paragraph consists of a series of parallel four-character phrases that state a condition (such as "dense"), add a connective ("thus," interpreted by most commentators as "then"),[8] and then repeat the first verbal adjective as a transitive verb with the explicit object "them." This works out to, "Dense, then dense them." Some commentators immediately understand this general formula as meaning, "If they are X, then treat them as X"—or, in other words, follow the prescription that states that the opposite should be employed to conquer them.[9] Thus, "if they are dense, then treat them as dense (by dispersing)." While this reflects basic orthodox theory—manipulate the enemy into a state in which one's forces in their present condition possess an advantage—even the latter interpretation for this sentence is questionable because one can disperse oneself or can force the enemy to disperse.[10] Moreover, this sort of reasoning works for conditions such as dense or dispersed, which are under voluntary control, but less so for masses and exhaustion, thereby suggesting that parallelism simply led the compilers astray.

A second possibility for this paragraph is as translated, following principles underlying Taoist thought and concretely expressed as a tactical principle in several of the military classics. Namely, whatever the enemy's condition or situation, destabilize them by further increasing or augmenting it; then apply military force in an appropriate manner to subjugate them. Chapter 36 of the traditional recension, the *Tao Te Ching*, states, "If you want to reduce something, you must certainly stretch it. If you want to weaken something, you must certainly strengthen it. If you want to abolish something, you must certainly make it flourish. If you want to grasp something, you must certainly give it away. This is referred to as subtle enlightenment. The pliant and weak will conquer the hard and strong."

Such thought is based upon the idea that extremes are unstable, that "reversal is the movement of the Tao."[11] The T'ai Kung employed this principle in his psychological warfare techniques and embraced it in "Three Doubts": "Now in order to attack the strong, you must nurture them to make them even stronger, and increase them to make them even more extensive. What is too strong will certainly break; what is too extended must have deficiencies. Attack the strong through his strength." Wei Liao-tzu noted, "Observing the enemy in front, one should employ their strength. If the enemy is white, then whiten them; if they are red, then redden them."[12] And the *Ssu-ma Fa* warned against making solid formations heavier, no doubt because they would become unmanageable and inflexible.[13] Accordingly, the formula would read as previously translated: "If they are dense, make them denser." In this particular case the objective could presumably be achieved by forcing the enemy into constricted areas where their density will render them unable to wield their weapons or implement any tactics. The only problematic term in this sort of reading is shortcut/roads, perhaps understandable as, "Force them onto even smaller paths."[14] Fortunately, in this chapter, despite the subsequent paragraphs that speak about mutual change, the conquest relationship is not stated as being symmetrical, even though the second term can be employed to overcome the first.

The third paragraph concretely elucidates the principle that a condition can change into its counterpart (or complementary condition). The dense can be dispersed, the dispersed brought together to become dense. The weary can be rested, the rested wearied. Those taking the shortest (or hastiest) route can be forced onto indirect ones. This accords with Sun-tzu's principle (and is applied in the chapter's last paragraph): "In military combat what is most difficult is turning the circuitous into the straight, turning adversity into advantage. Thus if you make the enemy's path circuitous and entice them with profit, although you set out after them you will arrive before them. This results from knowing the tactics of the circuitous and the straight."[15]

According to the next chapter, entitled "Unorthodox and Orthodox," the complementary conditions are in fact the basis of unorthodox tactics. The fundamental principle stated in the fourth paragraph—that identical attributes should not be used to oppose each other—is clearly asserted in one section of the next chapter: "Things that are the same are inadequate for conquering each other. Thus employ the different to create the unorthodox. Accordingly, take the quiet to be the unorthodox for movement, ease to be the unorthodox for weariness, satiety to be the unorthodox for hunger, order to be the unorthodox for chaos, and the numerous to be the unorthodox for the few." Furthermore, the fourth paragraph reflects Sun-tzu's and Sun Pin's basic philosophy of not opposing strength with strength, as discussed in previous chapters and their commentaries. Opposing the dense with the dense becomes a question of head-butting and attrition, a test of force and a waste of power rather than an effective tactic. Consistent with this tactical principle, in "The Questions of King Wei" and "Ten Deployments" Sun Pin outlined several formations and correlated a few of them with the deployments against which they would prove effective. While they are not invariably complementary conditions, they certainly never describe tactics that entail direct confrontations of strength.

The final portion concludes the chapter with statements of the form "If the enemy is X, then can X-complement." "If the enemy is dense, then one can be dispersed" would be the simplest and most direct understanding. However, the translated reading of "If the enemy is dense, then he can be made sparse" is coherent with Sun Pin's approach to manipulating the enemy and thereby creating an advantage that can be exploited by troops in a particular condition.[16] Obviously the initial situation defines the domain of possibilities: If one has fewer troops than the enemy, they cannot simply be multiplied. Instead, tactics must be employed to create localized, realizable advantages. Even when both sides are numerous and dense, orthodox frontal assaults are to be avoided. Rather, if a dense formation is to be employed, then the enemy should be forced to disperse somewhat, allowing deep penetration and a quick overwhelming of their troops. Tricks and ruses should be employed to compel them into movement, to divide and fragment them. Conversely, another unorthodox response would be to employ a diffuse formation to control and assault them. In implementing such tactics, the wisdom of the second paragraph becomes apparent: When employing a diffuse formation against a dense enemy, the more concentrated he may be, the more effective the diffuse methodology. Spread-out formations are employed to attack circular and square formations in Chapter 17, "Ten Questions," while other concrete analyses sustaining this argument are found scattered throughout the book. For example, Chapter 16 speaks about the diffuse providing a flexible response capability, whereas Chapter 14 suggests using the heavy to attack the light, a perfect example of employing the complementary (and thus unorthodox).

30

UNORTHODOX
AND ORTHODOX

The patterns of Heaven and Earth, reaching an extreme and then reversing, becoming full and then being overturned, these are [*yin* and *yang*].[1]

In turn flourishing, in turn declining, these are the four seasons.

Having those they conquer, having those they do not conquer, these are the five phases.[2]

Living and dying, these are the myriad things.

Being capable, being incapable, these are the myriad living things.

Having that which is surplus, having that which is insufficient, these are form and strategic power.[3]

Thus as for the disciples of form, there are none that cannot be named.[4] As for the disciples that are named, there are none that cannot be conquered. Thus the Sage conquers the myriad things with the myriad things;[5] therefore his conquering is not impoverished.

In warfare, those with form conquer each other. There are not any forms which cannot be conquered, but none know the form (disposition) by which one conquers.[6] The changes in the forms of conquest are coterminal with Heaven and Earth and are inexhaustible.

As for the forms of conquest, even the bamboo strips of Ch'u and Yüeh would be insufficient for writing them down.[7] Those that have form (dispositions) all conquer in accord with their mode of victory.[8] Employing one form of conquest to conquer the myriad forms (dispositions) is not possible. That by which one controls the form is singular; that by which one conquers cannot be single.

Thus when those who excel at warfare discern an enemy's strength, they know where he has a shortcoming. When they discern an enemy's insufficiency, they know where he has a surplus. They perceive victory as (easily) as seeing the sun and moon. Their measures for victory are like using water to conquer fire.[9]

When form is employed to respond to form, it is orthodox. When the formless controls the formed, it is unorthodox. That the unorthodox and orthodox are inexhaustible is due to differentiation. Differentiate according to unorthodox techniques, exercise control through the five phases, engage in combat with [three forces].[10]

..... Once differentiations have been determined, things take form. Once forms have been determined, they have names. .....

..... Things that are the same are inadequate for conquering each other.[11] Thus employ the different to create the unorthodox. Accordingly, take the quiet to be the unorthodox for movement, ease to be the unorthodox for weariness, satiety to be the unorthodox for hunger, order to be the unorthodox for chaos, and the numerous (masses) to be the unorthodox for the few.[12]

When (action) is initiated, it becomes the orthodox; what has not yet been initiated is the unorthodox.[13] When the unorthodox is initiated and is not responded to, then it will be victorious. One who has a surplus of the unorthodox will attain surpassing victories.[14]

Thus if when one joint hurts the hundred joints are not used, it (is because) they are the same body.[15] If when the front is defeated the rear is not employed, it is because they are the same form (disposition).

Thus (to realize) strategic power in warfare, large formations [should not be] severed; small formations [should not be] broken up. The rear should not encroach upon the front; the front should not trample the rear.[16] Those who are advancing should have a route out; those withdrawing should have a route for advancing.

If rewards have not yet been implemented and punishments not yet employed, but the people obey their commands, (it is because) the people are able to implement them. If rewards are high and punishments pervasive but the people do not obey their commands, it is because the people are not able to implement the commands. In spite of disadvantageous circumstances, to make people advance unto death without turning on their heels (is something) that even Meng Pen would find difficult; to require it of the people is like trying to make water flow contrary (to normal).[17]

Thus (to realize) strategic combat power, increase the victorious; alter the defeated; rest the weary; feed the hungry. Accordingly the people will see [the enemy's] men but not yet perceive death; they will tread on naked blades and not turn their heels.[18] Thus when one understands patterns of flowing water, he can float rocks and break boats.[19] When, in employing the people, one realizes their nature, then his commands will be implemented just like flowing water.

Commentary

Sun Pin's final chapter in the original reconstruction of the text is without doubt the paradigm expression and philosophical culmination of the strategy of the unorthodox and orthodox. Among the *Seven Military Classics,* only Sun-tzu's *Art of War* and the late *Questions and Replies* explicitly adopt these methods, although the *Six Secret Teachings* does advocate many tactics fundamental to actualizing the concepts. Sun-tzu himself might well be considered the concept's progenitor, at least insofar as extant materials allow attribution, and certainly its first proponent. However, in this chapter Sun Pin not only embraces the basic concepts but also elucidates and expands them, further systematizing and advancing them by integrating them with Sun-tzu's concepts of the formless and the *Tao Te Ching's* cosmogenic philosophy. While much of the material derives from Sun-tzu's *Art of War,* including some of the explanatory analogies, Sun Pin's comprehensive formulation transcends the original, and the chapter remains the most incisive discussion to be found among any of China's many military writings.

Although the entire *Art of War* is premised upon and largely reflects the implementation of the unorthodox and orthodox, Sun-tzu primarily advanced the concept in the chapter entitled "Strategic Military Power." The following paragraphs encapsulate his thought:

> What enables the masses of the Three Armies to invariably withstand the enemy without being defeated are the unorthodox and orthodox.

In general, in battle one engages with the orthodox and gains victory through the unorthodox. Thus one who excels at sending forth the unorthodox is as inexhaustible as Heaven, as unlimited as the Yangtze and Yellow rivers. What reach an end and begin again are the sun and moon. What die and are reborn are the four seasons. The notes do not exceed five, but the changes of the five notes can never be fully heard. The colors do not exceed five, but the changes of the five colors can never be completely seen. The flavors do not exceed five, but the changes of the five flavors can never be completely tasted. In warfare the strategic configurations of power do not exceed the unorthodox and orthodox, but the changes of the unorthodox and orthodox can never be completely exhausted. The unorthodox and orthodox mutually produce each other, just like an endless cycle. Who can exhaust them?

Clearly Sun Pin incorporated Sun-tzu's imagery as well, for the cyclic character of natural phenomena underlies the concept of the ever-evolving, ever-changing tactics of the unorthodox and orthodox.[20]

The conception's essence, as discussed briefly in the Historical Introduction and elsewhere in our writings,[21] is that engaging an enemy in battle in conventional, expected ways represents "orthodox" tactics, while employing unexpected, surprise attacks and movements is unorthodox. Naturally everything depends upon the enemy's assumptions and evaluations in a particular situation: If the enemy expects a flanking attack instead of the usual frontal advance in strength, the former then turns out to be "expected" and therefore by definition "orthodox" rather than "unorthodox," as initially conceived.[22] The question is thus reduced to a tactical one: how to create false expectations and how to exploit them. For the former the military writings are replete with techniques for confusing and befuddling the enemy; realizing the latter depends upon misdirection and the army's ability to maneuver, segment and recombine, and effect unexpected speed, such as through the employment of the cavalry arm appearing just after Sun Pin's era.[23]

Another concept echoed in Sun Pin's formulation of the unorthodox and orthodox probably stems from the *Tao Te Ching* but also appeared in other writings of the period, including (to a more limited extent) the *Art of War*.[24] The following famous words make up the opening lines of the *Tao Te Ching:* "The Tao that can be spoken of is not the constant Tao; The name that can be named is not a constant name. The nameless is the beginning of Heaven and Earth; The named is the mother of the myriad things." Embracing the terminology, if not necessarily the full philosophy of the *Tao Te Ching*, Sun Pin integrates Sun-tzu's concept of the formless and the Taoist perspective on names in this passage: "Thus as for the disciples of form, there are none that cannot be named. As for the disciples that are named, there are none that cannot be conquered." The key is the nature of the visible, of that which has attained form. Once something is visibly formed, it can be described; once describable, characteristics can be appended, predications can be made, and plans can be formulated. In the realm of concrete things—which for Sun Pin and the other military strategists includes military deployments or dispositions—any tangible form can be opposed and conquered by another. The normal penchant in the West, perhaps stemming from the Greek tradition, has been to oppose force with force, strength with strength. In contrast, Sun Pin here and in previous chapters advises against this wasteful and often futile approach, advocating instead consciously determining and employing the complementary position, the latter being identified as the "unorthodox." Examples of complementary dispositions found in the immediately preceding chapters (such as quiet/movement) are reiterated here, although now explicitly identified as unorthodox counterparts. The principle of employing the unorthodox should govern the tactics being implemented to achieve victory; this is the "singular" principle referred to by the sentence, "That by which one controls the form is singular; that by which one conquers cannot be single." Therefore, it is only necessary to develop the tactics appropriate to any particular battle, to find the "thing"—that is, the form or disposition—among the myriad things, whose strength will naturally counter and overwhelm the enemy.

In "Unorthodox and Orthodox" Sun Pin equally adheres to Sun-tzu's concept of the formless, both of them having absorbed the Taoist idea that the formless cannot be named, cannot be characterized. Sun-tzu himself stressed being formless in order to prevent the enemy from discerning one's intentions and disposition, thereby thwarting the development of effective tactics for attacking or defending. (This is to be accomplished through deception, numerous misleading techniques, and stealthy maneuvers.) In "Vacuity and Substance" he said, "Thus if I determine the enemy's disposition of forces while I have no perceptible form, I can concentrate my forces while the enemy is fragmented. The pinnacle of military deployment approaches the formless. If it is formless, then even the deepest spy cannot discern it or the wise make plans against it." But Sun Pin, while obviously embracing his doctrine, makes it even more comprehensive and explicit through the arguments developed in the several middle paragraphs, concluding that "when the unorthodox is initiated and is not responded to, then it will be victorious. One who has a surplus of the unorthodox will attain surpassing victories." This primarily applies to situations in which strengths are equal or one is outmatched rather than to those in which the enemy is outnumbered by a significant factor, and more "orthodox" tactics, such as convergent attacks, described in the other chapters can be employed.

The final few paragraphs venture into the area of motivation and the means for realizing strategic power. He insightfully focuses upon the need for organization and discipline in formations; otherwise any deployment will be fraught with the possibility of internally generated chaos. However, if the troops are treated appropriately and not compelled to undertake the impossible, a formidable force can be forged. Unexpressed but no doubt underlying Sun Pin's thought is Sun-tzu's concept of strategic power and its effects in making inevitable that men will perform as expected, that the army will prevail because conditions for the execution of tactics—the creation of an imbalance in power—are attained. As Sun-tzu said:

> One who excels at warfare seeks victory through the strategic configuration of power, not from reliance upon men. Thus he is able to select men and employ strategic power.
>
> One who employs strategic power commands men in battle as if he were rolling logs and stones. The nature of wood and stone is to be quiet when stable but to move when on precipitous ground. If they are square they stop, if round they tend to move. Thus the strategic power of one who excels at employing men in warfare is comparable to rolling round boulders down a thousand fathom mountain. Such is the strategic configuration of power.[25]

31
FIVE INSTRUCTIONS

Sun Pin said: "One who excels at instructing the fundamentals does not make changes when directing the army (in combat). Thus it is said that there are five instructions: instructions for controlling the state, instructions for arraying the lines, instructions for controlling the army, instructions for controlling deployments, and instructions for making combat advantageous when the armies are hidden and not mutually visible.

"What are the instructions for controlling the state ? . . . . . The filial, brotherly, and good, the five virtues, will warriors be without one of them? Then even though they can shoot (a bow), they should not mount a chariot. This being so, then those who excel at archery act as the left (of the chariot), those who excel at driving act as drivers, and those who lack both skills act as the right. Thus three men are emplaced on a chariot; five men are emplaced in the squad of five; ten men make a line; a hundred men make a company; a thousand men have a drum; ten thousand men act as a martial force, and the masses can be employed in great ways. The instructions for controlling the state are such.

"What are the instructions for arraying the lines like? The general's men must take responsibility even for broken down chariots and exhausted horses for they provide the means to efficiently . . . . . If the general establishes himself at ravines and strong points, they will provide the means to be respected . . . . . adequate. The instructions for arraying the lines are such.

What are the instructions for controlling the army like? The soldiers' leather armor and the wagons' metal layers are the implements (that make it possible) to deploy ..... in order to make those who excel glorious. Then the deployment will always be advantageous, and the formations will be substantial and abundant. The instructions for controlling the army are such.

"What are the instructions like for making combat advantageous when the forces are hidden and not mutually visible? .....

The five instructions ..... "

Commentary

This chapter appeared only with the 1985 revised reconstruction of the text, increasing the portions directly attributed to Sun Pin (due to the appearance of his name at the outset of the chapter) in the core material to sixteen chapters. A couple of subsequent modern Chinese versions incorporate it, including the SY edition, which is basically followed here, but for the most part the commonly available texts, whether in original or reprint, continue to present the fifteen-section format already seen in the first half of our translation.

Although the chapter is badly damaged, the basic theme—five instructions designed to ensure that the foundations of the military arts are well practiced and understood—remains clear. (The term *five instructions* also appears in the "Military Methods" chapter of the *Kuan-tzu,* but the five instructions are directed to the eyes, ears, feet, hands, and heart.) The overriding principle is that the commander and his troops should be at one, fully knowledgeable about each other, just as in the analogy previously cited of the mind and the four limbs. Thus, as Wu Ch'i and others pointed out, instruction in the basics should be strict; thereafter orders should not be changed. This allows a rational, deliberate approach to combat engagements that ensures the troops are capable of efficiently executing the chosen battle plan rather than floundering about in a flawed, disordered attempt to translate extemporaneous, even helter-skelter improvisations into survival tactics.

The principles behind the instructions' division and the exact contents of each category remain somewhat puzzling, partly as a result of the serious losses incurred over the centuries. However, the first topic, instructions for controlling the state, focuses not on government or administrative measures but instead on the need for ethical values, for the soldiers to be marked by characteristics that will make them reliable and exemplary. This is perhaps unique, for most of the military writings speak about such virtues being essential for rulers, generals, even the lower officers, but not soldiers in general, who were normally trained to immediate response and obedience, not the cultivation of Virtue. The paragraph may not be speaking about the masses of ordinary troops but perhaps just the selected few, the elite warriors, or even the officers (an older meaning for the term *shih*). Whatever the scope, the import is clear and merits noting.

This first paragraph also reflects the assignment of duties in chariot warfare as practiced from the Shang on downward: three men, with the archer (who was also the commander) on the left, the driver in the middle, and the shock-weapons specialist on the right to wield the halberd. The decade system of organization is also prominent, based on the squad of 5 but extending through companies of 100 and armies of 10,000 rather than the common pyramided sequence of 5, 25, 125, up to armies of 12,500. The decade-based system is thought to have characterized the peripheral states in contrast to the earlier Chou and the central plains states influenced by them and will be discussed in our other works.

The instructions for arraying or deploying the lines are unfortunately fragmented, but once again the importance of occupying ravines and exploiting strongpoints, as discussed in other chapters, is apparent. The need to conserve war materials, such as chariots and horses, reflects the Sun family emphasis upon the economics of warfare and also a new approach (at least in the extant written materials) to the practice of materials management, to the ensuring of adequate logistical means when required.

The fourth paragraph apparently would stress the virtues of armor in providing the army with a sense of security, in guaranteeing they would be willing to risk danger and fight. As Wu Ch'i said in "Controlling the Army," "If the weapons are sharp and the armor sturdy the men will easily engage in battle." Presumably Sun Pin constructs a bridge of effects in the missing portion, leading from the armor's sturdiness to the men manifesting an overwhelming confidence in their deployments, thereby realizing the army's strategic potential in combat.

The last two topics have been irretrievably lost. Of the two, the method for controlling an army when the enemy is not yet visible but known to be positioned at some discrete distance would be of great interest because it is not otherwise addressed in the *Military Methods* or the other military writings (in contrast to an enemy seen but whose capabilities and tactics, while not yet apparent, can be probed and otherwise subjected to analysis and evaluation).

32

EMPLOYING CAVALRY

(Supplement from the *T'ung Tien*)

Sun Pin said: "In employing the army there are ten (tactical objectives) for which (the cavalry) is advantageous:

"First, when moving to counter an enemy, to arrive first.

"Second, to exploit vacuities at the enemy's back.

"Third, to pursue the scattered and strike the chaotic.

"Fourth, when moving to counter an enemy, to strike their rear, forcing them to run off.

"Fifth, to intercept provisions and foodstuffs, to sever the army's roads.

"Sixth, to defeat (forces at) fords and passes, to open large and small bridges.

"Seventh, to surprise unprepared troops, to strike as-yet-unorganized brigades.

"Eighth, to attack lassitude and indolence, to go forth where not expected.

"Ninth, to incinerate accumulated stores and empty out market lanes.

"Tenth, to forage in the fields and countryside, to bind up their children. For these ten (tactical objectives) it is advantageous to employ the cavalry in warfare. Now the cavalry is able to separate and combine, able to disperse and assemble. A hundred *li* comprises a (marching) period; for a thousand *li* they travel forth, their going and coming unbroken. Thus they are termed 'the weapon of separating and combining.'"

Commentary

This highly coherent passage, preserved in *chüan* 149 of the *T'ung Tien*, explicitly discusses the circumstances in which it would be advantageous to exploit the swiftness and flexibility of the cavalry to attain tactical objectives. For centuries it was believed that this was the only surviving passage from Sun Pin's work, and it was taken as evidence of the latter's existence by those who continued to believe that both Sun-tzu and Sun Pin had penned original military works rather than Sun Pin perhaps having just edited the former's effort. As discussed in the Historical Introduction, since the titles were the same, confusion naturally ensued.

The chapter's subject matter coheres with that preserved in the *Art of War* and Sun Pin's *Military Methods*, lending credence to its attribution to the Sun family. Moreover, it parallels a chapter entitled "Cavalry in Battle" found in the *Six Secret Teachings*, as may be seen from the following excerpts reprising the conditions under which the cavalry might best be employed:

> When the enemy first arrives and their lines and deployment are not yet settled, the front and rear not yet united, then strike into their forward cavalry, attack the left and right flanks. The enemy will certainly flee.
>
> When the enemy's lines and deployment are well-ordered and solid, while their officers and troops want to fight, our cavalry should outflank them, but not go far off. Some should race away, some race forward. Their speed should be like the wind, their explosiveness like thunder, so that the daylight becomes as murky as dusk. Change our flags and pennants several times, also change our uniforms. Then their army can be conquered.
>
> When the enemy's lines and deployment are not solid, while their officers and troops will not fight, press upon them both front and rear, make sudden thrusts on their left and right. Outflank and strike them, and the enemy will certainly be afraid.
>
> When, at sunset, the enemy wants to return to camp, and their Three Armies are terrified, if we can outflank them on both sides, urgently

strike their rear, pressing the entrance to their fortifications, not allowing them to go in. The enemy will certainly be defeated.

When the enemy, although lacking the advantages of ravines and defiles for securing their defenses, has penetrated deeply and ranged widely into distant territory, if we sever their supply lines they will certainly be hungry.

When the land is level and easy, and we see enemy cavalry approaching from all four sides, if we have our chariots and cavalry strike into them, they will certainly become disordered.

When the enemy runs off in flight, their officers and troops scattered and in chaos, if some of our cavalry outflank them on both sides, while others obstruct them to the front and rear, their general can be captured.

When at dusk the enemy is turning back while his soldiers are extremely numerous, his lines and deployment will certainly become disordered. We should have our cavalry form platoons of ten, and regiments of a hundred; group the chariots into squads of five and companies of ten; and set out a great many flags and pennants, intermixed with strong crossbowmen. Some should strike their two flanks, others cut off the front and rear, and then the enemy's general can be taken prisoner. These are the ten [situations in which] the cavalry can be victorious.

The real question is whether the cavalry had already become an operational arm in Sun Pin's era, contrary to scholarly opinion (at least in the West) to date. The inclusion of certain statements in the *Military Methods* itself, unless inserted or amended by later editors, strongly suggests this possibility, as does the *T'ung Tien* remnant. However, the provenance and transmission of the latter are untraceable and, while highly congruent with Sun Pin's writings, may simply be a later formulation attributed to him solely to lend an air of authenticity, particularly as his work had disappeared from the imperial libraries. (This does not completely preclude its continued existence in secret libraries or families with military traditions but indicates it was no longer to be found in even limited public circulation.)

33

ATTACKING
THE HEART

(A *T'ai-p'ing Yü-lan* Fragment)

Sun Pin of Ch'i addressed the king of Ch'i saying: "Now in the Tao for attacking other states assaulting their hearts (*hsin*) is uppermost. Concentrate upon first causing their hearts (*hsin*) to submit. Now what Ch'in relies upon as its heart (*hsin*) is Yen and Chao. They are about to gather in Yen and Chao's authority. So if today you exercise your persuasion upon the rulers of Yen and Chao, do not use vacuous words and empty phrases. You must turn their hearts (*hsin*) with (the prospect of) substantial profits. This is what is referred to as attacking the heart (*hsin*)."

Commentary

This succinct discussion preserved in the *T'ai-p'ing Yü-lan* evinces a word play more characteristic of late Warring States period writings than of writings of Sun Pin's own era and may even have been composed many centuries after his death. However, it helped preserve his image as a historical figure and would seem to belong to the *Kuo-yü* genre.

The paragraph turns upon the dual meaning of the character *hsin*, "heart" and "mind." Obviously when physical location is intended, "heart" is immediately understood. However, beyond this circumscribed case, neither meaning ever completely excludes the other, just as when Sun Pin speaks about "assaulting their hearts" and "causing their hearts (*hsin*) to submit."

Notes to the Introduction, Translations, and Commentaries

Abbreviations for and Notes on Frequently Cited Books

CC   Chung-kuo Ping-shu Chi-ch'eng Pien-wei-hui, *Sun Pin ping-fa,* in *Chung-kuo ping-shu chi-ch'eng,* vol. 1 (Peking: Chieh-fang-chün ch'u-pan-she, 1987).

CL   Chang Chen-che, *Sun Pin ping-fa chiao-li* (Taipei: Ming-wen shu-chü, 1985).

CS   Huo Yin-chang, *Sun Pin ping-fa ch'ien-shuo* (Peking: Chieh-fang-chün ch'u-pan-she, 1986).

GSR   Bernhard Karlgren, *Grammata Serica Recensa* (Stockholm: Bulletin of the Museum of Far Eastern Antiquities, No. 29, 1957).

HW   Hsü Pei-ken and Wei Ju-lin, *Sun Pin ping-fa chu-shih* (Taipei: Li-ming wen-hua, 1976).

KK   *K'ao-ku*

KO   Kanaya Osamu, *Son Pin heihō* (Tokyo: Tokukan shoten, 1976).

MM   Murayama Makoto, *Son Pin heihō* (Tokyo: Tokukan shoten, 1976).

PH   Cheng Hsien-pin, ed., *Sun Pin ping-fa,* in *Pai-hua Chung-kuo ping-fa* (Ch'engtu: Ch'engtu ch'u-pan-she, 1992).

SP   Liu Hsin-chien, *Sun Pin ping-fa hsin-pien chu-shih* (Honan: Honan Ta-hsüeh ch'u-pan-she, 1989).

SPC   Shenyang Pu-tui Hou-ch'in-pu Sun Pin chuan pien-hsieh-tsu, *Sun Pin chuan* (Liaoning: Liaoning Jen-min ch'u-pan-she, 1978).

SY   Shenyang Pu-tui Sun Pin ping-fa chu-shih-tsu, *Sun Pin ping-fa chu-shih* (Liaoning: Liaoning Jen-min ch'u-pan-she, 1975).

TCT   Teng Che-tsung, *Sun Pin ping-fa chu-shih* (Peking: Chieh-fang-chün ch'u-pan-she, 1986).

WW   *Wen-wu*

References to various volumes in the *Seven Military Classics* are to our translation published by Westview Press in 1993. Titles are abbreviated as follows:

Sun-tzu's Art of War	*Art of War*
The Methods of the Ssu-ma	*Ssu-ma Fa*
Questions and Replies Between T'ang T'ai-tsung and Li Wei-kung	*Questions and Replies*
Three Strategies of Huang Shih-kung	*Three Strategies*
T'ai Kung's Six Secret Teachings	*Six Secret Teachings*

A single-volume *Art of War* with expanded discussion and historical analysis is also available from Westview Press (1994) and should be consulted for more extensive contextual material as indicated. This title is referred to as "single-volume *Art of War*." References to the "Military Classics" should be understood as comprising the first six of the *Seven Military Classics*, Sun Pin's *Military Methods*, and the sections on military theory, concepts, and even tactics found in such Warring States philosophical writings as the *Kuan-tzu*, *Huai-nan tzu*, *Hsün-tzu*, and *Mo-tzu*.

Introduction

1. General discussions and translations of these books may be found as follows: For Sun-tzu's *Art of War*, see Ralph Sawyer, *Art of War* (Boulder: Westview Press, 1994); or the translation and briefer introduction to the version found in *Seven Military Classics* (Boulder: Westview Press, 1993). For the *Wu-tzu*, see our translation and introduction in *Seven Military Classics*. For the *Book of Lord Shang*, see J. J. L. Duyvendak, *The Book of Lord Shang* (London: Arthur Probsthain, 1928).

2. First reports of the discovery appeared in China's cultural publication *Wen-wu*. Among the most important are Shantung Sheng Po-wu-kuan Lin-i Wen-wu-tsu, "Shantung Lin-i Hsi-Han-mu fa-hsien *Sun-tzu ping-fa* ho *Sun Pin ping-fa* teng chu-chien te chien-pao," WW, 1974 (2): 15–21; Hsü Ti, "Lüeh-t'an Lin-i Yin-ch'üeh-shan Han-mu ch'u-t'u te ku-tai ping-shu ts'an-chien," WW, 1974 (2): 27–31; and Lo Fu-i, "Lin-i Han-chien kai-shu," WW, 1974 (2): 32–35. In addition, all the works listed in the textual bibliography discuss the book's recovery to some extent.

3. For centuries scholars have even questioned Sun-tzu's existence because of the absence of contemporary references. For a discussion, see the introduction to our single-volume *Art of War*; and Jens Ostergard Petersen's subsequent article, "What's in a Name? On the Sources Concerning Sun Wu," *Asia Major* 1 (1992):1–32.

4. Opinions on the reliability of the *Chan-kuo Ts'e* vary considerably, ranging from those who assert it is a complete fabrication long after the fact to traditionalists who believe that the conversations are verbatim reports from court records. The truth probably falls somewhere in between; the extant book may well have a historical core, and perhaps many of the dialogues were once based upon recollections or notes, although certainly much polished and subsequently embellished. Unlike Sun-tzu, Sun Pin and T'ien Chi are mentioned in command of forces in the hereditary records of Wei and Ch'i, as well as in Sun Pin's biography. However, the *Shih Chi's* records should be internally consistent, having been penned by the great Ssu-ma Ch'ien himself.

5. Or the general of the Upper Army, if the tripart division and designation of Upper, Middle, and Lower are being employed. (The commanding gen-

eral would also normally command the central, or middle, army.) The quote differs somewhat from the original: "If you abandon your armor and heavy equipment to race forward day and night without encamping, covering two days normal distance at a time, marching forward a hundred *li* to contend for gain, the Three Armies' generals will be captured. The strong will be the first to arrive, while the exhausted will follow. With such tactics only one in ten will reach the battle site. If one contends for gain fifty *li* away, it will cause the general of the Upper Army to stumble, and by following such tactics half the men will reach the objective" (Chapter 7, "Military Combat").

6. The biographies of Sun-tzu, Sun Pin, and Wu Ch'i are combined in *chüan* 65 of the *Shih Chi*. The translation is based upon the three standard *Shih Chi* editions listed in the textual bibliography.

7. See SPC, pp. 12–13.

8. See CS, p. 2.

9. The Ch'en, T'ien, and Sun clans were all apparently interrelated. See TCT, p. 117 (citing Yang Po-chün).

10. A book entitled *Kuei Ku-tzu* in three *chüan* (chapters), Taoist in content and orientation, with many echoes and parallels with the *Tao Te Ching*, is found in the Taoist canon and other collections. It barely mentions military affairs, and its origins are obscure, although it may date back to the Han or Wei period, when various writings of the Horizontal and Vertical School, particularly those of Su Ch'in, were perhaps collected and attributed to Kuei Ku-tzu. (The school's famous proponents, Su Ch'in and Chang I, are traditionally said to have been his disciples.) While a *Kuei Ku-tzu* is noted in the Sui dynasty, whether it is this book or not remains an open question. Recently, numerous (contemporary) writings variously attributed to Kuei Ku-tzu have appeared, including several military texts, but they are clearly the modern products of enthusiasts. Scholarly consensus seems to incline toward identifying parts of the first two chapters as perhaps the writings of Su Ch'in, with the rest being later, post-Han additions and commentary. See, for example, *Hsü Wei-shu t'ung-k'ao* (Taipei: Hsüeh-sheng shu-chü, 1984) vol. 3, pp. 1663–1670.

11. See SPC, p. 10; CS, p. 2.

12. During the period 373 to 360 B.C., Ch'i was repeatedly attacked and often lost territory. Two invasions, those mounted by Chao and the lesser state of Wey, would have occurred in Sun Pin's native area. (See SPC, p. 9.) Sun Pin was born in what is now Shantung, within 100 *li* of Ma-ling, so he may have been familiar with the region. (See TCT, p. 117.)

13. This is somewhat problematic. If the traditional date of 307 B.C. for the introduction of the cavalry is accepted—and evidence seems to be building to the contrary—then these competitions would more likely have been chariot racing. This is especially plausible since teams of horses (*ssu,* roughly a "team of four horses," composed of the graph for "horse" combined with

that for "four") are mentioned. Note that some commentators understand the sentence as referring to betting on mounted archery, which would be extremely significant. However, here the term for "shooting a bow," *she*, seems to refer to placing a bet. Others (see, for example, CS, p. 3) would emphasize that horse riding and archery were becoming highly popular during the period and were another form of training for the cavalry. Accordingly, interpreting the characters as "racing and shooting a bow" would be appropriate. (Cf. SPC, p. 18.)

14. If the corrected dating for King Wei is employed, rather than the *Shih Chi* dates, King Hsüan did not ascend the throne until 320 B.C., long after the battle of Ma-ling. Accordingly, if T'ien Chi remained in exile until King Hsüan recalled him (whatever his reason for doing so), the earliest possible date would be about 319 B.C. His professed ignorance of the events at Ma-ling in Chapter 4 of the *Military Methods* is often cited as proving that he did not participate in the Ma-ling campaign. (However, this will be discussed later in the Historical Introduction.) Given Sun Pin's significant role in the Ma-ling effort, such an extended absence would require that he never accompanied T'ien Chi into exile. Alternatively, the reference to King Hsüan recalling him could be erroneous since the *Shih Chi* has the royal chronology wrong. King Wei may have summoned him back in the face of the growing crisis over Han, particularly if he felt that T'ien Chi's former threat had been directly solely toward Tsou Chi.

15. Which "three engagements" are unknown.

16. "The First Book of Ch'i," *Chan-kuo Ts'e*. The translation follows the *Chan-kuo Ts'e* (Taipei: Shih-chieh shu-chü, 1967), vol. 1, pp. 170–171.

17. This is the reason the *Three Strategies* advised, "Now once the masses have been brought together they cannot be hastily separated. Once the awesomeness of authority has been granted it cannot be suddenly shifted. Returning the forces and disbanding the armies are critical stages in preservation and loss. Thus weakening the commanding general through appointment to new positions, taking his authority by granting him a state, is referred to as a 'hegemon's strategy'" ("Middle Strategy").

18. If accurate, the event would probably date to 341 B.C. (See the "First Book of Ch'i.") The parallel *Shih Chi* passage adds further information: "When T'ien Chi heard about it he led his infantry to suddenly strike (the capital of) Lin-tzu seeking out Lord Ch'eng (Tsou Chi) but was not victorious and fled" (*chüan 46*).

19. The "First Book of Ch'i." Tsou Chi apparently feared T'ien Chi would persuade Ch'u's ruler to lend him the military forces to seize Wei, a reasonable possibility because he could probably generate considerable internal support. His logic was that if T'ien Chi was enfeoffed in Ch'u, he would have honor, position, and something to bind him and therefore be less inclined to foment trouble.

20. Perhaps in 319 B.C., based upon the corrected dates for King Wei. (See, for example, SPC, pp. 44–45.) The types of vegetation and preparation of bamboo strips mentioned in the *Military Methods* are taken as evidence that Sun Pin spent some time in Ch'u, presumably accompanying T'ien Chi in exile. (Details may be found in the chapter Notes.) However, it is unlikely that anyone in Sun Pin's era would have not known that bamboo flourishes in Ch'u and Yüeh and that the heavily used bamboo strips were produced there in great numbers.

21. Both are mentioned in the *Shih Chi* and other books of the *Chan-kuo Ts'e*, including Wei's own. See also SPC, pp. 45–46.

22. See SPC, p. 45; and TCT, p. 118 (citing Yang Po-chün's analysis). K'uang Chang and T'ien P'an, rather than T'ien Chi or Sun Pin, are recorded as being in command.

23. Unfortunately, there is no Western-language history of the Warring States period that might be consulted for additional information. The main Chinese sources—apart from the ancient chronicles such as the *Shih Chi* and *Chan-kuo Ts'e*—employed for this brief general background and numerous specialist articles are Yang K'uan, *Chan-kuo shih*, rev. ed. (Shanghai: Shanghai Jen-min ch'u-pan-she, 1980); and Chang Ch'i-yün's volumes on the Warring States period in his *Chung-hwa wu-ch'ien-nien-shih* (Taipei: Chung-kuo Wen-hua Ta-hsueh ch'u-pan-she, 1980). They are supplemented by the various military histories of varying quality on the period, the *Bamboo Annals*, and, of course, Ssu-ma Kuang's monumental overview of Chinese history, the *Tzu-chih t'ung-chien*. Further information may also be found in the introductory material included in our *Seven Military Classics* and the *Art of War*.

24. For further discussion of Wu Ch'i's life and tactics, as well as a complete translation of the extant *Wu-tzu*, see our *Seven Military Classics*.

25. It has been traditionally claimed that Li K'uei's codes continuously formed the basis for China's laws right through the Ch'ing dynasty. However, other views hold that his original writings were immediately lost and that the various codes were individually formulated thereafter, although on earlier precedent. Shang Yang, who had been serving in a minor capacity when Li K'uei was formulating his policies and the new laws, should have been familiar with them and may well have taken copies with him when he departed for Ch'in. See *Chan-kuo shih*, pp. 170–174, and *Chung-hwa wu-ch'ien-nien-shih*, vol. 6, *Chan-kuo shih* (hereafter *Chan-kuo shih*), pp. 16–17, 52. For an overview of Shang Yang's reforms, see Lin Ch'ing-ho, "Lun Shang Yang pien-fa," *Chung-kuo-shih yen-chiu* 1983 (3): 25–33.

26. For further discussion, see the introduction to the *Wu-tzu* in our *Seven Military Classics*. The *Wu-tzu* records Wu Ch'i's initial audience as having been with Marquis Wen; all subsequent ones are with Marquis Wu. Throughout the book the importance of selecting men and creating elite units is apparent.

27. Despite his brief tenure there, Wu Ch'i is credited with having had a great impact on Ch'u's laws and administrative structure, centralizing power under the ruler, and managing, with his death, to eliminate much of the obstinate nobility that had obstructed the ruler's efforts. The events are preserved in his biography. For a translation, see the introduction to the *Wu-tzu* in our *Seven Military Classics*. Also see *Chan-kuo shih*, pp. 176–179.

28. The historical records generally cohere in their depiction of Wei's defeats, particularly as Shang Yang tricked the Wei commander into carelessly meeting with him on the pretext of seeking harmonious grounds. However, whether the capital was shifted to Ta-liang because the defeat rendered the location untenable or whether the capital had previously been shifted in 361 B.C. remains at issue.

29. See the introduction to the *Six Secret Teachings* in our *Seven Military Classics*. Ch'i was regarded as a stronghold of military thought, and many military writings, including the *Ssu-ma Fa*, are associated with it.

30. These policies are advanced in what is generally believed to be the older portions of the *Kuan-tzu* (which is itself a very heterogeneous text compiled over a lengthy period) and otherwise noted in the writings chronicling the Spring and Autumn period. For a translation of the *Kuan-tzu*, see W. Allyn Rickett, *Guanzi* (Princeton: Princeton University Press, 1985).

31. Such as are found in Book I, "Benevolence the Foundation."

32. King Wei's dates are variously given as 379–343 B.C. and 357–320 B.C. The former is based on the *Shih Chi*, although it sometimes appears with a one-year variation, yielding 378 B.C. for the actual enthronement; the latter, upon the *Bamboo Annals*. For example, Chang Ch'i-yün accepts the former in his *Chan-kuo shih* (p. 39), while D. C. Lau advances the second set in his *Mencius* (Hong Kong: Chinese University Press, 1984), pp. 309–310. The dates of 357–320 B.C. are used throughout our works.

A story recorded in the *Chan-kuo Ts'e* depicts the youthful king as idly passing nine years in debauchery until suddenly awakening to his duties. Apart from the fantastic aspects of such stories—how could the state have survived?—it is impossible to integrate it with any known chronology. Nine months, however, would have been possible.

33. He is also credited with patronizing the Chi-hsia Academy, which was probably established by his father but was associated more with King Hsüan and King Min late in the century. The academy provided a forum at Chi-hsia for scholars to express a diverse spectrum of viewpoints. Military theorists were prominent among them. For an overview, see Yüan Te-chin, "Chi-hsia ping-chia chi ch'i che-hsüeh ssu-hsiang ch'u-t'an," *Chung-kuo che-hsüeh-shih yen-chiu* 1988 (2):21–28. See also Yang, *Chan-kuo shih*, pp. 182–184; and Chang, *Chan-kuo shih*, p. 40. Similar academies were designed to provide the monarch with innovative policies, and solutions were established by other rulers, including King Hui in Wei, who tried to imitate Ch'i's visible success.

34. This was also the period when states undertook extensive wall building for defensive (and sometimes offensive) purposes. King Wei of Ch'i was particularly energetic in this regard, but the state of Wei also constructed extended fortifications about Ta-liang, while Chao built a protective wall to the north. See Yang, *Chan-kuo shih*, pp. 295–301, for a summary.

35. Probably more out of fear than respect, although the latter is not precluded.

36. "T'ien Ching-chung Wan shih-chia," *chüan* 46. The story is also preserved in the *Chan-kuo Ts'e*.

37. See "Han, Chao, and Wei," in Li Xueqin, *Eastern Zhou and Qin Civilizations,* trans. K. C. Chang (New Haven: Yale University Press, 1985).

38. For example, see Su Ch'in's persuasion of the king of Chao at the beginning of "Second Book of Chao" in the *Chan-kuo Ts'e*.

39. King Wu-ling is traditionally credited with being the first to have his soldiers wear barbarian-styled clothes in order to facilitate the cavalry's employment and with being the first to utilize cavalry in combat, which went on to defeat the contiguous northern barbarians (called the Lin Hu, or "Forest Hu") and annex their lands.

40. For a discussion of Shen Pu-hai and his philosophy, see Herrlee G. Creel, *Shen Pu-hai* (Chicago: University of Chicago Press, 1974).

41. See *Eastern Zhou and Qin Civilizations*, pp. 59–64.

42. Han's forces were comparatively fewer but still significant, perhaps consisting of a standing force of 200,000 men, more than adequate to protect the state's comparatively smaller area. Their lack of success against Wei's 100,000-man expeditionary force is therefore surprising. Han is noted as a source for strong crossbows in "First Book of Yen," the *Chan-kuo Ts'e*.

43. Insofar as our objective is providing sufficient contextual material to facilitate understanding of the translation and its concepts, this section provides only an extremely brief overview of salient topics. Further material may be found in our *Seven Military Classics,* and our forthcoming *History of Warfare in China: The Ancient Period*.

44. See especially "Planning for the State." However, Wu Ch'i was not alone in advocating the creation of elite forces to spearhead the army. For example, see "Selecting Warriors" in the *Six Secret Teachings*.

45. In Chapter 14 Sun Pin speaks about executing prisoners.

46. See the discussion of the Kuei-ling campaign that follows.

47. Based on Su Ch'in's persuasion of King Min of Ch'i, which constitutes Ch'i's entire fifth book in the *Chan-kuo Ts'e*. The speaker was probably Su Ch'in's brother rather than Su Ch'in himself.

48. See Yang, *Chan-kuo shih*, pp. 285–286. In another persuasion Su Ch'in analyzes the economic and military potential of Lin-tzu, Ch'i's capital, concluding that it could raise 210,000 soldiers by itself from among a populace of 70,000 families. (See "First Book of Ch'i" in the *Chan-kuo Ts'e*.) The fig-

ures provided for the chariots are either wildly underestimated, or the numbers must have declined precipitously from the Spring and Autumn period. (King Hui of Wei even refers to Ch'i as a state of 10,000 chariots in the passage already translated comparing jewels and talents.)

49. See *Chung-kuo li-tai chan-cheng-shih* (Taipei: Li-ming wen-hua, 1980), vol. 2, p. 112.

50. "First Book of Ch'i," *Chan-kuo Ts'e*. The translation is based upon the *Chan-kuo ts'e,* ed. Yang Chia-lo (Taipei: Shih-chieh shu-chü, 1966), vol. 1, pp. 168–169. Throughout the passage "Marquis T'ien" has been changed to "King Wei" for consistency in our discussions.

This is one of three similar discussions in which Tsou Chi opposes rescuing an embattled enemy state (first Chao, later Han). As Yao Ming-ta pointed out in 1918, there are chronological problems with the sequence of passages and also Tsou Chi's apparent stupidity. See "*Shih-chi* T'ien Ching-chung Wan shih-chia chung Tsou Chi te san-tuan-hua," reprinted in *Ku-shih pien,* ed. Ku Chieh-kang (Shanghai: Shang-hai Ku-chi ch'u-pan-she, 1982), vol. 2, pp. 118–126. The identity of Tuan-han Lun is also questioned, some suggesting he was an alter ego for Sun Pin. In the *Shih Chi* biography and other materials, Sun Pin always suggests concrete, indirect tactics rather than the general principle of waiting for the enemy to exhaust themselves.

51. The attacks are also mentioned in the T'ien family annals, "The Hereditary Biography of T'ien 'Ching-chung' Wan," *chüan* 46; and the various *Bamboo Annals.* There are some inconsistencies about the date of the actual attack with, for example, "Hereditary Biographies of the House of Wei," indicating that "the feudal lords besieged us at Hsiang-ling" the year after the battle at Kuei-ling. The *Ku-pen chu-shu chi-nien,* however, gives the sequence as an attack on Hsiang-ling followed by an attack directed toward Ta-liang and—despite this account's own questionable reliability—is taken as the basis for our discussions. However, see also the section on tactical evaluation that follows.

52. See *Chung-kuo chün-shih-shih,* vol. 2, *Ping-lüeh* (hereafter *Ping-lüeh*) (Peking: Chieh-fang-chün ch'u-pan-she, 1986), p. 102.

53. Chao apparently was not unaware of being manipulated and may have rejected Ch'u's aid for exactly this reason. However, they had no alternative. (See *Ping-lüeh,* p. 102.) Ch'u held a court discussion much like Ch'i's about whether to assist Chao and similarly decided to send only a token force in order to stiffen them and preserve their own options to exploit the unfolding situation. (See the "First Book of Ch'u" in the *Chan-kuo Ts'e*.)

54. In another version of these court discussions preserved in *chüan* 46 of the *Shih Chi,* Tuan-kan P'eng clearly asserts that failing to respond to Chao's request would be unrighteous, a criticism Ch'i much wished to avoid. (Further discussion will be found in the tactical evaluation section.)

55. As Chao Chün-yao points out, the *Lü-shih Ch'un-ch'iu* states, "Wei's army besieged Han-tan for three years but was not able to take it." He concludes that Han-tan never fell and that Ch'i deliberately implemented a strategy designed to rescue it after the combatants had suitably exhausted themselves. See "*Tzu-chih t'ung-chien*, Wei-Wei chiu-Chao, Han-tan tsun-chiang-k'ao," *Chung-kuo-shih yen-chiu* (January 1990):144–145. However, the "Hereditary Records of Chao" in the *Shih Chi* states that Wei "took" Han-tan and then returned it two years later when an alliance was concluded at Chang-shui. In contrast, the editors of the SPC edition (p. 29) argue that King Hui was unwilling to abandon the attack on Han-tan and therefore split off some of P'ang Chüan's forces to respond to the threat, proof that Han-tan had not yet fallen. Their argument overlooks the need to deploy an occupation force if the city had been conquered but not butchered.

56. "Planning Offensives," the *Art of War.*

57. Chapter 5, "Responding to Change."

58. Brief discussions of the psychology and role of *ch'i* may be found in the *Art of War* and the *Seven Military Classics,* as well as the commentaries and Notes for several of the *Military Methods* topics. See especially Chapter 13, "Expanding *Ch'i*."

59. See, for example, CS, p. 6; and CL, p. 7, following the original reconstruction commentaries. In "The Hereditary Biography of T'ien 'Ching-chung' Wan," *chüan* 46, the scheming Tsou Chi is advised to suggest that T'ien Chi attack Hsiang-ling and does so. While either position can be effectively argued, it appears that given the detailed description of the attack on P'ing-ling in the *Military Methods,* somewhat greater credence should be given to this possibility. The records may be confused about the battle of Hsiang-ling, about which there are date inconsistencies, but it also seems to have been widely known and clearly part of the planning process. Sun Pin's ruse, however, was clearly undertaken while on the campaign, not a premeditated decision as the product of court discussions, and resulted in defeat. Perhaps because it was a minor action, it was simply eclipsed by the fame and scope of the greater picture.

60. "Army Orders, II," the *Wei Liao-tzu.*

61. Whether Ta-liang had become Wei's designated capital in 361 or subsequently in 340 B.C. after it became untenable due to Ch'in's power and invasions remains irresolvable. However, without doubt Ta-liang was the focal point for Wei's eastern portion and therefore critical to exercising command over the region, as will be discussed later.

62. For basic descriptions, see Chapter 4, *Eastern Zhou and Qin Civilizations,* and the articles referred to therein. In the last twenty years numerous ancient cities have been discovered, pushing the threshold of fortified towns back many centuries into the Hsia period and earlier. However, their develop-

ment as significant economic centers did not begin until well into the Spring and Autumn period, eventually causing them to command attention as preferred military targets in the Warring States period.

63. In a persuasion preserved in the "First Book of Wei" of the *Chan-kuo Ts'e*, the famous strategist Chang I analyzed Wei's power and relative vulnerability, focusing on Ta-liang's exposed position as follows: "Wei's territory does not reach a thousand square *li*; your troops do not exceed 300,000 men; your land is level in all four directions, penetrating to the feudal lords in four directions, just like the spokes of a wheel converging. You lack obstacles such as famous mountains or great rivers."

64. See the analysis in *Chung-kuo li-tai chan-cheng-shih*, vol. 2, pp. 112–113 and Mu Chung-yüeh and Wu Kuo-ch'ing, *Chung-kuo chan-cheng-shih* (Peking: Chin-ch'eng ch'u-pan-she, 1992), vol. 1, pp. 266–268. The distances involved are actually not extensive. For example, from Han-tan to Ta-liang was said by Chang I (see preceding note) to be only 100 *li*; modern maps indicate a distance of about 150 miles or more, depending upon how circuitous the route might actually be. Naturally, with a maximum average sustained marching speed of 30 *li* per day, four days' minimum would be required to speed from Han-tan to Kuei-ling unless P'ang Chüan wanted to try for the 100 *li* a day strongly condemned in the *Art of War*.

65. One hundred years later Hsün-tzu observed that "Ch'i's skillful attacks are inadequate to oppose Wei's martial troops" ("Discussion of the Military"). A generation earlier Wu Ch'i had characterized Ch'i's battle array and their states as follows: "Although Ch'i's battle array is dense in number, it is not solid. Ch'i's character is hard; their country is prosperous; the ruler and ministers are extravagant and insulting to the common people. The government is expansive but salaries are inequitable. Each formation is of two minds, with the front being heavy and the rear light. Thus while they are dense, they are not stable. The Way to attack them is to divide them into three, harrying and pursuing the left and right, coercing and following them for then their formation can be destroyed" (Chapter 2, "Evaluating the Enemy"). Presumably King Wei, Sun Pin, and T'ien Chi rectified the problems that Wu Ch'i had observed roughly fifty years earlier.

About the Three Chin Wu Ch'i also said, "The battle arrays of the Three Chin are well controlled, but they prove useless. The Three Chin are central countries. Their character is harmonious and their government equitable. The populace is weary from battle but experienced in arms, and they have little regard for their generals. Salaries are meager, and as their officers have no commitment to fight to the death, they are ordered but useless. The Way to attack them is to press [points in] their formations and when large numbers appear oppose them. When they turn back, pursue them in order to wear them out" (Chapter 2, "Evaluating the Enemy").

66. Echoing Sun-tzu's position throughout the *Art of War,* Sun Pin in Chapter 7 states, "If he perceives victory, he engages in battle; if he does not perceive it, he remains quiet."

67. Chapter 3, "Planning Offensives," the *Art of War.*

68. As cited in note 53.

69. See *Chung-kuo chan-cheng-shih,* vol. 1, p. 266.

70. For the seventeenth year, as noted, the *Ku-pen chu-shu chi-nien* has Sun and Wey joining Ch'i to besiege Hsiang-ling, followed by Ch'i's attack toward Ta-liang in the same (or seventeenth) year and then Wei, with Han's aid, defeating the feudal lords at Hsiang-ling the next year. However, a minor possibility exists that if the other accounts are followed, and Han-tan was still partially under siege even after the defeat at Kuei-ling, passages that assert the feudal lords attacked Hsiang-ling in the succeeding year would be plausible and therefore would account for Wei retuning Han-tan to Chao thereafter.

71. For example, "Explosive Warfare," in the *Six Secret Teachings;* and "Tactical Balance of Power in Defense," in the *Wei Liao-tzu.*

72. Accounts clash over whether the incident occurred twelve or fifteen years later. For example, the *Tzu-chih t'ung-chien* indicates fifteen years, and the *Shih Chi,* twelve, but the latter has variations in the records of the individual states. For a summary, see CS, p. 11. There is also much disagreement as to whether the conflict began in 342 or 341 B.C. For a detailed discussion, see Wang Huan-ch'un, "Ma-ling-chih-chan hsin-chieh," *Hsien-Ch'in, Ch'in-Han-shih* 1991 (5): 49–54.

73. See *Ping-lüeh,* vol. 1, pp. 105–106. The assembly's date is also disputed. Yang suggests it occurred in 344 B.C. (*Chan-kuo-shih,* p. 318); others suggest 342 B.C., the year Wei attacked Han. (In the *Shih Chi,* see the "Ch'in-pen-chi," Ch'in Hsiao-kung twentieth year, and "Liu-kuo nien-piao." However, the "Chou-pen-chi" has Chou Hsien-wang, twenty-fifth year.)

74. The sources disagree as to whether King Wei had died or was still on the throne. We follow the corrected dating given by the *Bamboo Annals,* yielding a reign period of 357–320 B.C. For the *Shih Chi,* see Ch'i's annals, *chüan* 46.

75. There is some disagreement as to whether Tsou Chi had already died and T'ien Chi been banished to Ch'u at this time. The historical disputants can best be understood as representing positions rather than particular individuals. However, see the *Chan-kuo Ts'e* version that follows. The SPC edition (pp. 23–24) postulates that T'ien Chi, T'ien Ch'i, and even T'ien P'an were all the same individual, the names being pronounced similarly. Others have advanced similar theories to account for one man apparently having different identities in different contexts.

76. Corrected from King Hsüan to King Wei in accord with note 74.

77. Presumably he had been forced to flee to Ch'u after falling victim to Tsou Chi's schemes. However, others claim he never left, or never returned, and that the identification is erroneous.

78. Correcting from Chao to Ch'i.

79. *Shih Chi, chüan* 46.

80. For the benefit of readers not conversant with the original sources, we provide a translation of the similar *Chan-kuo Ts'e* account in order to furnish an example of the difficulties faced in reconstructing the history of the period: "When the difficulty at Nan-liang arose, Han requested that Ch'i rescue them. King Wei summoned his high ministers to make plans and asked: 'Would it be more conducive to rescue them early on or later?' Chang Kai replied: 'If we rescue them later, Han will have been destroyed and absorbed by Wei. It would be better to effect their rescue early on.' T'ien Ch'en-ssu said: 'It is not possible. Now if we rescue them when the armies of Han and Wei are not yet exhausted, we will be standing in for Han, subjecting ourselves to Wei's anger, and—conversely—obeying Han's commands! Moreover, Wei is determined to destroy Han. When Han sees it is about to perish, they will certainly advise us. We can then secretly conclude an alliance of friendship with Han and, late in the conflict, take advantage of Wei's exhaustion. Then the state will be made weightier, profits will be gained, and our name will be respected.' King Wei said: 'Excellent.' Thereupon he secretly informed Han's emissary (of their support) and sent him forth. Han thought it had the concentrated support of the state of Ch'i and so fought five battles with Wei but failed to be victorious five times. They came east to inform Ch'i. Ch'i then mobilized its army, struck Wei, and severely destroyed them at Ma-ling. Wei having been destroyed and Han weakened, the rulers of Han and Wei, in accord with T'ien Ying's (advice), faced north and acknowledged Ch'i's sovereignty" ("First Book of Ch'i").

81. For a typical view, see "Ma-ling-chih-chan hsin-chieh," pp. 51–52.

82. The "Book of Sung and Wey" in the *Chan-kuo Ts'e;* also found in Wei's annals in the *Shih Chi*. Sung was a weak state, easily encroached upon by both Wei and Ch'i and frequently crossed by their troops en route to engagements in third states. The discussion about the prince's campaign army going east, crossing the Yellow River, is generally seen as evidence that the site of Ma-ling was in Ch'i. See, for example, Wang Huan-ch'un, "Wei Ch'i Ma-ling-chih-chan tsai T'an-ch'eng Ma-ling-shan," *Chung-kuo-shih yen-chiu* 1985 (1):163; and "Ma-ling-chih-chan hsin-chieh," p. 51.

83. "Second Book of Wei," the *Chan-kuo Ts'e*. Several writers assert that T'ien P'an, rather than T'ien Chi, commanded the troops. See, among others, "Ma-ling-chih-chan hsin-chieh," p. 52.

84. "Collected Biographies of Lu Chung-lien and Tsou Yang," *chüan* 23.

85. In a famous passage in the *Mencius* King Hui says, "There was no one stronger than Chin, as you well know. However, during my own rule we have

been defeated by Ch'i in the east where my eldest son died; in the west we have lost 700 *li* of land; and to the south we have been insulted by Ch'u" (IA5). The death is confirmed again in Book VII (VIIB1). Whenever describing the battle or associated events, the *Shih Chi* indicates that the prince was captured. However, the *Chan-kuo Ts'e* records are contradictory: Ch'i's annals state he was captured; Wei's and Sung's, that he died and never returned to his native state. Furthermore, the *Lü-shih Ch'un-ch'iu*, a work composed about a century after Ma-ling, cites the prince as one example of men who did not know themselves and therefore perished. See Part 3, "Self-knowledge," *chüan* 24. See also "Ma-ling-chih-chan hsin-chieh," pp. 51–52.

86. For example, see *Chung-kuo li-tai chan-cheng-shih*, vol. 2, p. 120; and *Chung-kuo chan-cheng-shih*, vol. 1, p. 285.

87. Chapter 7, "Military Combat," the *Art of War*. In Chapter 26, "The General's Losses," Sun Pin identifies a number of errors on the general's part that can certainly be said to have marked P'ang Chüan's disposition of his troops in the rush to Ma-ling. However, the events that unfolded at Ma-ling may also be interpreted somewhat differently: Probably the defeat was not due solely to an exhausted, disorganized force confronting an entrenched, well-rested one but rather to light, mobile elements not being able to contend with entrenched, heavy forces, particularly on constricted terrain. This will be discussed in our *History of Warfare in China: The Ancient Period*.

88. Chapter 3, "Controlling the Army," the *Wu-tzu*.

89. Book II, "Obligations of the Son of Heaven," the *Ssu-ma Fa*. For a translation and discussion of the text, see our *Seven Military Classics*.

90. Chapter 7, "Military Combat," the *Art of War*.

91. In Chapter 12, "Incendiary Attacks," Sun-tzu states, "Now if someone is victorious in battle and succeeds in attack but does not exploit the achievement, it is disastrous, and his fate should be termed 'wasteful and tarrying.'"

92. See "Ma-ling-chih-chan hsin-chieh," pp. 49–51; SPC, pp. 145–146; and TCT, pp. 145–146.

93. A number of scholars, even entire committees, have trekked through the countryside in search of historical materials (such as steeles and local records) marking Sun Pin's life and also investigated the possible locations for various places identified in the records, such as Kuei-ling, Ma-ling, P'ing-ling, and Hsiang-ling. In conjunction with ancient maps and textual notes, their efforts have yielded—perhaps "forced"—probable locations for all of them, and these results have been incorporated in our maps and analyses, although without reproducing the lengthy, detailed justifications. However, of particular interest are similar descriptions of the Ma-ling road (even by those who disagree as to its exact location and identification) as winding through uneven terrain that even today retains extensive vegetation. For various locations and similar details, see, for example, SPC, pp. 145–146; "Ma-ling-chih-

chan hsin-chieh," pp. 51–52; and "Wei Ch'i Ma-ling-chih-chan tsai T'an-ch'eng Ma-ling-shan," pp. 163–164. The third of these identifies the famous battle site as the road to Mount Ma-ling in T'an-ch'eng in the the southern part of modern Shantung. (As it seems probable, it has been used for our reconstructed maps.) Other views, such as that of SPC, place the battle at Fan-hsin just north of the Yellow River; however, this does not seem to cohere with presently available data.

94. For example, see Chapter 4, "Strict Positions," in the *Ssu-ma Fa*.

95. See Chapter 14 of the *Military Methods*.

96. This is disputed, with many claiming that the move occurred in 361 B.C. However, many analysts who focus upon military history conclude it took place in 340 B.C., the year after the defeat at Ma-ling. See, for example, *Chung-kuo li-tai chan-cheng-shih*, vol. 2, pp. 118–119: and *Chung-kuo chan-cheng-shih*, vol. 1, p. 284. However, the latter contradicts itself, giving a date of 361 B.C. in its historical state-by-state overview on p. 34. The *Ping-lüeh* authors subscribe to 361 B.C. (p. 100), as does Yang (*Chan-kuo-shih*, p. 276). *Tzu-chih t'ung-chien* records the move as occurring after the great defeat inflicted by Ch'in under the command of Shang Yang (Lord Shang).

97. Questions of priority are irrelevant to any meaningful study but naturally evoke interest by themselves. If the first real employment of a "subtracted" or "tactical" reserve is attributed to Hannibal in the Second Punic War (at the battles of Cannae in 216 B.C. and Zama in 202 B.C.), then clearly Sun Pin's concept of keeping up to two-thirds of his force in reserve (which falls in the middle to late fourth century) predates Hannibal by a century or more. However, if the changes in organization and composition being witnessed early in the fourth century B.C. in Greece and Rome are considered as manifestations of the concept of a fighting reserve, however limited, then the battle of Leuctra in 371 B.C. would obviously predate Sun Pin.

98. In Chapter 6, "Lunar Warfare," Sun Pin implies that there is a hierarchy of results by which the success of a battle may be judged: numbers killed, troop commanders captured, encampments taken, commanding general not captured, and the enemy's army destroyed and their general killed.

99. In Chapter 1, "Planning for the State."

100. See Chapters 17, 18, 28.

101. For an extensive discussion of the nature and concept of *shih*, "strategic power," see the introduction to our translation of the *Art of War*.

102. In Chapter 9, "Preparation of Strategic Power," Sun Pin states, "Yi made bows and crossbows and imagized strategic power on them. . . . How do we know that bows and crossbows constituted (the basis for) strategic power? Released from between the shoulders, they kill a man beyond a hundred paces without him realizing the arrow's path. Thus it is said that bows and crossbows are strategic power." Earlier Sun-tzu likened strategic power to pent-up water and the fully drawn crossbow: "The strategic configuration of power is visible in the onrush of pent-up water tumbling stones along. The effect of constraints is visible in the onrush of a bird of prey breaking the

bones of its target. Thus the strategic configuration of power of those that excel in warfare is sharply focused, their constraints are precise. Their strategic configuration of power is like a fully drawn crossbow, their constraints like the release of the trigger" (Chapter 5, "Strategic Military Power").

103. Despite appearing in the first part of the book, the entire eighth chapter incorporates what can only be termed miscellaneous material loosely organized around concepts of conquest and *yin-yang* relationships. As will be discussed in the section on the text and its date, since five-phase thought apparently became amalgamated with *yin-yang* categorization only after Tsou Yen, who postdated Sun Pin by some decades, the chapter obviously was compiled by Sun Pin's disciples.

104. The basis for the latter, apart from the rigid application of an inappropriate descriptive system, is not readily visible.

105. Of course, the general must also comprehend the factors of Heaven, including climate and weather. For an overview of such factors by an experienced contemporary military officer, see Karl W. Eikenberry, "Sun Bin and His *Art of War*," Military Review (March 1991):52–57.

106. The concept of imbalance of power, or *ch'üan*, prominent in the *Art of War* is only briefly mentioned in the *Military Methods*. For a discussion of the term, see the introduction to our translation of the *Art of War*.

107. In "Military Combat" Sun-tzu states, "Do not obstruct an army retreating homeward. If you besiege an army you must leave an outlet. Do not press an exhausted invader." In "Nine Terrains" Sun-tzu describes how casting men onto fatal terrain will elicit their desperate commitment to fight to the death.

108. For example, see Chapter 17 of the *Military Methods*. The concept of a startling employment contravening all normal expectations, behavior that can only (and appropriately) be termed *berserk*, should not be confused with that for warriors resigned to die and therefore fighting to the death, which appears in several of the other military writings. However, in the *Six Secret Teachings* the T'ai Kung, to emphasize the need for swift action and total commitment, states, "Advance as if suddenly startled; employ your troops as if deranged" ("The Army's Strategic Power").

109. The section entitled "Canine Secret Teaching" contains several passages on the general theme of training the soldiers; however, an even better explication of his method (as understood by Li Ching) appears in Book II of the *Questions and Replies*.

110. For a discussion of the general's rising importance, see the introduction to our translation of the *Art of War*, where a comparative list of characteristics may be found.

111. The necessary independence of the general once he had been commissioned and dispatched to take command of his troops was particularly emphasized by Sun-tzu, no doubt in reaction to the habit of rulers (who had previously exercised field command themselves) constantly interfering with

the responsibilities of the newly risen *professional* general. For a discussion, see the introduction to our translation of Sun-tzu's *Art of War.*

112. By which would be meant not only the individual virtues but also the cultivation of personal *Te,* or "power."

113. Compare the views expressed in Chapters 3 and 5. Sun Pin identifies "authority" as the means to assemble the people.

114. See the chapter commentary and Notes for further discussion of this critical concept.

115. The initial reports of the tomb's discovery and opening have already been cited in note 2. Unlike many others, this tomb had not been plundered; therefore the grave goods were still intact and remained in their original positions.

116. It was customary to inter objects that had particular meaning for the deceased, as well as utensils and vessels, for use in the afterlife. However, entombing so many texts on military subjects, apart from indicating the deceased must have been a prominent military official, was unusual. Perhaps the time had come when they could no longer be safely possessed, or his family simply wanted to be rid of them now that the old general had died but could not bear to burn them. (Their possession could have been dangerous if discovered by Han imperial authorities.) Or possibly the deceased had instructed that his precious books be buried with him so that he might continue to peruse them in the afterlife (or simply did not want anyone else in the family to have them, feeling that they did not deserve them).

117. Published by *Wen-wu* in a limited quantity, they are entitled *Yin-ch'üeh-shan Han-mu chu-chien (1): Sun Pin ping-fa* and *(2): I-shu ts'ung-ts'an.*

118. The original edition of the *Military Methods* incorporated a total of 364 strips, many damaged or fragmented, from among a much larger number.

119. The tomb is located in what was once the ancient state of Chü. Chü had been annexed by Ch'i and subsequently became famous as the last bastion from which T'ien Tan revived the nearly vanquished state of Ch'i. (This conflict is extensively analyzed in our forthcoming *History of Warfare in China: The Ancient Period*). Possibly the deceased had some connection with the famous Sun family and thus had access to the family writings, although it is more likely he was simply a high official entrusted with military affairs who had managed to acquire the text.

120. Huo Yin-chang (CS, pp. 14–15) has pointed out that Ssu-ma Ch'ien (the author of the *Shih Chi*) refers to Sun Pin in his personal introduction and also states that his "*ping-fa* has been transmitted for generations." Furthermore, Huo also cites a reference in the *Lü-shih Ch'un-ch'iu* that couples the *Sun Pin ping-fa* with the *Mo-tzu* and the *Lao-tzu* as widely known books. However, in both cases it is not invariably true that *ping-fa* refers to the

book's title; it may simply mean Sun Pin's tactics or military methods. Hsü and Wei, whose translations are somewhat unreliable but who bring a military perspective and frequently provide important insights, contend—no doubt correctly—that neither Ssu-ma Ch'ien nor Pan Ku, the main author of the *History of the Han*, ever saw the text (HW, pp. 9–12).

121. See HW, pp. 12–13. Ts'ao Ts'ao was excoriated by later scholars for supposedly butchering the eighty-two-*pien Art of War* when he edited the diverse materials and appended his commentary.

122. See Huo Yin-chang, CS, p. 11.

123. This is Hsü and Wei's conclusion (HW, p. 11). However, Huo Yin-chang asserts the latest possible date for the entombment is 118 B.C. (CS, p. 11).

124. See CL, preface, p. 2. However, Chang assumes that people privately making copies would bother to observe the taboos, which may be doubtful.

125. See the end of Chapter 3, "The Questions of King Wei."

126. The Tao of the Sun family is mentioned in Chapter 4, "Ch'en Chi Inquires About Fortifications." This chapter is thought by many to be a discussion of the tactics employed at Ma-ling.

127. See, for example, Huo Yin-chang, CS, pp. 12–14.

128. See Yang Po-chün's analysis of a passage reflecting the later amalgamation of *yin-yang* and five-phase thought in *Hsü-Wei-shu t'ung-k'ao*, vol. 3, pp. 1629–1670. Hsü Pao-lin, a widely acknowledged expert on military writings, views the entire text as a compilation from numerous hands. See *Chung-kuo ping-shu t'ung-lan* (Peking: Chieh-fang-chün ch'u-pan-she, 1990), pp. 108–112.

129. Robin D.S. Yates draws a similar conclusion in "New Light on Ancient Chinese Military Texts: Notes on Their Nature and Evolution, and the Development of Military Specialization in Warring States China," *T'oung Pao* 74 (1988):224. Yates points out it was probably composed about the same time as *Mencius* and the *Six Secret Teachings*, to which we would add parts of the *Wei Liao-tzu* as well.

Chapter 1

1. As noted in the Historical Introduction, the marquis of Wei became known as King Hui late in his rule when both he and the ruler of Ch'i (King Wei) assumed the regal title. Moreover, his state was referred to as Liang only after he moved the capital from An-i to Ta-liang. (The controversy over the date of the actual move is examined in the Historical Introduction.) Thus, this chapter was penned subsequent to these events, when such honorific forms of reference had become common, although still possibly within Sun Pin's own lifetime.

2. Some commentators suggest a different pronunciation for the city's name: Ch'a-ch'iu rather than Ch'ih-ch'iu. See, for example, SY, p. 26; CL, p.

5; and TCT, pp. 2–3. Chang suggests it was located in Wey's territory near the border with Ch'i, about 200 *li* north of Ta-liang (CL, p. 5).

3. The army's size is historically significant. If the fighting troops numbered 80,000, a minimum support force of 20,000 would be expected, bringing the army up to at least 100,000. Apparently both sides could easily field such numbers from their standing forces by this time, suggesting that military expenses must have imposed great burdens on their respective states.

4. Unfortunately, the name of the city is no longer legible, although several possibilities are suggested by the commentators. (See, for example, CL, p. 6.) Chang's text also has "seized" following attack, for "attacked and seized," although he stands alone in deciphering this character from the illegible strip.

As discussed in the Historical Introduction, Wey was a minor state dominated by the major powers. Forces from Ta-liang would have had to pass through Wey to strike at Han-tan, so this would have been a requisite preliminary move to ensure an unobstructed route (even though Wey could hardly have refused passage to mighty Liang). Moreover, since Chao had seized Han-tan from Wey in an earlier conflict, Wey would hardly have been sympathetic to Chao's plight.

5. Rather than make them doubtful, as some editions suggest.

6. Apparently a minor city in Wei or Wey. See CL, p. 8; and CS, p. 29. The former is more likely because it poses a threat, contrary to Chang's identification of it as being in Wey. However, it might be an abbreviated writing for the name of Wey's capital, Ti-ch'iu.

7. As Chang points out, Ch'i tended to use *tu* (essentially metropolitan centers with attached administrative areas) when other states were implementing the *hsien* (district) system. (See CL, pp. 8–9.) Within Ch'i, especially near the borders, these administrative regions would have their own standing military forces under local command. The *ta-fu* (high-ranking functionaries) with appropriate areas of responsibility also commanded the troops, despite the increased specialization of the period, although normally as part of a component force whenever major actions were entailed. (In some instances the officials responsible for military affairs were specialists rather than simply overall administrators. The extent and degree of specialization deserve further study.)

8. It is suggested that T'ien P'an was the *ta-fu* in charge of the defense of Kao-t'ang. (See CC, p. 96; and CL, p. 9.) However, while he might have served there at one time, he obviously was not killed in this battle and was generally known as a competent and feared commander, not a dolt or idiot.

9. Chang provides an extensive note on the names and possible locations of these two cities (CL, pp. 10–12).

10. There is considerable disagreement as to what the characters *huan t'u* (translated as "regional roads") actually mean. The original commentators suggest they might be a place-name or a general's name. Chang provides an

enlightening note (partly based upon earlier suggestions) that *huan* referred to regional roads capable of accommodating seven chariots abreast and *t'u,* to roads of five-chariot width. These roads presumably extended 200 *li* from the capital, making the area accessible from Ta-liang. See CL, p. 12; CS, pp. 29–30; SY, p. 26; and CC, p. 196.

11. Translating *mo-chia* as "vanguard" (or "elite front") in accord with Sun Pin's use of the term in Chapter 16, "Ten Deployments," and the context. (Under Sun Pin's plan the main force, with its elite front, would present a formidable target, whereas the two *ta-fu* would present isolated, easy targets of opportunity.) It is also suggested that the term refers to the rear guard (cf. CC, p. 97; CS, p. 30; SY, p. 26; and CL, p. 13). In terms of the overall tactics, whether the term refers to a vanguard or a rear guard is rather unimportant; the army's readiness and integrity coerce Wei into directing its attack toward the two sacrificial lambs.

12. Chang believes these sentences record P'ang Chüan's tactical analysis of the situation (CL, pp. 9–10). However, others, such as Huo Yin-chang (CS, p. 32), concur with the present translation.

13. "Flurried assault," literally like ants, echoing Sun-tzu's famous image in "Planning Offensives": "If the general cannot overcome his impatience but instead launches an assault wherein his men swarm over the walls like ants, he will kill one-third of his officers and troops, and the city still will not be taken."

14. Here and in the chapter's title the term is *ch'in,* which means "to capture" rather than simply defeat, giving rise to tortuous efforts to reconcile his capture at the battle of Kuei-ling with his subsequent death at Ma-ling (as discussed in the Historical Introduction).

Chapter 2

1. The term *shih,* which is translated throughout as "strategic configuration of power" (or shortened to "strategic power"), inherently entails the integrated concepts of power and positional advantage. One of the critical concepts of the ancient military strategists, it plays a prominent role in Sun-tzu's *Art of War* (where it was first articulated) and Sun Pin's *Military Methods.* Apart from the Historical Introduction, for further discussion see the introduction to our *Art of War* translations.

2. The *Wei Liao-tzu,* probably a fairly late Warring States work, observes that if a besieged city has ample supplies of grain (and some prospect of being rescued), the people will be stalwart in their defense. See "Tactical Balance of Power in Defense." Other chapters in the *Wei Liao-tzu,* such as "Combat Awesomeness," also stress the need for adequate material reserves.

3. Note that warriors who are not "good" or lack certain other virtues are not qualified to serve in chariots according to Chapter 31, "Five Instructions."

4. Yao was a semilegendary paragon of Virtue, a great Sage Emperor, totemic figure, and probable tribal leader.

5. "Central states" was a concept of Sun Pin's time, not of antiquity, and thus refers to the central plains area. Since the text only accounts for six peoples, a phrase must have dropped out.

6. The legendary Shen Nung (given as Shen Jung in the strips), one of the Three August Ones, is commonly honored as the progenitor of agriculture and father of herbal medicine. According to Teng Che-tsung (TCT, p. 9), the Fu and Sui were two tribes with whom Shen Nung did battle, marking the earliest conflict mentioned as occurring in Chinese history. Other commentators take Fu-sui as a tribal or state name (cf. SY, p. 32; CL, p. 24; and CS, p. 39).

7. The Yellow Emperor, another legendary Sage Emperor and totemic figure, is traditionally credited with numerous cultural inventions and essential technical creations that constitute the foundations of civilization itself, including swords. (See Chapter 9, "Preparation of Strategic Power.") Apart from Teng's view as discussed in the previous note, the Yellow Emperor's battle with Ch'ih Yu has long been taken as marking the start of armed conflict in China, although historical and archaeological materials prove conclusively that combat originated many thousands of years earlier. (This battle is extensively analyzed in our *History of Warfare in China: The Ancient Period*.) The site for this famous battle has always been identified as Chuo-li rather than Shu-lü, based upon a *Chan-kuo Ts'e* passage. (See CC, p. 113.)

8. Kung Kung, apparently the leader of an unsubmissive tribal people, is otherwise unknown.

9. *Ch'e* is a tentative pronunciation for an obscure, previously unknown character. It may be a tribal name or a leader's name. (Cf. TCT, p. 10.)

10. The Miao were an important early people of southwest China mainly active in modern Honan, Hunan, and Chianghsi. "Three Miao" perhaps refers to three tribes within them, a single group, or even a place-name. (Interpretations vary.)

11. The single-character *kuan*—possibly a place-name—follows a broken fragment.

12. King T'ang, the founding ruler of the Shang dynasty, forcibly overthrew the last evil ruler of the Hsia dynasty, King Chieh, reportedly through a devastating military campaign.

13. King Wu, claiming the Mandate of Heaven, led the peripheral state of Chou to overthrow the debauched, tyrannical King Chou, last ruler of the Shang dynasty, and establish its own dynasty in the epoch-making battle at Mu-yeh around 1045 B.C. For an extensive discussion, see the historical introduction to the *Six Secret Teachings* in our *Seven Military Classics*.

14. The Duke of Chou, who served as regent for his youthful nephew after King Wu died, commanded the campaign that suppressed the revolt of vari-

ous Shang peoples and remnant states (in alliance with two of King Wu's other brothers) subsequent to the king's death. Although in this conflict he was forced to slay his own brothers to preserve the lineal succession and the integrity of the Chou house, Confucius canonized him as a paragon of Virtue and raised him as an awesome inspiration for all men of moral courage and rectitude. (The translation follows the many commentators who emend the original character *ti*, "emperor," to "Shang." See TCT, p. 10; CS, p. 40; SY, p. 33; and CL, p. 24.)

15. Various lists are given for the Five Emperors, although most include the Yellow Emperor, Yao, and Shun. As Huo Yin-chang notes, Sun Pin apparently intended to include Shen Nung among them, which would be unusual (CS, p. 40).

16. Here the "Three Kings" are the founding kings of the successive dynasties of the Hsia, Shang, and Chou.

17. "Initial Estimations."

18. As noted in Chapter 9, "Preparation of Strategic Power." Similar statements are also found in the famous *I Ching*.

19. That Sun Pin's criticism is directed toward the Confucians is obvious from his description of the (effete) bureaucrat in his flowing robes, committed to practicing the rites and music as well as cultivating benevolence and righteousness. (Sun Pin concurs with the latter two, as discussed in the commentary and seen in this chapter.)

20. Sun-tzu primarily advised against lengthy campaigns (which even by then already entailed several battles) in "Waging War." Others, such as Wu Ch'i (in "Planning for the State") and Wei Liao-tzu (in "Discussion of Regulations"), warned against the pernicious effects of frequent battles.

21. In "Lunar Warfare."

22. In Chapter 23, "The General's Righteousness."

23. In "Five Instructions."

24. "Planning for the State," the *Wu-tzu*.

25. For further discussion, see the introductory material in our *Seven Military Classics*.

Chapter 3

1. King Wu inquires about this exact situation in Chapters 36, "Approaching the Border," and 37, "Movement and Rest," in the *Six Secret Teachings*. However, the T'ai Kung suggests radically different tactics.

2. Wu-tzu also suggested employing this technique to probe unknown generals and thereby determine character and preferred tactics from their responses in Chapter 4, "The Tao of the General." He states, "Order some courageous men from the lower ranks to lead some light shock troops to test him. [When the enemy responds] they should concentrate on running off instead of trying to gain some objective." "Light troops" could also be under-

stood as "light companies," as some commentators suggest. However, insofar as a few hundred men would clearly be inadequate even for reconnaissance efforts, the indefinite "troops" seems more appropriate in this context.

3. In Chapter 3, "Planning Offensives," Sun-tzu discusses tactics appropriate to various relative strength levels. However, the ancient strategists were mainly troubled by finding themselves outnumbered, and virtually all the theoretical situations discussed in the Military Classics are premised upon extricating oneself from a disadvantageous condition. King Wei phrases the question in terms of the "Tao" of the military, equivalent to "methods" here.

4. Following Chang (CL, p. 31, n. 5), who understands *tsan* in the sense of "to induce," "to draw (out)," rather than in the basic sense of "to assist," "to support," or even "to manifest." Hsü and Wei (HW, p. 46) understand it in the latter sense, as a "supportive army." Since the basic tactic requires displaying apparent disorder to induce the enemy to venture forth and commit its forces, the former is clearly more appropriate.

The deliberate creation of apparent disorder was apparently a common technique in antiquity. For example, Sun-tzu repeatedly warned against becoming entrapped, while he advanced the idea that simulated disorder, which stems from compete control and discipline, can be used to entice an enemy into desirable actions. (See Chapter 5, "Strategic Military Power.") Sun-tzu also warned that interference by an ignorant ruler would lead to confusion and equally draw the enemy on to victory. (See Chapter 3, "Planning Offensives.") Chapter 27, "The Unorthodox Army," of *Six Secret Teachings*, similarly suggests manifesting an appearance of chaos, and other writings (including Chapter 3, "Determining Rank," the *Ssu-ma Fa*) warn against being manipulated through such techniques.

5. The problems of confronting a numerically superior enemy were much remarked upon in the Military Classics. For example, in the *Ssu-ma Fa* states, "With a small force it is advantageous to harass the enemy" (Chapter 5, "Employing Masses"). Wu-tzu said, "If his troops are numerous and yours few, then use improvised measures to harry him, never giving him any rest" (Chapter 5, "Responding to Change"). Chapter 49, "The Few and the Many," the *Six Secret Teachings,* among others, focuses on this challenge.

6. This apparently means that the army's full strength, however limited, should not be exposed in a direct frontal confrontation or displayed to the enemy. Rather, the army should "yield to the enemy's awesomeness." (Cf. CL, p. 32.) However, Hsü and Wei believe it refers to having part of the army effect a nominal advance, withdrawing to draw the enemy into an ambush (HW, p. 47). Since the chapter's first tactic already encompasses this maneuver, such repetition seems unlikely.

7. The commentators differ over the probable missing character—whether the short weapons should be "in support" (CL, p. 33) or "in the rear" (HW, p. 47). Chapter 2 of the *Ssu-ma Fa* states, "When the five weapons are not in-

termixed it will not be advantageous. Long weapons are for protection; short weapons are for defending." The short weapons would be used in the close fighting that results when trying to contain enemy breakthroughs. Sun Pin wisely employed formations that retained an empty middle ground designed for this purpose. (For example, see the discussion in the next chapter.)

8. This same situation is analyzed in Chapters 45 and 46 of the *Six Secret Teachings*, where action, rather than nonaction, is advised. The two missing characters at the start of the sentence are almost certainly *chu chün*, "main army." Most editions gloss "capabilities" (*neng*) as *pi*, "exhausted," which would yield, "Wait for the enemy to become exhausted." (See, for example, SY, p. 38; SPC, p. 77; and TCT, p. 15.) However, insofar as no action has been taken to manipulate and exhaust the enemy—Sun-tzu's and Sun Pin's normal method—the gloss is unwarranted. (Note also CL, p. 33.)

9. From "Dangerous Completion" to the end of the paragraph the translation follows Chang Chen-che's reconstruction obtained by incorporating fragments previously found at the end of the chapter (CL, p. 33). The 1985 edition follows his suggestion and adds the characters that fill out the passage. (See SY, p. 23.) On the basis of these sentences, Chang explains "Dangerous Completion" as describing the situation of committing one-third of one's troops to countering a frontal action, the remaining two-thirds (or two forces) then effecting a flanking assault once the enemy has committed its troops and exposed itself. (The "danger" stems from the weakness of the single unit directly confronting the enemy; the "completion," from the segmented forces mounting [unorthodox] strikes to achieve victory. See CL, p. 33.) The tactic of dividing into three, rather than some larger and perhaps uncontrollable (or too extensively fragmented) number, is found in many of the military writings. For example, see Chapters 38 and 40 of the *Six Secret Teachings*.

10. Understanding *cheng* as referring to "being well ordered," as commonly found in the Military Classics, rather than launching a "frontal assault" as Chang suggests (CL, p. 33).

11. Probable characters are supplemented by the translators.

12. Probable characters are supplemented by the translators.

13. The term *exhausted invaders* appears in Chapter 7 of "Military Combat," the *Art of War*. Sun-tzu generally advises not pressing an enemy into a desperate situation as it will then fight with fervor, and in Chapter 7 states, "Do not obstruct an army retreating homeward. If you besiege an army you must leave an outlet. Do not press an exhausted invader." Other works, such as the *Ssu-ma Fa* and the *Six Secret Teachings*, address the same theme, advising aggressive actions that preserve, or at least appear to allow, an escape route for the enemy. The *Wei Liao-tzu* also exploits this perception in Chapter 22, "Military Instructions, II." Thus, Sun Pin apparently suggests that the defenders can wait for the invaders to determine a "route to life," an es-

cape route, attacking when they have turned their attention toward retreating and are neither focused nor prepared for combat. (Hsü and Wei believe the term *exhausted invader* refers either to a vanquished enemy or to one defending a city to the point of death. Cf. HW, p. 47; and note SY, p. 38.)

14. Fragmenting the enemy's forces while concentrating one's own troops in a decisive strike was one of Sun-tzu's central principles. See, for example, Chapter 6, "Vacuity and Substance."

15. Wu Ch'i said, "It is said for one to attack ten nothing is better than a narrow defile. For ten to attack a hundred nothing is better than a deep ravine. For a thousand to attack ten thousand nothing is better than a dangerous pass. When using small numbers concentrate upon naturally confined terrain" (Chapter 5, "Responding to Change"). A similar view is advanced in Chapter 27, "The Unorthodox Army," in the *Six Secret Teachings*. Sun-tzu discussed combat on constricted terrain in Chapter 10, "Configurations of Terrain," and Sun Pin advocated their exploitation in these circumstances in Chapters 17, 20, and 27.

16. This sentence, found in "Initial Estimations," the first chapter of Sun-tzu's *Art of War*, is among Sun Wu's most famous dicta, one frequently quoted in other military writings.

17. Accordingly, the T'ai Kung and others advocated selecting skilled, valiant soldiers to make up elite units that would occupy the forward position, providing a powerful thrusting force to forge an opening for the main body. See, for example, Chapter 10, "Configurations of Terrain," in the *Art of War*, where Sun-tzu comments on this problem; and such chapters in the *Six Secret Teachings* as "Selecting Warriors," "Martial Chariot Warriors," and "Martial Cavalry Warriors." Also compare "Planning for the State" and "Evaluating the Enemy," the *Wu-tzu;* and "Combat Awesomeness," in the *Wei Liao-tzu*. Wu Ch'i even calls these elite units the "army's fate." Sun Pin stresses the need for a "weighted front" throughout the *Military Methods,* particularly in Chapter 9. This query may be understood with reference to the enemy or in general to oneself. It is either the final question in the series focusing on the enemy or the initial one on general topics.

18. Sun-tzu states, "If orders are consistently implemented to instruct the people, then the people will submit" (Chapter 9, "Maneuvering the Army"). The creation and development of credibility, and thus unquestioned obedience, were a major concern for the military writers and such famous Legalist theorists as Lord Shang. See, for example, the extensive discussions in the *Wei Liao-tzu,* including Chapter 3, "Discussion of Regulations."

19. The importance of exploiting ravines when numerically outnumbered has already been raised in note 15. See also note 26.

20. Wetlands have always posed severe problems for campaign armies, and most of the Military Classics advocated policies in concord with Sun-tzu's dictum to rapidly distance oneself from them. (See, for example, Chapter 9,

"Maneuvering the Army.") Apart from the physical difficulties created by muddy and watery terrain, which especially bogged down armored infantry and the chariots (and later the cavalry), marshlands also afforded the enemy excellent cover for ambushes. In several chapters of the *Six Secret Teachings,* the T'ai Kung focused upon avoiding such hazards as well as not being caught when still partially enmired. Remarkably, although such dangers were frequently remarked upon, far less was ever said about exploiting the opportunities they presented. However, Wu Ch'i did briefly mention the possibility of flooding or incinerating an enemy force tarrying within them (Chapter 4, "The Tao of the General") and also of attacking an enemy in the midst of fording rivers or crossing wetlands (Chapter 2, "Evaluating the Enemy").

21. Because of the lost characters, the question and thus the exact content of the reply are unknown, and Sun Pin's response is variously interpreted. Among other possibilities are, "Drum the advance and probe them; destroy them; assume a kneeling posture." (Cf. CS, p. 51; and HW, pp. 48–49.)

22. Four concepts central to Sun-tzu's *Art of War*. Authority (*ch'üan*), a term that originates in the concept of weighing, came to mean a "strategic imbalance in power," although here it clearly refers to political authority, the power of the ruler over the people. *Shih*, "strategic configuration of power" (or "strategic power"), encompassed the advantages conveyed by the integration of power (including military composition, total numbers, and firepower) with those of terrain. Plans should be detailed and complete before military action is initiated. And "deception," a fundamental, essential technique, became inextricably identified with Sun-tzu, who said, "Warfare is the Tao of deception" (Chapter 1, "Initial Estimations"). Sun-tzu consistently advocated implementing tactics designed to deceive the enemy and thereby gain victory with a minimum expenditure of force and loss of men and material. For an extensive discussion of these concepts, refer to the introduction and notes to our translation of the *Art of War*.

23. See Sun-tzu, "Strategic Military Power"; and Chapters 19 and 30 of the *Military Methods.*

24. Several additional principles from Sun-tzu. See, for example, Chapters 7, "Military Combat," and 10, "Configuration of Terrain," on the importance of determining the configuration of the terrain, and Chapter 1, "Initial Estimations," on evaluating the enemy. Evaluating terrain is a common theme in most military writings, including Chapter 3, "Determining Rank," the *Ssu-ma Fa;* and the *Wu-tzu.*

There is some disagreement about the meaning of the four characters invariably/attack/not/defend. Hsü and Wei interpret it as, "Invariably attack, do not defend," explicitly commenting that aggressive, rather than defensive, measures are to be chosen, in contradistinction to Sun-tzu, who equally advised employing defensive measures as appropriate (HW, p. 49, n. 43. See also SY, p. 42; and SP, p. 28. This reading is favored by most commentators).

However, as Huo Yin-chang points out, this sentence is equivalent to Sun-tzu's principle of "attacking where the enemy does not defend, striking where they do not expect it," also expressed as "to ensure taking positions in an attack, strike positions that are undefended." Furthermore, according to Huo, while such aggressive actions are properly directed toward realizing the army's ultimate objective of vanquishing the enemy, Sun Pin also discussed defensive and temporizing measures even in this chapter. (See CS, pp. 51, 57–62. See also TCT, p. 20. The *Mou-lüeh k'u* contains a commentary upon this sentence largely derived from Huo's notes, but with an additional emphatic denial that it could ever be understood as "not defending" [Tzu Yü-ch'iu, ed. (Peking: Lan-t'ien ch'u-pan-she, 1990), pp. 100–102].) In addition, while Hsü and Wei's interpretation is not impossible, it is less likely than as translated, for the negative "not" is not the more emphatic "do not" (*wu*), which Sun Pin frequently employs in his writings.

25. Note that Sun-tzu said, "If I do not want to engage in combat, even though I merely draw a line upon the ground and defend it, they will not be able to engage me in battle because we thwart his movements" (Chapter 6, "Vacuity and Substance").

26. As noted, exploiting ravines was an important tactical concept in ancient military thought. Conversely, a force that chose to occupy them rather than engage in battle was considered to be afraid, lacking in combative spirit. Chapter 5 of the *Wei Liao-tzu* states, "One who occupies ravines lacks the mind to do battle." When in difficulty or outnumbered, deploying in such strategic terrain was still the most advisable course, as Sun Pin here suggests. Sun-tzu also warned against an enemy occupying ravines in Chapter 9, "Maneuvering the Army." Since ravines could also be exploited to establish a constricted killing zone by occupying the heights, the tactic was inherently risky.

27. "Be greedy," following Hsü and Wei's suggestion, HW p. 49, n. 46. However, the 1985 edition suggests the partially illegible characters are *li ch'ien* (advantage/to the fore), which would yield, "you must not seek advantage to the fore." This would be understood as meaning that one should not pursue a seeming advantage to the front as it may simply be an enticement meant to draw one out from a secure position and make one's forces vulnerable. However, while the partial for the first character might allow "advantage," the second seems rather skewed. (Cf. SP, p. 24.)

28. This brief sentence occasions considerable disagreement among commentators. In summary, there are two questions: Should the character here understood as "augment" mean "reduce"? To whom does "expand determination" refer? According to the original commentary, the general should deliberately have the walls "made lower" in order to show that he and his troops are unafraid and thereby expand the defender's determination. However, this sort of foolhardy action seems extremely dubious as a general tactic because no commander would waste energy to reduce his fortifications and thereby

expose his forces to greater danger. While such bravado might serve to stimulate the troops and result in victory in the exceptional case, in general strategists concurred with the T'ai Kung, who reputedly said, "Deep moats, high ramparts, and large reserves of supplies are the means by which to sustain your position for a long time" (Chapter 27, "The Unorthodox Army," the *Six Secret Teachings*). The *Wei Liao-tzu* also notes the importance of high ramparts in Chapter 1, "Heavenly Offices," and Chapter 4, "Combat Awesomeness," and in Chapter 22 states, "When the general is light, the fortifications low, and the people's minds unstable, they can be attacked." As both Chang Chen-che and Hsü and Wei point out, the character obviously means to "increase" or "augment," perhaps in terms of width as compared with height (since the latter was advocated in the previous passage. See CL, pp. 37–38; and HW, p. 49). Chang, however, believes "determination" refers to the general's own resolve (which seems unlikely since the military works normally discuss actions generals take with regard to their men and commands), arguing that it contrasts with ordering the masses. Nevertheless, it seems more likely to refer to the elite soldiers (or even officers) as compared with the masses. (This is the view of the original commentators and of Hsü and Wei. See HW, p. 49. The CS edition simply follows Chang's commentary, CL, p. 52.) Also observe that Sun Pin advises relying on fortifications for security and defense. (See, for example, Chapter 19.) Some editions interpret *kuang chih* (expand/determination) as referring to setting out flags. (See SY, p. 43; and SP, p. 31.)

29. Repeating Sun-tzu's famous dictum. See note 16.

30. Sun Pin is cautioning against being hasty and trying to force a precipitous conclusion. Although Sun-tzu strongly warned against the dangers of protracted warfare (in "Waging War"), he equally condemned foolhardy actions. Chang differs from the other commentators in understanding the sentence as, "You must take note of this," which seems unlikely (CL, p. 38).

31. The Awl and Wild Geese formations are more extensively described in Chapter 16.

32. A topic consistently addressed by the Military Classics, including the *Six Secret Teachings*. (See note 17.) The translation preserves the older distinction, still somewhat seen in the Warring States period, between *shih* as "officers" and *tsu* as "ordinary troops." *Shih* gradually came to simply mean "warriors" and then "soldiers" as the original hereditary trappings eroded.

33. Although Chang offers an extended note 42 (CL, pp. 39–40) justifying a reading of "racing along (in chariots) and firing"—rather than understanding as "rapidly firing," as most commentators do—it is unlikely that crossbows would have been extensively employed on chariots because of the difficulty of recocking. However, infantry troops could brace the crossbow against firm ground when pulling the string back and thus could shoot and run. Conversely, an archer with a long bow or a compound bow could

achieve a high rate of fire even from a racing chariot. (Cf. SY, p. 44; CL, p. 56; and TCT, p. 17.)

34. Supplying *tuo pien* ("many changes") as the two missing characters based upon Chapter 16's elucidation of the quick archery response.

35. Following Chang, who argues convincingly against any characters being missing in these sentences. (Cf. CL, p. 41; HW, p. 45; and CS, p. 48.) Note that Chapter 5 of the *Wei Liao-tzu* observes, "Now, in general, one who presumes upon righteousness to engage in warfare values initiating the conflict."

36. Because Ch'i did in fact precipitously decline and approach virtual extinction after the reigns of King Wei and his immediate successor, King Hsüan, some historians claim the book must have been compiled by Sun Pin's disciples subsequent to the extensive military defeats suffered under King Min. However, Sun Pin may have been perspicacious, or "three" might merely signify an indefinite "few" generations. (Cf. CL, p. 41; and CS, p. 53.) Similarly, even discussing Sun Pin in the third person does not really preclude him from having been the author—in which case either he was using the device to impart a sense of objective history to the moment and his self-portrayal, or his disciples applied some heavy-handed editing.

37. Several fragments follow, the most comprehensible of which are translated to give a sense of the topics involved.

38. The partially obliterated characters seem to be those for "Confucius."

Chapter 4

1. The original surname of the T'ien family, before fleeing from the state of Ch'en to Ch'i in the Spring and Autumn period, was Ch'en. Later descendants were generally known by T'ien, but apparently Ch'en was also employed somewhat interchangeably. Ch'en Chi is thus T'ien Chi, the inquirer throughout the chapter. It is also thought that the characters were pronounced similarly. (Cf. CL, p. 46; and SY, p. 46.)

2. The portion from just after "Our troops" to the note number is supplemented from the 1985 edition. (See SY, p. 32.)

3. Some twenty or more characters are missing; the probable intent as derived from the chapter's content is summarized within the brackets. "Fortifications" encompass not only permanent, enclosed structures, such as forts or walled cities, but also various types of field entrenchments and ramparts, as well as the disposition of troops within them. As the commentary will discuss, the chapter is thought to describe the temporary field defenses erected at Maling.

4. Supplementing from the chapter's contents. Hsü and Wei suggest "worry the enemy"; however, compare CS, p. 65.

5. Sun-tzu discusses "fatal terrain" in Chapters 8 and 11 of the *Art of War*.

6. The use of "took," which already connoted great ease in the act of seizing in the *Art of War*, is clearly pejorative. Furthermore, Wang Huan-ch'un asserts it means "to kill," in contradistinction with *ch'in*, "to capture." (See "Ma-ling-chih-chan hsin-chieh," p. 51.)

7. *Hsing*, translated here as "form," is a technical term in the *Art of War* and thereafter, comprising the concepts of shape and disposition, as well as the intangible but integrated principles governing the conduct of the battle. (Although simpler, "circumstances" would thus be an inadequate translation.) T'ien Chi's statement is seen as evidence that he did not personally participate in the campaign, although other interpretations are possible, such as not recalling or not knowing exactly how the battle was shaped.

8. The specific terms for "parapets" and "battlements" occasion much commentary. Basically the former, consisting of the chariot walls, enclosures, and coverings, present a solid appearance, while the battlements, formed from the various large and small shields, would naturally have gaps. (Cf. CS, p. 66; CL, pp. 47–48; and HW, p. 62.)

9. Following Hsü and Wei (p. 63), who understand *t'ou chi*, literally "throwing machines," as "machines for throwing stones" (trebuchets), a reading that seems appropriate because Sun Pin appears to be comparing the role of the crossbowmen with that of stone hurlers (using trebuchets) inside the city walls. (The chapter also states that the top of the fortifications should be split equally between crossbows and spear-tipped halberds, and in static city defenses trebuchets may also have been used on the walls, presumably for greater height and more accurate sighting.) Robin D.S. Yates, "Siege Engines and Late Zhou Military Technology," in *Explorations in the History of Science and Technology in China* (Shanghai: Shanghai Chinese Classics Publishing House, 1982), pp. 414 ff., notes that this is probably the second-oldest reference to such devices in the ancient Chinese military corpus. (See p. 416. However, we differ with his reading of the text—"Sun Pin recommends that crossbows be used against throwing machines"—because the chapter throughout employs *tang* in the sense of "employed as" or "substitute for," as in "caltraps are employed as ditches and moats," rather than "against" or "oppose." He also notes that the Mohists apparently emplaced trebuchets on top of the walls. See p. 418.)

However, Chang Chen-che specifically rejects this possibility, asserting that through the Han dynasty *t'ou chi* was not the term used for trebuchets. (See CL, pp. 48–49; and also note "Siege Engines," especially p. 418, for the opposite conclusion.) In his view the term should be interpreted in the sense of "springing a trap" or ensnaring, citing various texts in support, including a *Wu-tzu* sentence that he apparently understands as, "Establish an ambush and spring the trap." (However, the sentence cited may also be translated as, "Establish ambushes and take advantage of the moment," as we have in our *Seven Military Classics*.) Furthermore, he notes, in the *Six Secret Teachings* and *Huai-nan Tzu* the release or firing of a crossbow trigger (*chi*) is employed as an analogy for speed and swiftness, and in the context of a discussion about using the firepower of crossbows to swiftly and rapidly strike the enemy, lends support for the idea of springing a trap. Thus, the crossbowmen, placed inside, would overwhelm the enemy forces (which would likely be entangled in the caltraps). In this interpretation he is also followed by Huo Yin-chang (CS, pp. 66, 68). For further discussion of trebuchets in China, see Joseph Needham, *Science and Civilisation in China*, vol. 5, *Chemistry and Chemical Technology* (Cambridge: Cambridge University Press, 1986), Part 7, pp. 276 ff.

10. See the commentary for a discussion of Chang's alternative interpretation and an appropriate translation.

11. The enemy's entrance into the caltraps as a condition for firing seems to be implied by the "them." (Cf. HW, p. 64.)

12. This amazingly simple sentence—"high," "then," "square," "them"—is open to many interpretations besides the one given in the translation. Among those mentioned by various commentators, in rapidly descending probability, include "If the enemy is superior then use a square flag," " If the enemy is coming from above use a square flag," and "Make the upper part (of the lookout tower) square." (The translation partially follows SY, p. 49. Cf. SP, p. 34, for a completely different understanding; CL, p. 50; and CS, p. 66.) A square deployment of sentinels could be accomplished by positioning men at the four corners, assuming they maintain mutual lines of sight. A circular perimeter defense would require far greater numbers.

13. Hsün Hsi and Sun Chen were active in the Spring and Autumn period. The former was especially noted for his righteousness, the latter for his victories, but why they, among many far more famous and distinguished generals, should have been singled out for comment is unknown. For a convenient summary of the historical material, see CL, pp. 50–53.

14. Wu and Yüeh were major forces during the Spring and Autumn period, Yüeh eventually destroying the formerly great power of Wu, which had attained its position of dominance through Sun-tzu's efforts. Ch'i provided the stage for Sun Pin's activities. For a discussion of the historical Sun-tzu, see the introduction to our single-volume *Art of War*.

15. See CS, pp. 44–49.

16. See CL, pp. 69–72.

17. For a summary of the evolution of warfare in ancient China, see the introduction to our translation of the *Seven Military Classics*.

18. A third possibility, depending upon the interpretation of the character "fill" and likely reconstruction of the virtually illegible ones that follow, might be "fill with provisions and foodstuffs." (Chang believes the left side shows the traces of the "tree radical," and postulates the character as "tree." However, it might equally be the "chariot" or "rice" radical and the charac-

ters "baggage train" or "foodstuffs," respectively. See CL, p. 49.) Alternatively, emphasizing the aspect of "filling" in a dynamic situation, the character "to fill" could be understood as "to supplement (weakness)," thus lending support to allowing the unrestricted movement of reinforcements across a space without men, as suggested in our commentary. Later texts, such as the *Questions and Replies,* extensively discuss creating and exploiting the empty spaces within formations to facilitate the deployment of troops in unorthodox tactics. While Sun-tzu already elaborated the concept of the unorthodox and orthodox, and Sun Pin utilizes it, it does not seem intended here.

19. See CL, pp. 44–49; and CS, pp. 69–72.

Chapter 5

1. The character translated as "regulations" (connoting both military regulations and discipline) is also variously understood as meaning "commands" or "instructions" (proceeding from the ruler) and as indicating a tight organization with strict discipline. (Cf. CL, p. 56; and HW, p. 68.)

2. *Hsin* encompasses the concepts of good faith, trust, sincerity, and—with respect to rewards and punishments—credibility and certitude. ("Trust" is employed in the translation because of its appearance later in the chapter. When the context is clearly focused upon rewards and punishments, we usually translate the term as "credibility" or "certitude," emphasizing that the system must be believable and believed in.) Here it refers to the words and commands of the general; the troops must trust them completely to enthusiastically implement them. This is one of the fundamental perceptions underlying the psychology of rewards and punishments prominently developed by both the Legalists and the military strategists. As Chang points out, when men believe in the rewards, the army will be "sharp," spirited, and motivated (CL, p. 56; CS, p. 74. All the commentators, Chang included, understand *li* as "sharp"; however, another meaning for the character [apart from "profit"] in the military writings is "advantage." Cf. SY, p. 53). However, Hsü and Wei apparently believe that this trust refers to the mutual trust and good faith that must exist among all the ranks for affairs to proceed expeditiously, "advantageously" (HW, p. 68).

3. The elusive term *Tao* embraces a wide range of disparate meanings, encompassing many philosophical orientations, metaphysical assumptions, and even religious beliefs. In the military writings it generally has a somewhat more circumscribed scope, frequently referring to either the "Tao of the military," which would be the science of all military arts, or the "Tao of government." The Way (Tao) is thus often equated with the specific knowledge and techniques for command and governing, generally on the basis of a saying in Sun-tzu or another ancient text, but there is no need to so delimit it. In fact, in the earlier *Art of War* the term is never really defined, although in his first

chapter Sun-tzu touches upon its scope and role, saying, "The Tao causes the people to be fully in accord with the ruler." In Chapter 7, "Eight Formations," the *Military Methods* (as Chang points out, CL, p. 56), the Tao is synonymous with knowledge of the Way of Heaven, the patterns of Earth, the psychology of the people, the nature of the enemy, and the methods of deployment. (For Sun Pin, based upon his use of the term throughout the text, Tao must be even more extensive than that, including methods of command, tactics, and strategy.)

Tê, here translated as "power," also means "virtue," in the sense both of moral virtue and a person's or object's innate power. The latter is primarily intended here, but the sentence is fraught with echoes of Taoist concepts too complex and extensive to explore in a note.

4. This accords with the fundamental assumption of the military theorists: A ruler who practices Virtue attracts the willing allegiance of his people (as discussed in the commentary). The latter constitutes the army's greatest resource. (For example, see the doctrines summarized in Chapter 1, "King Wen's Teacher," the *Six Secret Teachings,* and expanded in the subsequent chapters of "Civil Secret Teaching.")

5. This simple sentence unfortunately does not make good sense. It would seem that something like "Trust is the way to make rewards clear to the army" should be intended.

6. Most of the military writings warn against becoming enthralled with warfare and vociferously condemn rulers who take pleasure in it. Sun-tzu also condemned protracted campaigns because they exhaust the resources of the state, although by the end of the Warring States period, massive undertakings resulted in states fielding armies that reputedly approached 1 million men in strength. Sun Pin clearly continues Sun-tzu's thoughts here, noting that the ruler who detests warfare is the one who will most effectively commit the army to battle. Furthermore, in this chapter he has already warned against engaging in frequent conflicts and noted that an army's strength derives from a rested populace. Wu Ch'i voiced similar thoughts: "Being victorious in battle is easy, but preserving the results of victory is difficult. Those who have conquered the world through numerous victories are extremely rare, while those who thereby perished are many" (Chapter 1, "Planning for the State"). However, note that the commentators offer other interpretations, such as, "Those who detest warfare are the army's greatest assets," understood as referring to generals and the troops, or "Detesting warfare is the army's greatest implement." (Cf. SY, p. 54; CS, p. 74; and SP, p. 38.)

7. Supplementing the missing characters based on HW, p. 68. The military theorists generally agreed that gaining the willing support of the people was essential to creating a prosperous state and mounting a successful campaign. As Sun Pin states in the next paragraph, "One who gains the masses will be victorious." Note the difference in terms: In the former case it is *ch'ü,* "to

take," rather than *te,* "to obtain," "to gain," or "to get." In the initial chapter of the *Six Secret Teachings,* the T'ai Kung also speaks about "taking the people" and "taking the realm."

8. With the rise of military specialization and a gradual shift away from the Shang and Early Chou practice of rulers normally commanding in person and toward entrusting the army's actual command to professional generals, the importance of granting unhampered authority to the field commander became obvious. Sun-tzu apparently first raised the issue in Chapter 3 of the *Art of War,* although for another century there would still be men, such as the famous Wu Ch'i, who combined the talents of administrators and generals. Thereafter virtually all the military writings voiced the need for such independence, and in the ceremony empowering the commanding general described in Chapter 21, "Appointing the General," of the *Six Secret Teachings,* sole authority is explicitly conferred. Sun Pin perceived that any field general suffering under the yoke of external influences would be doomed.

9. "Terrain" is literally "ravines." The entire sentence closely follows one found in Sun-tzu's Chapter 10, "Configurations of Terrain": "Analyzing the enemy, taking control of victory, estimating ravines and defiles, the distant and near, is the Tao of the superior general."

10. The commentators take pains to gloss "perverse" as meaning "distanced" or "estranged" from his troops, but it need not be so restricted. (Cf. CL, p. 58; HW, p. 69, "not in harmony"; and SY, p. 55.) "Perverse" should be defined with respect to the Tao, for someone who is not in accord with the Tao (of Heaven, man, Earth, command, and so forth) has thus distanced himself from the normative world.

11. Following Sun-tzu's thrust as expressed in his infamous chapter, "Employing Spies."

12. A number of other words could equally well be supplied for the missing character posited as "trust": "Knowledge," "loyalty," and "strength" are three commonly found in the military writings.

13. Following CL, p. 58; and SY, p. 58. The character *pao* means "shield," "screen," "fan," or even "umbrella," with Hsü and Wei interpreting it as referring to King Wu's "standard" or "battle objectives" (HW, p. 69). King Wu, literally the "Martial King," led the house of Chou in overthrowing the tyrannical and debauched last ruler of the Shang dynasty. The king, who inherited a prosperous, well-governed state from his brother, King Wen, founded the famous Chou dynasty through a lightning military campaign and an epoch-making victory over vastly superior Shang forces at Mu-yeh. For a discussion, see the introductory material to our translation of the *Seven Military Classics.*

14. Literally, those who "do not excel," the "not good."

15. Rising military specialization and professionalism (see note 8) mandated the granting of sole authority to the commanding general. Transfer-

ring such power immediately created a potential threat to the ruler himself, for only a truly loyal general would be able to resist the subsequent temptation to usurp the throne.

16. The point being that if the rewards are not trusted, they will not be taken seriously, and they will fail to motivate the soldiers.

In the Early Chou period the "hundred surnames" had been the members of the aristocracy; only the royal and other noble clans had surnames. With the passage of time, decline of the old aristocracy, rise of talented lower family members, and ongoing establishment of numerous minor collateral lines, many more people were granted or assumed surnames. In the Warring States period, families, and thus individuals, with surnames were still members of the privileged class, but the distinction continued to be eroded over the centuries until the term became effectively synonymous with "the common people." For Sun Pin, the "hundred surnames" probably referred to all members of the nobility, including the *shih,* or lowest-ranked members, who largely furnished the officer corps and many of the most spirited warriors; various freemen with landholdings, financial position, and political power; and perhaps even government officials. Most of the commentators confine the scope of the reference here to the officers and soldiers. (See, for example, CL, pp. 58–59; CS, p. 75; and SY, p. 56.)

17. Huo Yin-chang, no doubt following Marxist guidelines, equates the masses with the soldiers rather than the populace at large (CS, p. 83). However, while the term often refers to troops, especially infantry troops later on, in the context of the general political theory espoused by most of the military thinkers, such as encompassed by the *Ssu-ma Fa,* "masses" should be understood as referring to the populace at large.

18. As mentioned in note 6, many of the military writings speak out against war not just because of the expense and danger incurred but also because of the inherent misery and loss of life. At the same time, skill in warfare was becoming essential if a state were to survive, and martial studies and preparations could be ignored only by the doomed. Sun-tzu said, "Warfare is the greatest affair of state, the basis of life and death, the Tao to survival or extinction" ("Initial Estimations"). And he also said, "Unless endangered do not engage in warfare. The ruler cannot mobilize the army out of personal anger. The general cannot engage in battle because of personal frustration. When it is advantageous move, when not advantageous stop. Anger can revert to happiness, annoyance can revert to joy, but a vanquished state cannot be revived, the dead cannot be brought back to life" ("Incendiary Attacks").

19. For example, Sun-tzu said, "The general encompasses wisdom, credibility, benevolence, courage, and strictness" ("Initial Estimations"; see also "Nine Changes"). The *Six Secret Teachings* devotes three chapters to the character and qualifications of generals (19, "Discussion of Generals"; 20, "Selecting Generals"; and 23, "Encouraging the Army"). Wu-tzu's chapter,

"The Tao of the General," also touches upon them, while Sun Pin's *Military Methods* enumerates an extensive array of virtues and vices in Chapters 23–26.

Chapter 6

1. The precise meaning of the title is unclear. It is generally understood as referring to the phases or influences of the moon, which are cited in the chapter itself to explain unexpectedly numerous victories, but dissident voices (wrongly) suggest it might refer to night battles. (See SY, p. 57; and CS, p. 87.) It may simply have been appended by a later editor or disciple as a convenient rubric for remembering the main contents.

2. Supplementing the four missing characters.

3. Mencius, roughly contemporary with Sun Pin, is noted for having stated, "The seasons of Heaven are not as good as the advantages of Earth; advantages of Earth are not as good as harmony among men" (*Mencius,* II:B2). A virtually identical statement is found in Chapter 4, "Combat Awesomeness," in the *Wei Liao-tzu,* and echoed in other writings of the late Warring States period such as the *Hsün-tzu.* Here Sun Pin advises, as would most of the military thinkers, that while men are the most important element, all three—Heaven, Earth, Men—must be relied upon. (The translation follows CL, p. 63, understanding the missing character as *i,* "to take," rather than *hou,* "after." However, note Hsü and Wei's interpretation of the entire sentence [HW, p. 74], resulting in a markedly different translation: "For this reason one must first suffer insult, and only then engage in battle." This would seem to be an inappropriately modern reading of the characters.)

4. All the commentators connect these two clauses, no doubt due to their continuous positioning on a single bamboo strip and the less likely probability that a sentence would begin with the negative *pu.* However, while it seems that the second clause reflects Sun Pin's avowed principle that war should be avoided unless necessary and appropriate conditions have been realized, the common view is that even when these conditions are not realized, one must still fight. This seems totally discordant with the thrust of Sun-tzu's and Sun Pin's thought and therefore inappropriate. Moreover, a famous passage in Chapter 31 of the *Tao Te Ching* employs similar phrasing to state that "weapons are inauspicious implements; they are not the implements of the perfected man. Only when it is unavoidable are they employed."

5. Within the overall context of the chapter and the passage to this point, it seems obvious that *shih,* "time" or "season," is intended in the latter sense. However, while some of the commentators agree (CL, p. 63; CS, p. 88; KO, pp. 178–179) and even supply correlative material from other texts (such as CL, pp. 60–62; and CS, p. 88), others take it as referring to "time" in the sense of "timeliness" or the "moment." (See, for example, SY, p. 58; and SP, p. 41.) This would yield a translation something like, "If you grasp the moment to engage in warfare. . . . " Still another, radically different interpreta-

tion is offered by Hsü and Wei (HW, pp. 74–75), who gloss *fu* (GSR 103p) as *wu*, "without," essentially deleting the hand radical. Consequently, their version would read, "If you engage in battle without regard to the seasons you will not be able to again employ the people." Thus, they embrace the concept of seasonal restrictions on warfare but through a negative adjunctive. In Chapter 2, "Waging War," Sun-tzu condemned lengthy campaigns, concluding, "No country has ever profited from protracted warfare." Furthermore, he specifically noted, "One who excels in employing the military does not conscript the people twice, nor transport provisions a third time."

6. Literally "calendrical influences"—the changes of the sun (such as angle and length of day), seasons, phases of the moon, and temporal position among the twenty-four seasonal segments into which the year is divided, as modified by their mutual influences and interactions. (The translation follows CL, p. 63; and CS, p. 60. As usual, Hsü and Wei interpret dramatically differently: "If one lacks the advantages of terrain but engages in battle and gains a small victory, it will be a waste of resources" [HW, p. 74].)

7. This sentence presumably continues the thought of the previous lines— that is, the general is acting without any basis. (See the discussion in the commentary.)

8. As discussed in the commentary, the moon stood as the paramount symbol and embodiment of *yin*. Therefore, with its waxing and waning, military activities would be more or less auspiciously undertaken. Gaining a number of victories without any basis is thus interpreted as the army's actions having fortuitously coincided with the moon's ascendant influence.

9. In Sun Pin's view, numerous battles, even if producing nothing but victories, will inevitably spawn disaster. (There may also be some sense that because *yin* is peaking within its cycle, reversal—which is a constant of the Tao—is both to be expected and unavoidable in military affairs.)

10. In Sun-tzu's *Art of War* (as Robin D. S. Yates has noted, "New Light on Ancient Chinese Military Texts," p. 219), the connectives frequently appear to have been supplied by the compilers even though there is not any inherent connection, logical or otherwise, between the conjoined sentences. Consequently, in the *Art of War* "therefore" and "thus" may be essentially empty bridging words. This appears to be equally true here, with Sun Pin proceeding to list some general cases (which may well be classified as incomplete victories but need not necessarily be so). Clearly, five examples of incomplete victories cannot be forced out of the possibilities that follow. (However, see CL, p. 64; CS, pp. 90, 92; and the translation provided in note 18.)

11. While it appears that the idea carries over from the conclusion of the first paragraph, this section being an expansion upon the theme of astrological influences, the second section is found on another bamboo strip headed by a concluding connective particle. Therefore, although its placement seems sequentially appropriate, it may be erroneous.

12. *Analects,* XI:12.

13. Chapter 13, "Employing Spies."

14. The incident is recorded in his *Shih Chi* biography; the *Han-shih wai-chuan;* and Book III, *Questions and Replies,* where Li Ching stresses that military affairs are a "matter of human effort." The latter in fact derives from the opening sequence of the *Wei Liao-tzu* (which is mentioned immediately following in the commentary).

15. Unfortunately, this issue is too complex to discuss here, and even basic studies of these questions are lacking.

16. As discussed in note 10.

17. Compare CS, p. 92. In Chapter 3 of the *Art of War,* Sun-tzu establishes the principle that emerging victorious while preserving the enemy intact is primary, while decimating the enemy can claim only second best. "For this reason attaining a hundred victories in a hundred battles is not the pinnacle of excellence. Subjugating the enemy's army without fighting is the true pinnacle of excellence."

18. Based upon the interpretations in CL and CS, this paragraph might be translated as follows: "There are five factors which preclude a victory from being considered a (complete) victory. If a single one of the five is present, it will not be a (complete) victory. Thus in the Tao of warfare, if many men are killed but the troop commanders are not captured, (it is not a complete victory). If the troop commanders are captured but the encampment is not taken, (it is not a complete victory). If the encampment is taken but the commanding general is not captured, (it is not a complete victory). If the army is destroyed and the commanding general killed, (it is a complete victory). Thus if one realizes the Tao, even though the enemy wants to live, they cannot." (Cf. CL, p. 64; and CS, p. 90.)

Chapter 7

1. Literally, "self-reliant." The commander clearly suffers from overconfidence in his limited abilities.

2. Literally, "self-expanding." *Kuang* is perhaps a loan for "courage." (See CL, p. 66.) Lacking courage, the commander apparently displays a false front, "bravado." However, Hsü and Wei understand it rather differently as meaning the commander wants to thereby augment his awesomeness (HW, pp. 78–79). Note that several analysts interpret *kuang* in the sense of broadening the ruler's territory. (Cf. SY, p. 61; and SP, p. 45.)

3. The phrase "number of battles inadequate" is perfectly clear in itself; the problem arises as to its implicit referents. We understand it in the general sense of the commander's experience being inadequate but equally that he simply has not grasped the Tao of the military because he has not had enough experience to penetrate it. (Note CL, p. 94.) It might also mean that several battles are inadequate for him to be victorious, so the army is relying on luck alone. (Cf. HW, pp. 78–79; SY, p. 61; and SP, p. 45.)

4. Only the two or three largest of the original seven states in the Warring States period could field, or even maintain, 10,000 chariots. Most of the others could deploy 2,000 or 3,000 at best. The *Wei Liao-tzu* states, "A state of ten thousand chariots [concentrates upon] both agriculture and warfare" ("Martial Plans").

5. Following Chang's gloss of "glory" for "expand" (CL, p. 66) rather than "expand the kingship of a state of ten thousand chariots." (Cf. SY, p. 61; CS, p. 95; and SP, p. 45.)

6. Reading *cheng*, "to remonstrate," as *ching*, "still" or "quiet." (Cf. CC, p. 177; and SY, p. 62.) Several commentators explain the former as embracing the idea that a good general will "remonstrate" or "struggle" with his ruler if he feels victory cannot be attained. However, apart from the likelihood that *cheng* was a simple substitution, the sentence as translated expresses a theme commonly found in the military writings, such as Sun-tzu's *Art of War*. (See the chapter commentary for further discussion and quotations.) Moreover, once the army had embarked on its campaign, the ruler would no longer have any voice, nor would there be time for direct consultations. Accordingly, such remonstrance would be feasible only in court discussions of strategy—such as before Ma-ling—not actual battlefield decisionmaking. (Cf. SP, p. 45; and CL, p. 76.)

7. Configurations of terrain are discussed in the chapter commentary and the Historical Introduction.

8. These classifications were first advanced by Sun-tzu in the *Art of War*. "Easy" terrain was primarily defined with respect to chariot requirements; therefore it refers to reasonably flat and unobstructed ground. "Difficult," whose Chinese character *hsien* has the basic meaning of "ravine" and an extended one of "precipitous" (and is used in both senses in the *Art of War*), here probably represents a transitional terrain too irregular or otherwise restricted to easily accommodate the passage of chariots but still accessible to horsemen. For a discussion of the probable origins and evolution of the chariots and cavalry in China, see the appendices to our translation of the *Seven Military Classics*; or the revised view in our forthcoming *History of Warfare in China: The Ancient Period*.

9. The crossbow began to play an increasingly important role in the Warring States period, providing unprecedented firepower while creating the possibility of long-range killing zones, such as employed in the battle at Ma-ling. "Constricted terrain" would confine the enemy's mobility and reduce its options, preventing enemy forces from easily breaking out. The heights of ravines and valleys would not only offer an advantageous angle of fire but also retard the progress of any counterattack mounted against massed crossbowmen. (For a brief history of the crossbow in China, see the appendix referred to in note 9. The term translated as "constricted" is a different character than the one that appears in Sun-tzu's Chapter 10. Both refer to narrow valleys, terrain that is severely constricted.)

10. "Fatal" and "tenable" terrains are discussed in the commentary.

11. "Configurations of Terrain."

12. "Incendiary Attacks."

13. "The Army's Strategic Power."

14. "Military Discussions."

15. See, for example, CL, pp. 68–71; and HW, pp. 80ff. Huo Yin-chang and others suggest that "eight formations" simply refers to military formations in general rather than to any specific eight. (See, for example, CS, pp. 93, 96–97; and SY, p. 61.)

16. Historical formations are extensively analyzed in the first two books in the *Questions and Replies*. The SP edition of the *Military Methods* includes a proposed reconstruction of the eight formations in terms of a nine square deployment arrayed as three lines of three units. While the nine block array discussed in the later military works provides great flexibility, the SP formation is simply the standard square array proven effective both in the East and West throughout history and still requires an extensive explanation as to how the "eight formations" are subsumed within it. Since Sun Pin discusses a square formation in other chapters, however imaginative the reconstruction, its validity is questionable. (The proposed reconstruction has the force components numbered so that the sum of any three horizontal, vertical, or diagonal blocks is always fifteen, constituting what is known as a "magic square." [The authors fail to point out this aspect of their diagram.] There is no need or justification for this numbering.)

17. "Configurations of Terrain."

18. "Nine Terrains."

19. Ibid.

20. "Nine Changes."

21. "Nine Terrains."

Chapter 8

1. As translated, the title follows the interpretation of *pao* (GSR 1057f), "screen" or "fan" as standing for *pao* (GSR 1059a), "treasure" rather than *pao* (GSR 1057a), "preserve," "stronghold," for which it is sometimes a loan word. (Cf. HW, p. 87; CL, p. 73; CS, p. 99; and SP, p. 49.) The chapter discusses various principles for exploiting "Earth," incorporating important observations on the relative values of different types of terrain. Thus, it continues Sun-tzu's ideas found in "Configurations of Terrain," justifying the translation of "treasures" rather than "strongholds" and the choice of "terrain" rather than "Earth." (If there had been a chapter entitled "Treasures of Heaven," then the title would clearly have to be "Treasures of Earth" to correspond.)

2. Note that this is not a simple identification: "*Yang* is the exterior." *Yang* is generally identified with the exterior and *yin*, the interior, in ancient writings, similarly being extended even to government and behavior, as in Chapter 13 of the *Six Secret Teachings*. In Chapter 23 of the *Wei Liao-tzu*, the fol-

lowing interesting passage appears: "The military takes the martial as its trunk, and takes the civil as its seed. It makes the martial its exterior, and the civil the interior. One who can investigate and fathom the two will know victory and defeat. The civil is the means to discern benefit and harm, discriminate security and danger. The martial is the means to contravene a strong enemy, to forcefully attack and defend." Also note the discussion in Book II of *Questions and Replies,* which is too extensive for inclusion here. *Yang* as the exterior is the visible and thus equally associated with orthodox tactics; *yin* as the interior, secluded, dark, and not visible is associated with unorthodox tactics. This principle is essential to comprehending the sentences that follow.

3, 4. Some commentators believe that the "direct" (*chih*) both here and in the conjoined sentences that follow refer to "main roads," while *shu,* translated as "techniques" here, refers to the "indirect" or small byways. (See SY, p. 65; SP, p. 50; and TCT, p. 34, which notes a similar usage in the first chapter.) However, Chang Chen-che identifies "straight" as flat land that allows direct passage, and *shu* (our "techniques") as referring to a road with steep sides. Thus, the former would be easily traversed, while the latter, not easily negotiated, would be a land of "half death." (See CL, p. 74.) Hsü and Wei, analyzing the two sentences, believe "direct" and "techniques" refer to the orthodox and unorthodox. Accordingly, unorthodox techniques, which (as already discussed in note 2) are associated with *yin,* the secret, and the difficult, would appropriately exploit terrain otherwise shunned because the foliage is half-dead. (See HW, p. 88.) Our translation assumes that a contrast between luxuriant vegetation and semibarren lands is intended rather than that the terrain itself is "half-dead," as Chang apparently believes, and that Hsü and Wei's explanation is basically sound. (*Shu* usually refers to "techniques" or "tactical methods" throughout the ancient military writings.) Note that the *Ch'ien-shuo* contrasts the vegetation found on flat land (the "straight") and that on contorted terrain, where ambushes are possible (CS, p. 100).

5. Although Sun Pin's work clearly contains vestiges of astrological thought, the physical sun, rather than auspicious and inauspicious days, seems to have been intended. Sun-tzu had previously advised avoiding the dark and occupying sunny areas (including the heights) in "Maneuvering the Army," whereas Sun Pin apparently focuses upon assuming positions that would take advantage of the angle of the sun's rays by orienting the sun directly behind the army.

Even though the ancient strategists often advocated deploying on a particular side of a mountain or river, the appropriateness of any actual disposition at the moment of attack would depend upon the time of day and even the season. For example, assuming a position that places the army to the east of an enemy (especially if on the west side of a mountain) would effectively force one's opponents to stare into the sun when advancing (assuming the height of the mountain did not completely block it). Unfortunately, no additional concrete statements or discussions are preserved in the extant *Seven Military Classics*. (Note that sunset attacks, mounted when men think of encamping and have lost their spirit, are frequently suggested as unorthodox techniques, as in Chapter 7 of the *Art of War*, and Chapters 49 and 58 of the *Six Secret Teachings*.)

6. The "eight winds" correspond to the eight directions. As with the sun, it is preferable to have the wind at one's back when making an attack. Thus, Wu-tzu said, "When about to engage in combat determine the wind's direction. If favorable, yell and follow it; if contrary, assume a solid formation and await the enemy" ("Controlling the Army"). And the *Ssu-ma Fa* states, "Keep the wind to your back" ("Employing Masses"). Advancing into a strong wind would not only hinder progress but also present the enemy with an opportunity to employ dust, smoke, and even fire. For example, the *Six Secret Teachings* states, "High winds and heavy rains are the means by which to implement unorthodox plans" ("The Unorthodox Army"). Chapter 36 advises stirring up dust to fatigue the enemy, making it difficult for him to fathom plans and determine dispositions, while Chapter 41 discusses extricating oneself from a situation in which the enemy ignites fires to take advantage of strong prevailing winds.

7. Sun-tzu earlier discussed the dangers of rivers: "After crossing rivers you must distance yourself from them. If the enemy is fording a river to advance, do not confront them in the water. When half their forces have crossed, it will be advantageous to strike them. If you want to engage the enemy in battle, do not array your forces near the river to confront the invader, but look for tenable ground and occupy the heights" ("Maneuvering the Army"). Wu-tzu said, "When fording rivers and only half of them have crossed, they can be attacked" ("Evaluating the Enemy").

8. Sun-tzu said, "Do not approach high mountains, do not confront those that have hills behind them" (Chapter 7, "Military Combat").

9. Sun-tzu said, "Do not confront the current's flow" (Chapter 9, "Maneuvering the Army").

10. Five "killing grounds" are defined at the end of the chapter. See note 20.

11. Forests ("masses of trees") not only concealed troops in ambush (as Sun-tzu had warned in "Maneuvering the Army") but also offered protection to the archers and crossbowmen, who were just becoming an important element of firepower in Sun Pin's era.

12. Chang and many others would understand the characters *chün chü*, translated as "all these I have mentioned," as "in all these cases depart," glossing *chü* as "to go away" (based on a *Ch'u ts'u* commentary. See, for ex-

ample, CL, p. 75; and SY, p. 65. This would be more appropriate if the clause concluded the sentence and included something like "you must," thereby advising a course of action). Huo Yin-chang understands these characters as "cases of equal strategic power" (CS, p. 100). Other variants include the odd TCT explanation of *chü* meaning "attack" and the fragment as "will be attacked" (TCT, p. 35). Our translation is based upon the meaning of "raise" or "mention" (GSR 75a) and happens to cohere with HW, p. 88.

13. Hsü and Wei provide the most plausible explanation for this sentence, noting that battles generally commenced in the morning. Therefore, deploying on the southern slope would ensure being in the light without facing into the rays. Deploying on the eastern slope—the side of the mountain facing the east—would require confronting the rays as the enemy approached from the east and possibly staring into the sun even though presumably looking downward. (See HW, p. 88.) However, as suggested in note 5, deploying to the west—and therefore facing out westward—would seem to be even more advantageous as the advancing enemy would be staring directly into the sun's rays as they advanced eastward up the mountain (assuming the peak did not block the rays). Furthermore, the basis for such sweeping statements is unclear, particularly as the military thinkers emphasized flexibility in determining the terrain to be occupied. Chang Chen-che suggests that the sentences should be read differently and in conjunction: "(If) mountains arrayed to the south are mountains of life, (then) those arrayed to the east (and north and west) are mountains of death" (CL, p. 75). Others advance the theory that it is the mountains that are "arrayed" in the east, meaning north to south, and to the south, meaning east to west. (See SP, p. 57; TCT, p. 35; and SY, p. 65.) Naturally the same arguments against this sort of reading apply: How could all the mountains in a major region prove to be untenable? (See Chang's ruminations, CL, p. 75.)

It is possible that this sentence has been removed from a far different context, having originally been part of a discussion devoted to a specific situation. The *Six Secret Teachings*, for example, discusses defensive measures to be taken when deploying on either side of a mountain, whether *yin* or *yang*, in Chapter 47, "Crow and Cloud Formation in the Mountains."

14. Again Hsü and Wei provide the most likely explanation, suggesting that since Sun Pin lived in the Shantung-Honan region, a reasonably level area, rivers flowing to the east would be going to the sea and therefore naturally associated with life. Rivers and streams flowing in other directions would likely end in still bodies of water, such as wetlands and marshes—terrain fraught with death (HW, pp. 88–89).

15. Clearly a concrete expansion of the military maxim earlier expressed by Sun-tzu, "The army likes heights and abhors low areas," and found in the *Six Secret Teachings*, "Occupying high ground is the means by which to be alert

and assume a defensive posture" (Chapter 27, "The Unorthodox Army"). In each case a relative height advantage can be exploited by long-range weapons and the onrush of suddenly descending forces.

16. Or possibly "rushes." In this case the relative hierarchy is from stronger, greater vegetative obstacles to smaller, more easily penetrated ones. However, the formula has changed because the word *conquers* is absent. The five are simply enumerated and the nature of their relationship implied from the context and initial phrase.

17. This passage reflects the five-phase correlative thought becoming influential in Sun Pin's era and already, although briefly, apparent in the *Art of War.* (As noted in the Historical Introduction, the amalgamation of five-phase and *yin-yang* thought marks the chapter as probably having been compiled some decades after Sun Pin's death.) While this is not the only possible arrangement, as ordered this is the conquest cycle found later in the *Huai-nan tzu* and other works. However, the implications are not simplistic: A position on yellow soil will not invariably result in overcoming those on red soil, if in fact such idealized configurations could even be realized. Nor would deployments that consciously employ the directions associated with the colors invariably produce victory, although that would seem to be the inevitable conclusion. How it might be determined that a particular soil is red rather than yellow (contrary to appearance), why these relationships have been tied to soil rather than being simply abstract, and how they may be exploited all require further research. For a general discussion of five-phase theory, see Joseph Needham, *Science and Civilisation in China,* vol. 2, *History of Scientific Thought* (Cambridge: Cambridge University Press, 1962), pp. 232–269. For further discussion, including the role of Tsou Yen, see the following: Kanaya Osamu, "Yin-yang wu-hsing-shuo te chuang-li," *Chung-kuo che-hsüeh-shih yen-chiu* 1988 (3):22–27; Hsin Ch'i, "Tsou Yen ssu-hsiang te chuan-pien chi ch'i yin-yang wu-hsing hsüeh-shuo te i-i fa-wei," *Chung-kuo che-hsüeh-shih yen-chiu* 1988 (3):28–32; Yin Nan-keng, "Wu-hsing pen-i suo-chieh," *Chung-kuo che-hsüeh-shih yen-chiu* 1988 (3):17–22; and Chao Chi-ping, "Yin-yang wu-hsing hsüeh-pai te tai-piao—Tsou Yen," *Chung-kuo che-hsüeh-shih yen-chiu* 1985 (2):57–67.

18. Although the character is obliterated, "valleys" appears to have been the original word (CL, p. 76).

19. As previously noted, most of the military thinkers advised against occupying low-lying, potentially submersible terrain, particularly as it would likely be characterized by wet areas that could enmire an army. The progression is from radically defined, precipitous terrain, generally marked by water, to increasingly expansive, level ground.

20. The five killing grounds derive from Sun-tzu's classification and terminology in Chapter 9, "Maneuvering the Army." (Wu-tzu also discussed a few

dramatically defined configurations, such as Heaven's Furnace, in "Controlling the Army.") These are terrains on which an army can become entangled and bottled up and that easily create a killing ground that becomes a virtual funeral mound for the careless. Naturally, they could and should be exploited by the wise commander who strives to lure or compel the enemy onto them, following the principle asserted in the preceding chapter: "Attack (an enemy) on fatal ground."

21. Heaven's Well is generally understood as a significant depression, including valleys, surrounded on all four sides by hills or mountains. It presents a surrounding force with many lethal possibilities, such as directing sustained crossbow fire downward from an unassailable position and diverting streams to inundate the enemy.

22. The commentators generally identify *t'ien yüan*, "Heaven's Bowl" or "Heaven's Basin" (a translation that assumes GSR 260b is a loan for 260o), with Sun-tzu's term, "Heaven's Jail," perhaps to distinguish it from Heaven's Well. (For Sun-tzu the latter is generally a valley enclosed on three sides with limited access.) However, it is more likely that Heaven's Well has more precipitous sides and therefore resembles a well, while Heaven's Bowl, although still a significant depression, is less sharply defined. (Chang suggests that *yüan* refers to a dry well as distinguished from a well with water in it. See CL, p. 77. Compare HW, p. 89; and CS, pp. 101–102.)

23. Although the term is *t'ien li*, "Heaven's Departure," the commentators all agree that it is identical with Sun-tzu's term, Heaven's Net. The defining characteristics for this terrain are dense growth and entangling vegetation that naturally impede the passage of chariots and infantry.

24. Another of Sun-tzu's terms, it suggests terrain that appears like a fissure in the earth. Accordingly, it should probably describe long, narrow passages constricting the army's passage, rendering troops an easy target.

25. *T'ien chao* is generally taken by the commentators as identical to Sun-tzu's "Heaven's Pit," particularly as the tomb text of the *Art of War* has another variant of *chao* for the traditional text's *hsien*. In the *Art of War* it apparently refers to soft, probably muddy terrain, which will enmire the vehicles and retard the infantry's progress. However, Chang believes that *chao* is the proper term and glosses it as "target," as in an archery target. Therefore, the term would then be "Heaven's Target," suggesting an obvious focal point for archery and crossbow fire. (See CL, p. 77.)

26. Various explanations are concocted to explain this dictum rationally rather than consigning it to the realm of superstition. They generally focus upon the luxuriant growth and high moisture levels found in the spring and the dryness and decay of fall. Consequently, in the spring it would be preferable to remain in the highlands, as there would be adequate plants to sustain the army's animals, while the dampness of the lowlands could be avoided.

(Note that Sun-tzu said, "Now the army likes heights and abhors low areas, esteems the sunny [*yang*] and disdains the shady [*yin*]. It nourishes life and occupies the substantial. An army that avoids the hundred illnesses is said to be certain of victory" ["Maneuvering the Army"].) Correspondingly, come fall the heights would be drier and more susceptible to burning, potentially trapping and endangering anyone occupying them. However, this is highly speculative. (Cf. HW, p. 89; CL, p. 78; and CS, p. 102.)

27. This sentence is generally interpreted as reflecting the belief that an army should keep heights to its right rear and water to its left front. For example, in "Maneuvering the Army" Sun-tzu said, "On level plains deploy on easy terrain with the right flank positioned with high ground to the rear, fatal terrain to the fore, and tenable terrain to the rear," and "where there are hills and embankments you must occupy the *yang* side, keeping them to the right rear. This is to the army's advantage and (exploits the natural) assistance of the terrain." The *Ssu-ma Fa* states, "In general, as for warfare: keep the wind to your back, the mountains behind you, heights on the right, and defiles on the left" ("Employing Masses"). Consequently, the modern commentators would translate it as, "Neither the army nor any deployment should confront (such configurations of terrain) to the front right," reading *cheng* (GSR 833r) as *cheng* (GSR 833j), "upright" or "straight" and therefore forward. (See CL, p. 78; and CS, p. 102.) However, Hsü and Wei understand it to mean "to encamp" and, by extension, "to attack" (HW, p. 89), which achieves the translation we have provided but through a dubious course. Rather, it is more likely that *cheng* is a loan for *cheng* (GSR 833o), meaning "to attack." Therefore one should not attack to the right front. While Sun Pin does not provide any rational basis for this tactical principle, if the commentators' views have any basis, it is probably because the enemy will have tried to assume a position that would put entrapping terrain to the left front, thereby presenting effective obstacles to advancing forces launching an attack to their own (i.e., the attacker's) right. (Attacking to the right would follow the normal tendency, noted in ancient Greek phalanx warfare as well, for an army to be stronger to the right side, and thus drift to the right, due to the weapons being wielded by the right hand.) Note that this contravenes the commentators' explanations, which incorrectly would put mountains to an enemy's front rather than to one's own rear.

28. The commentators continue the line of understanding discussed in the previous note for this sentence as well, assuming that it refers to deployments or advances that are either sited without obstacles to the right, or keep the obstacles to the right. While our translation reflects a more static interpretation for *chou*, "circle" or "perimeter," it might also mean, "Do not circle to the right; circle to the left." Although it more likely refers (perhaps as a continuation of the previous sentence) to establishing fixed positions (against the ene-

my's obstacles to prevent an unorthodox attack), it may simply assert that movement to the right should be avoided. (Cf. CL, p. 78; CS, p. 102; and HW, p. 89.)

Chapter 9

1. As discussed in the commentary, "strategic power" (*shih*), changes, and the "strategic imbalance of power" (*ch'üan*, a character often having the meaning of "authority") are all concepts originally appearing in Sun-tzu's *Art of War*. While Sun Pin discussed several types of formations and their employment in Chapter 3, the extant *Art of War* is silent. For definitions and discussions of these concepts, consult the introductions to this translation and our *Art of War*.

2. *Ping*, the character translated as "military," also has the primary meaning of "weapons." While "military" is clearly appropriate here, strong connotations of the sense of "weapons" are invariably present.

3. The analogy emphasizes that the army constantly practices deploying into formation but rarely engages in actual combat.

4. As discussed in Chapter 5, "Selecting the Troops," Sun Pin and the later military theorists emphasized the need for an elite front to penetrate the enemy's ranks. (Also note the last sentence of the chapter in this regard.)

5. Again, as discussed in note 2, the character for military affairs is *ping*, carrying the connotation of "weapons," thereby echoing the pair "sword—weapon."

6. Following suggestions found in CL, pp. 82–83. In terms of content the chapter's last sentence would seem appropriate here; however, the layout of the text on the actual bamboo strips precludes this possibility.

7. Following Chang's suggestion to insert this fragment here (CL, p. 80).

8. Most of the commentators understand the character *chu*, translated here as "ruler," as referring to the general (in his role as "master" of the army). However, in our view it is in fact the ruler whose fame grows; were the general intended, the term for general (*chiang*) would certainly have been employed. (Cf. SP, p. 58. Also note the last sentence of Chapter 10.)

9. The two sentences from notes 7–9 are inserted based upon the 1985 revised edition. (See SP, p. 54.)

10. These two sentences originally appeared at the end of the chapter and have been moved here in accord with the 1985 revised edition.

11. In Chapter 7, "Military Combat," Sun-tzu said, "Gongs, drums, pennants, and flags are the means to unify the men's ears and eyes. When the men have been unified the courageous will not be able to advance alone, the fearful will not be able to retreat alone. This is the method for employing large numbers. Thus in night battles make the fires and drums numerous, and in daylight battles make the flags and pennants numerous, in order to change the men's eyes and ears."

12. Chang (CL, pp. 83–84) interprets the character "to send" as "to attain," as in attaining or realizing a fighting spirit. This would be consonant with Sun-tzu's purpose for multiplying the drums and flags but seems to be an unnecessary explication. (However, compare SY, p. 72.)

13. As discussed in the commentary that follows.

14. Traditional belief asserted that the Yellow Emperor created bows and the crossbow, even though Yi was recognized as the paradigm archer of antiquity. For some additional cultural dimensions, see Book II of *Questions and Replies*.

15. The *Huai-nan tzu* was probably authored by a group of scholars under the auspices, and possibly active editorship, of Liu An, the king of Huai-nan, shortly after the middle of the second century B.C. This is the initial passage from Chapter 15, "Military Strategy." The translation follows the *Huai-nan tzu chu*, ed. Yang Chia-lo (Kao Yu, commentator) (Taipei: Shih-chieh shu-chü, 1969), p. 251. A partial, although dated, translation of *Huai-nan tzu* is Evan Morgan, *Tao: The Great Luminant*, rpt. (Taipei: Ch'eng-wen, 1966). Charles Le Blanc is noted for his study of this text, one of his works being *Huai Nan Tzu* (Hong Kong: Hong Kong University Press, 1985).

16. The *Lü-shih Ch'un-ch'iu* was compiled about the middle of the third century B.C. The translation follows Ch'en Ch'i-yu, *Lü-shih ch'un-ch'iu chiao-shih* (Shanghai: Hsüeh-lin ch'u-pan-she, 1984), vol. 2, p. 1321.

17. Hsün-tzu (340–245 B.C.?) was the last outstanding pre-Ch'in philosopher, as well as a prolific teacher and writer. While generally identified with the Confucian school, much of his thought is far more complex and inclined to the systematic imposition of behavioral controls. His analysis of human nature cannot be reduced to the simplistic position that he is invariably identified with—namely, that man's nature is evil. However, in contrast with Mencius (who essentially believed man's nature was inherently good or, more properly, tended to the good), Hsün-tzu embraced the view that men are innately selfish and that their desires stimulate antisocial behavior.

18. The translation is based upon Wang Hsien-ch'ien, *Hsün-tzu chi-chieh*, ed. Yang Chia-lo (Taipei: Shih-chieh shu-chü, 1974), p. 231.

19. "Benevolence the Foundation."

20. The *Three Strategies* particularly espouses this view.

21. This sentence, which also appears in the *Six Secret Teachings*, is from Chapter 31 of the *Tao Te Ching*. For an analytical discussion of Taoism and its relationship with military thought, see Christopher C. Rand, "Chinese Military Thought and Philosophical Taoism," *Monumenta Serica* 34 (1979–1980):171–218.

22. Book III, "Inferior Strategy," the *Three Strategies*.

23. The clearest statement of the necessity for a military program oriented toward rescuing the people, to be founded upon surpassing moral worth, appears in the earliest chapters of the *Six Secret Teachings*.

24. Chapter 2, "Waging War."

25. In contrast, theories about the invention of civilization's key activities and artifacts found in the Appendices to the famous *I Ching* describe the creation process as first envisioning an image, a hexagram, and then proceeding to fashion the concrete item. For example, boats and oars were based upon the hexagram Huan, "Dispersion," and bows and arrows upon K'uei, "Opposition."

26. Swords, whether short or long, did not actually appear in China until long after the Yellow Emperor's legendary existence, gradually evolving from single-edged to double-edged weapons. For a discussion, see the appendix on weapons to our translation of the *Seven Military Classics*.

27. Sun-tzu formulated the concept of the strategic configuration of power in Chapter 5, "Strategic Military Power."

28. Change, as conceived by Sun-tzu, refers primarily to the changes in tactics from orthodox to unorthodox but also to "segmenting and reuniting."

29. *Ch'üan*, always translated as "strategic imbalance of power" in the *Military Methods*, is closely related to *shih*, "strategic configuration of power," and also temporal parameters. In the context of ever-evolving battlefield situations, "tactical imbalance of power" more closely approximates the situation. However, in the greater perspective, when armies and their strategic advantages are being weighed, it is a strategic imbalance that can be exploited. For a discussion of the term, see the introduction and notes to our *Art of War* translation.

Chapter 10

1. The title is supplied by the editors as none was found for this chapter.

2. Note that the arrows have feathers rather than vanes or simply being bolts (as employed later).

3. Literally, "they are rhinoceroslike and excel at running." The commentators generally gloss "rhinoceros" (*hsi*) as "sharp"; however, in antiquity the image of the rhinoceros was one of power and ferocity. For example, in an initial interview with Marquis Wen, Wu Ch'i remarked upon Wei's obvious military preparations, noting, "Right now, throughout the four seasons, you have the skins of slaughtered animals covered with vermilion lacquer, painted with variegated colors, and embellished with glistening images of rhinoceros and elephants" (Chapter 1, "Planning for the State," the *Wu-tzu*). Chapter 8 of the "Martial Plans," *Wei Liao-tzu*, also mentions the toughness of rhinoceros hide for armor. Clearly anything "rhinoceroslike" must be "powerful." (Cf. CL, p. 86; HW, p. 101; CS, p. 115; SY, p. 75; and SP, p. 61.)

4. The original bamboo strip character translated here as "well ordered" appears to be *pien* (GSR 219b), "to distinguish" or "to discriminate

among." Hsü and Wei believe it is a loan for *pien* (GSR 219e), "to argue" or "to dispute," and thus means that the troops do not obey their orders (HW, p. 100; there is the additional possibility that the original character is GSR 219a, which has the primary meaning of "to wrangle" or "to argue" and appears in Chapter 11). However, most of the commentators, following the original notes citing usage in the *Kuan-tzu* and other early writings, understand it in the sense of "to manage," with an extended meaning of "to complete," thereby establishing a contrast between the ease of deploying in normal times and the results in combat. Although this requires assuming the "but" in the translation, it coheres with the degree of training and expectations at that time. Moreover, the *Wu-tzu* contains a passage in which Wu Ch'i describes a number of failings that develop when the battle commences: "Although Ch'i's battle array is dense in number it is not solid. That of Ch'u is dispersed, with the soldiers preferring to fight individually. Ch'u's formations have good order, but they cannot long maintain their positions. Yen's formations are adept at defense, but they are not mobile. The battle arrays of the Three Chin are well controlled, but they prove useless" ("Evaluating the Enemy"). Perhaps Sun Pin was thinking of the concrete cases of Ch'i and Ch'u. (Cf. CC, p. 202; CL, p. 86; and CS, p. 115, which follows the former.)

5. Understanding *ping*, which means "handle," as referring to the stock of the crossbow. (However, compare CL, p. 86; CS, p. 116; SY, p. 75; SP, p. 61; and HW, p. 100, where *ping* is taken as referring to the "bowman.") Consistent with appearances in earlier chapters, *cheng*, "upright" or "correct," is read in the sense of "straight."

6. That is, the four just enumerated in the preceding sentence: the arrow's lightness and heaviness correct, front and rear appropriate, the crossbow drawn straight, and the archer correct. However, the text is somewhat problematic insofar as only three factors (considering the first two as really one) are enumerated, and the "four" and subsequent "three" do not dovetail very well. (However, the SP edition identifies the "four" as the lightness and heaviness being appropriate, the stock straight, the arms at one, and the archer correct. See p. 75.)

7. Book 4, "Strict Positions."

8. Book 3, "Determining Rank."

9. "Strategic Balance of Power in Attacks."

10. Chapter 21, "Appointing the General."

11. "Superior Strategy."

12. Those without personal archery experience may find the following classic works illuminating: Robert Hardy, *Longbow*, 3d ed. rev. (London: Bois d'Arc Press, 1992); Ralph Payne-Gallwey, *The Crossbow*, rpt. (London: Holland Press, 1990); and George M. Stevens, *Crossbows*, rev. ed. (Cornville, Ariz.: Desert Publications, 1985).

Chapter 11

1. The character translated as "affect" (*i,* "to move" or "to shift") frequently refers to government measures to deliberately transfer population, as well as to people voluntarily emigrating. Some commentators assume it refers to making the people susceptible to being employed by the state; others, that it refers to the people turning their allegiance to the ruler. However, within the chapter's context it would appear to mean "affect" the people, particularly through the employment of rewards and punishments. (Cf. HW, p. 103; CL, p. 89; SY, p. 78, "use the people"; and SP, p. 63, attract immigrants.)

As discussed in the commentary, most commentators understand this sentence as simply, "The Tao for employing the military and affecting the people is (the employment of) the steelyards." They take *ch'üan heng* as referring to two types of steelyards or to the concept of "steelyards" in general. (Cf. HW, p. 103; and CS, p. 121.) However, while *ch'üan heng* appears with this meaning in such early works as the *Book of Lord Shang, Chuang-tzu,* and the *Han Fei-tzu, ch'üan* is a critical concept in the military writings of the period, including the *Military Methods,* where it designates a strategic (or occasionally tactical) imbalance in power, as previously discussed. It is also found with the primary meaning of authority, and was a concept well developed by the Legalists, including Lord Shang (discussed later) and Han Fei-tzu. For the reasons cited in our commentary, it appears that "authority," even in the context of a discussion about steelyards, remains preferable. (However, compare SY, p. 78; and SP, p. 64.)

2. The original compilation editors indicated that the text suffers a break in the middle of this sentence, but Chang Chen-che effectively argues that the sentence, although requiring some emendations to understand, is a single one. (In this he is followed by later editions, such as SP, pp. 63, 64.) The translation substantially follows his suggestions (CL, p. 90). However, Hsü and Wei would understand it as meaning, "Correct the balances, and then apply the accumulated rewards. . . . to those who have exhausted themselves in loyal (service)" (HW, p. 104).

3. Chang Chen-che would understand *hsiang* as a loan for a similar character meaning "to feast," drawing attention to the sixth chapter in the *Wu-tzu* ("Stimulating the Officers"), and the sentence's thrust as "determine (achievements) by weighing with the standard and feast (synonymous with 'reward') them in conspicuous fashion." (See CL, pp. 90–91; CS, p. 122; and TCT, p. 45.) Our translation largely accords with HW, p. 104; and KO, pp. 201–202.

4. In this period wealth would have been only minimally held by individuals and then largely concentrated in the hands of the great families and aristocrats, while state wealth was essentially identical with that of the extended ruling family. Chapter 3 of the *Wei Liao-tzu* states, "We should employ all the resources under Heaven for our own use." In the *Six Secret Teachings* the T'ai

Kung is portrayed as ɔdvising King Wen to employ all his fiscal resources to attain kingship over the realm, on the premise that once he takes the realm, they will become one, while the *Ssu-ma Fa* states, "When the masses have material resources, the state has them" ("Determining Rank").

The character translated as "wealth" (*ts'ai*, GSR 943h) is glossed by many commentators as "ability" or "talent" (GSR 943j). While they are correct in pointing out that *huo* is generally used by Sun Pin for material wealth, the gloss seems unnecessary and requires further interpretation to produce a translation such as, "Whether a person is a private individual or a (member of the) ruling family, their talent should be treated in the same fashion." (Cf. CL, p. 91; and CS, pp. 122, 124.) Within the context of the immediately following discussion, "wealth" seems more appropriate.

5. As the CC edition points out (p. 208), Sun-tzu framed one of his discussions in terms of wealth and longevity. (See the passage quoted in note 18.) This section is taken as meaning that people desire whatever they find insufficient. (See, for example, TCT, p. 45; SP, p. 64; and SY, p. 78.)

6. Only an enlightened king can implement policies designed to exploit the imbalances in the distribution of goods and longevity and thereby retain the people in government service, in battle, or perhaps in the state (although the era of greatest mobility and disaffection had not yet dawned. Cf. CL, p. 91; HW, p. 104; and CS, p. 122, which emphasizes military service, including combat. Another interpretation glosses *liu*, "to retain," as meaning "to take hold of" or "to control" with reference to the Tao. See SY, p. 79; TCT, pp. 45–46; SP, p. 64.) The T'ai Kung is recorded as speaking extensively about employing fragrant bait (salary) to catch men, even the righteous, for service. For further discussion of this sentence, see the commentary.

7. Proposed referents for this sentence vary. "The dead will not find it odious" presumably refers to death in military service. Those who have the misfortune to die in combat will not resent or begrudge their deaths because they voluntarily, even fervently, participate, as opposed to being impressed and virtually enslaved by an oppressive ruler. (Cf. CS, p. 122.) Similar views are found throughout the military writings, with various means and methods suggested for motivating the soldiers sufficiently that they will view death as if "returning home." Lord Shang, who advocated a draconian system in which rewards would be granted only for military and agricultural service, stated that under optimal conditions "the dead will feel no regret, while the living will concentrate upon exerting themselves" (Chapter 17, "Rewards and Punishments," the *Book of Lord Shang*).

The second part, "those from whom it is taken," presents somewhat greater questions. For example, Huo Yin-chang believes it refers to officials who have been demoted, expelled, or otherwise had their positions "taken away." Because the implementation of rewards and punishments is equitable, they apparently do not resent it (CS, p. 122). Others take it more generally, as those from whom wealth and riches are confiscated (no doubt as punishment) do not resent it, do not bear grudges. (Cf. KO, p. 202.)

Hsü and Wei, as noted in the commentary, interpret the entire sentence as referring to the ruler's policy of taking the excess riches of wealthy individuals upon their deaths in order to supplement (through rewards and government incentives for service?) the needs of the impoverished. (Cf. HW, pp. 104–106.) While Sun Pin was not the only thinker to advocate such policies in antiquity, it was hardly a common view and requires further investigation. Note that Chapter 4 of the *Wei Liao-tzu* states, "The state of a true king enriches the people."

8. Following the suggestions of the commentators (CL, p. 92; CS, p. 122; SP, 65). Those holding power "near" the ruler will not presume upon their positions to commit thievery, while those far off will not assume that they are too remote to be noticed and become lazy. (This requires assuming that the verb *able* should be read to mean either "dilatory" or "exhausted." Possibly the copyist failed to include a pivotal character or two after *neng*, thereby necessitating such forced interpretations.) Hsü and Wei, however, understand it passively—that those near "will not be robbed" (HW, p. 104). A minority view is found in the TCT edition, where the modern Chinese translation reads something like, "Neighboring states will not dare make incursions, while distant states will lack the strength to invade" (p. 46).

Tsei, translated here as "thievery," generally designated a person who committed a serious crime, one normally accompanied by violence, or an offense against the established order. It referred to individuals of position or rank rather than the common people and can be translated as "brigand" or "murderer." The Legalists, such as Lord Shang and Han Fei-tzu, warned against the danger of high officials usurping authority and acting corruptly, thereby lessening the ruler's power and imperiling the state. Even the military writings, focusing upon the government's integrity, depicted the dangers, as in Chapter 9, "Honoring the Worthy," of the *Six Secret Teachings;* and extensively in the initial book, "Superior Strategy," of the *Huang Shih-kung.*

9. Understanding *pien* as "contentious," consistent with possibilities raised in the Notes to the previous chapter. However, the commentators vary in their understanding. Surprisingly, Hsü and Wei regard it in a common use as a loan for *pien* (GSR 246b), "everywhere" or "universally" (HW, p. 104); Huo, as "becoming estranged from" (CS, p. 123); and Chang, as "to complete," affairs being managed successfully (CL, p. 92). The SP commentators think that "plentiful material goods" refers to rewards. (See p. 65; and the discussion in the commentary.) A more radical interpretation equates the term *huo*, translated as "material goods," as the taxes imposed on the people. Therefore, when they are heavy, the people suffer injury (*pien*). See SP, p. 65.

10. Following Chang's suggestion that the largely obliterated character is *hsiang*. It would therefore appear to echo its use in the first paragraph, al-

though Chang glosses it differently there (as discussed in note 3). Here he understands *hsiang* as "to incline toward." (Cf. CL, p. 92.)

11. "State treasure" is supplemented from the 1985 edition.

12. Such as for evaluating officials and generals, as in Chapters 6 and 20 of the *Six Secret Teachings.*

13. Although the *Ssu-ma Fa* states that "within the army there should be standards," and standards for action and conduct were explicitly promulgated in great detail to armies in the Warring States period, no body of theory ever developed to sustain the concept.

14. Yang Chia-lo, ed., *Shang-chün-shu chieh-ku ting-pen* (Taipei: Shih-chieh shu-chü, 1973), p. 50. As discussed in the Historical Introduction, Lord Shang was active at approximately the same time as Sun Pin, dying in 338 B.C. Both had sought positions in Wei and were probably influenced by its heritage, including the thought of Wu Ch'i. The book attributed to Lord Shang clearly contains chapters from different hands and dates but probably preserves the essentials of his thought and likely was created shortly after his demise.

15. While the Legalists are most closely identified with methods to ensure conformance between position and performance, the military thinkers also pondered the almost insurmountable problem, although in less systematic and conceptualized fashion. For example in Chapter 10 of the *Six Secret Teachings,* the T'ai Kung is recorded as advising, "Your general and chancellor should divide the responsibility, each of them selecting men based upon the names of the positions. In accord with the name of the position, they will assess the substance required. In selecting men, they will evaluate their abilities, making the reality of their talents match the name of the position. When the name matches the reality you will have realized the Tao for advancing the Worthy." The concept of matching name and reality apparently originated with Shen Pu-hai, subsequently being integrated and elaborated by Han Fei-tzu in his lengthy work. For an introductory discussion, see Herrlee G. Creel, "The Meaning of Hsing-ming," *Studia Serica Bernhard Karlgren Dedicata,* Copenhagen, 1959, pp. 199–211.

16. Through retaining the sole power to implement rewards and punishments. This concept was embraced by the military thinkers as well, for they recognized that the commander's awesomeness stems from his power over life and death. As quoted in the *Three Strategies:* "The ruler cannot be without awesomeness, for if he lacks awesomeness he will lose his authority" ("Middle Strategy").

17. This is the initial paragraph of Chapter 14, "Cultivating Authority," in the *Book of Lord Shang.* (The translation follows *Shang-chün-shu chieh-ku ting-pen,* p. 49.) As with the Sun Pin text, there is considerable disagreement as to whether *ch'üan,* which appears in the title of the chapter as well as the text, should be understood as "authority," as we have translated, or "stan-

dard," as J. J. Duyvendak translates in *The Book of Lord Shang*, rpt. (Chicago: University of Chicago Press, 1963), pp. 260–261. The concept of employing standards, of weights and balances, and thereby determining the reality of a situation was further expanded by Han Fei-tzu. See, for example, Chapter 6, "On Having Measure," of the *Han Fei-tzu*.

18. Sun-tzu said, "If our soldiers do not have excessive wealth it isn't because they detest material goods. If they don't live long lives, it isn't because they abhor longevity. On the day that the orders are issued the tears of the soldiers who are sitting will soak their sleeves, while the tears of those lying down will roll down their cheeks. However, if you throw them into a hopeless situation they will have the courage of a Chu or Kuei" (Chapter 11, "Nine Terrains," the *Art of War*).

19. See HW, pp. 105–106.

20. "Virtue" not as a moral quality but as synonymous with the ruler's beneficence—that is, the people regard his rewards positively.

21. The commentators have long—and, we believe, unnecessarily— emended "rewards" to "ruler" both in this work and the *Han Fei-tzu*. (See *Shang-chün-shu chieh-ku ting-pen*, p. 48.)

22. From Chapter 13, *Shang-chün-shu chieh-ku ting-pen*, p. 48. It should be noted that the extant text of the *Book of Lord Shang* preserves somewhat contradictory views on whether rewards should be generous or meager, numerous or few. For example, while Chapter 5 contains a statement describing what Lord Shang apparently perceived as the interrelationship of rewards and punishments—"When punishments are numerous then rewards are esteemed; when rewards are few then punishments are esteemed"—in Chapter 22 he is found advising that rewards for combat should be numerous. In general the military writings tended toward the latter view, seeking to thereby motivate men to action.

Chapter 12

1. The title is extensively discussed in the commentary.

2. This whole paragraph is deleted in the 1985 edition.

3. This passage is added in the 1985 edition. Wu Ch'i was particularly known for inquiring about the health of his soldiers. See his biography in the introduction to our *Wu-tzu* translation in the *Seven Military Classics*.

4. For further discussion, see the notes to our translation of the *Six Secret Teachings*.

5. Cf. CS, pp. 127–128.

6. "Rewards and Punishments."

7. "Discussion of Regulations." Also see "The Source of Offices."

8. For an extensive discussion, consult the notes to our *Wei Liao-tzu* translation in the *Seven Military Classics*.

9. Cf. CS, pp. 127–128.

10. "Martial Plans." The original passage appears in "The General's Awesomeness" in the *Six Secret Teachings.*

11. Cf. CL, p. 94; and CS, p. 127.

12. "Incendiary Attacks."

13. See, for example, "Military Discussions" in the *Wei Liao-tzu.* The *Six Secret Teachings* states that "in military affairs nothing is more important than certain victory" ("The Army's Strategic Power").

14. "Military Disposition."

15. "Incendiary Attacks." For further discussion of the various interpretations and implications of this passage, see the notes to our *Art of War* translation.

16. "Combat Awesomeness."

17. See, for example, "Encouraging the Army" in the *Six Secret Teachings.*

Chapter 13

1. Supplementing in accord with the commentators' unanimous view. See CL, p. 96; CS, p. 129; HW, p. 109; and SY, p. 83.

2. *Ping,* translated as "soldiers," also has the basic meaning of "weapons" and a secondary one of "army." Sun Pin generally uses *chün* for "army"; therefore the choice should be between "weapons," favored by many of the commentators, and "soldiers." The latter seems preferable because the topic is reassembling the army and developing its *ch'i.* (Cf. CL, p. 96; CS, p. 130; HW, p. 110; and SP, p. 70.)

3. The predetermined assembly point for campaign armies was frequently the enemy's border. (See, for example, Chapter 5, "Tactical Balance of Power in Attacks," of the *Wei Liao-tzu.*) Apart from its tactical significance, the border was psychologically critical because the troops would clearly realize they were leaving the relative security of their homeland and would naturally feel uneasy, if not frightened. Thus, as discussed in the commentary, crossing the border required implementing measures to boost morale and toughen their attitude.

4. "Expand their *ch'i,*" *yen* (GSR 203a) *ch'i.* The character *yen* is generally understood as meaning "to extend and make longer (in duration)."

5. Or possibly "the commanding general's orders." Obviously, the following two or three sentences have been lost. "Stimulating *ch'i,*" *chi* (GSR 1162e) *ch'i.*

6. "Sharpen *ch'i,*" *li* (GSR 519a) *ch'i.*

7. "Hone *ch'i,*" *li* (GSR 340a) *ch'i.* The character *li* basically signifies a whetstone, with the extended meaning of "to grind" or "to polish."

Several texts exhort the commanding general not to wear a robe in the cold or use an umbrella in the heat, among other measures designed to visibly stimulate the soldier's morale while also providing the commanding general with personal knowledge of the soldiers' hardship and experience. (See Chapter 23 of the *Six Secret Teachings;* Chapter 4 of the *Wei Liao-tzu;* and the "Superior Strategy" in the *Three Strategies.*) Sun Pin is apparently providing an explanation as to why the soldiers wear the short coat or tunic. (The SP editors envision the short tunic simply as facilitating combat, missing the point of the whole chapter. See SP, p. 71.)

8. Carrying only minimal rations had logistic, economic, and psychological effects. Logistically, fewer rations require fewer vehicles and support troops, freeing more men for actual combat service (assuming human resources are not unlimited). Economically, Sun-tzu emphasized minimizing the hardship imposed upon the state by having the army forage for its food once in enemy territory (Chapter 2, "Waging War"). Psychologically, "Employing Masses," the *Ssu-ma Fa,* states, "Writing letters of final farewell is referred to as 'breaking off all thoughts of life.' Selecting the elite and ranking the weapons is termed 'increasing the strength of the men.' Casting aside the implements of office and carrying only minimal rations is termed 'opening the men's thoughts.' From antiquity this has been the rule." (Note that in "Orders for the Vanguard" in the *Wei Liao-tzu,* the vanguard carries three days of prepared food.)

9. Supplementing from the 1985 edition. (See SP, p. 69.) The intent is obviously to preserve secrecy and probably to create an aura of finality before the onset of battle.

10. "Easy terrain" is another concept first recorded in Sun-tzu's *Art of War,* although such a simple classification probably dates back to the earliest stages of analytic military thought. For example, Sun-tzu said, "On level plains deploy on easy terrain with the right flank positioned with high ground to the rear, fatal terrain to the fore, and tenable terrain to the rear" (Chapter 9, "Maneuvering the Army"). It was well known that easy, open terrain would require large numbers and a brave spirit, as Sun Pin asserts in this fragment. The *Six Secret Teachings* states, "(Deploying) on clear, open ground without any concealment is the means by which to fight with strength and courage" (Chapter 27, "The Unorthodox Army"). And Wu-tzu said, "When employing larger numbers concentrate upon easy terrain, when using small numbers concentrate upon naturally confined terrain" (Chapter 5, "Responding to Change"). In this Wu Ch'i echoes instructions found in the *Ssu-ma Fa:* "When you employ a large mass they must be well ordered. . . . When employing a large mass advance and stop" ("Employing Masses"). Sun Pin himself discusses "easy" terrain in similar terms in Chapters 7, 14, and 17. (The TCT edition interprets *ying* not as "encamping" [as translated] but glossed as "confuse [the enemy]," and *i* ["easy"] not as terrain but as "slight [the enemy]." This yields an understanding for the sentence something like: "Confuse the enemy so that they will treat their opponents lightly." See SP, p. 71; and compare CS, p. 131, for a more orthodox interpretation.)

11. *Chuo,* "awkward," "stupid," "inept," here translated as "plodding" in concord with the following sentences that expand upon the sense of dispirited slowness. In the *Art of War* Sun-tzu discussed the dire effects of protracted warfare, concluding, "Thus in military campaigns I have heard of awkward (*chuo*) speed but have never seen any skill in lengthy campaigns. No country has ever profited from protracted warfare" ("Waging War").

12. Augmenting the original text from one single "mass" to "masses, masses" in accord with the 1985 edition. (See SP, p. 69.) This whole passage is closely related to the third paragraph in the main portion, but the physical configuration of the strips does not allow for simply integrating them. (See CC, p. 217.) (From note 12 to the end of the paragraph is supplemented based upon the 1985 edition.)

13. Emending *po* to *chüan.*

14. Augmenting based upon the 1985 edition.

15. Augmenting based upon the 1985 edition.

16. This would seem to summarize regulations current in their time (and seen prominently in Lord Shang's writings) for punishing the remaining squad members who have had one of their five captured or perhaps a larger unit that has had its commander captured. Numerous such examples are found in the extant *Wei Liao-tzu,* particularly in the later portions that may preserve Ch'in's military regulations and organization. For instance, Chapter 16, "Orders for Binding the Squads of Five," states, "Five men comprise the squad of five. They collectively receive a tally from command headquarters. If (in battle) they lose men but capture (or kill) an equivalent number of the enemy, they negate each other. If they capture members of an enemy squad without losing anyone themselves they will be rewarded. If they lose members without capturing (or killing) equal numbers of the enemy, they will be killed and their families exterminated." Similar punishments applied for losing a squad leader, platoon commander, company colonel, and army general.

17. Book IV, "Strict Positions."

18. Chapter 4, "Combat Awesomeness." Sun Pin similarly notes in Chapters 26 and 27 that a decline in the *ch'i* leads to defeat.

19. Particularly as discussed in Chapters 6 and 7, "Vacuity and Substance" and "Military Combat." In the former Sun-tzu states, "In general, whoever occupies the battleground first and awaits the enemy will be at ease; whoever occupies the battleground afterward and must race to the conflict will be fatigued. Thus one who excels at warfare compels men and is not compelled by other men."

20. "Military Combat," the *Art of War.*

21. Chapter 2, "Evaluating the Enemy."

22. For examples, see Chapter 52, "Military Vanguard," the *Six Secret Teachings;* Chapter 4, "Strict Positions," the *Ssu-ma Fa;* and Chapter 2, "Evaluating the Enemy," in the *Wu-tzu.* Sun-tzu not only advises attacking

the exhausted but also advocates manipulating armies so that they will become tired. Conversely, he equally cautions against armies and states becoming worn-out and thereby offering an easy conquest opportunity to their enemies. (See Chapter 2, "Waging War.") As discussed in the Historical Introduction, Sun Pin equally embraces the concept of exploiting an enemy's exhaustion.

23. Chapter 26, "The Army's Strategic Power," in the *Six Secret Teachings*. In the same chapter the T'ai Kung also states, "One who excels in warfare will not lose an advantage when he sees it, nor be doubtful when he meets the moment. One who loses an advantage or lags behind the time for action will, on the contrary, suffer from disaster. Thus the wise follow the time and do not lose an advantage. The skillful make decisions and have no doubts." The *Wei Liao-tzu* states, "When doubts arise defeat is certain" (Chapter 18, "Orders for Restraining the Troops"). And Wu-tzu, among others, said, "The greatest harm that can befall an army stems from hesitation, while the disasters that befall an army will be born in doubt" (Chapter 3, "Controlling the Army"). Sun Pin advises attacking doubts in Chapter 17 and notes that doubt marks an enemy that can be defeated in Chapter 26.

24. For example, the *Ssu-ma Fa* states, "Move to observe if they have doubts. Mount a surprise attack and observe their discipline. Mount a sudden strike on their doubts. Attack their haste" (Chapter 5, "Employing Masses").

25. Chapter 52, "Military Vanguard," the *Six Secret Teachings*.

26. Chapter 27, "Unorthodox Army," the *Six Secret Teachings*.

27. Chapter 4, "Strict Positions," the *Ssu-ma Fa*. Other chapters offer the same advice.

28. See Chapter 2, "Evaluating the Enemy," and Chapter 5, "Responding to Change," in the *Wu-tzu*.

29. While fear and punishment were generally emphasized by the Legalist political thinkers, the military writers often advised offering generous rewards as incentives to spur men into the most dangerous situations. However, in general, as in the West until the last century, men fought more out of fear of punishment than other motives. A passage in the *Wei Liao-tzu* summarizes the general trend of thought: "Now the people do not have two things they fear equally. If they fear us then they will despise the enemy; if they fear the enemy they will despise us. The one who is despised will be defeated; the one who establishes his awesomeness will be victorious." Awesomeness is of course established through fear and fear through the threat and implementation of strong punishments. (See Chapter 5, "Tactical Balance of Power in Attacks.")

30. The *Ssu-ma Fa* states, "In antiquity the form and spirit governing civilian affairs would not be found in the military realm; those appropriate to the

military realm would not be found in the civilian sphere. . . . In the civilian sphere words are cultivated and speech languid. In court one is respectful and courteous, and cultivates himself to serve others. Unsummoned he does not step forth; unquestioned, he does not speak. It is difficult to advance but easy to withdraw. In the military realm one speaks directly and stands firm. When deployed in formation one focuses on duty and acts decisively. Those wearing battle armor do not bow; those in war chariots need not observe the forms of propriety; those manning fortifications do not scurry. Thus the civilian forms of behavior and military standards are like inside and outside; the civil and martial are like left and right" (Chapter 2, "Obligations of the Son of Heaven").

31. Chapter 4, "Strict Positions," the *Ssu-ma Fa*.

32. Ibid.

33. Ibid.

34. Chapter 6, "Stimulating the Officers," the *Wu-tzu*.

35. Chapter 3, "Discussion of Regulations." Note that Sun Pin also speaks about acting as if deranged in certain situations. (See Chapter 18.)

36. Chapter 22, "Military Instructions, II."

37. Perhaps the most famous historical example is found in the *Wei Liao-tzu:* "When Wu Ch'i engaged Ch'in in battle, before the armies clashed one man, unable to overcome his courage, went forth to slay two of the enemy and return with their heads. Wu Ch'i immediately ordered his decapitation. An army commander remonstrated with him, saying, 'This is a skilled warrior. You cannot execute him.' Wu Ch'i said, 'There is no question that he is a skilled warrior. But it's not what I ordered.' He had him executed" (Chapter 8, "Martial Plans").

38. Chapter 4, "Strict Positions."

39. As discussed in note 37.

40. Chapter 12, "Tactical Balance of Power in Warfare."

41. For example, Wu Ch'i said, "In general to govern the state and order the army you must instruct them with the forms of propriety, stimulate them with righteousness, and cause them to have a sense of shame. For when men have a sense of shame, in the greatest degree it will be sufficient to wage war, while in the least degree it will suffice to preserve the state" (Chapter 1, "Planning for the State").

Chapter 14

1. Although in Chapter 9, "Maneuvering the Army," Sun-tzu speaks about *ch'u chün*, "deploying the army," in similar terms, the reading here for *ch'u tsu* appears to be more general, "to command troops." Similarly, commentators who suggest "deploying companies" interpret *tsu*, which was also

employed to refer to a company of 100 men, too narrowly. (Cf. HW, p. 116; CS, p. 139; TCT, p. 52; SP, p. 74; and SY, p. 85.)

2. Understanding *t'i*, "body," in the sense of "to make a unified body," "to unify," or "to systematize," rather than as glossed by Chang in the sense of "to divide" or "to distinguish" (CL, p. 102). Many commentators understand *t'i* in the sense of "to control." (See, for example, SY, p. 85; and TCT, p. 52.) The term for "mailed soldiers" derives from characters originally meaning "armor" and "weapons," stimulating the SP editors to believe the sentence refers to properly distributing weapons among the troops (SP, p. 74).

3. The scope of the term *offices* is the key to this opening paragraph and to whether in an extended meaning it can encompass the disparate contents of the entire chapter. As this is a major issue, it is discussed in the commentary. (The title was found on the back of a bamboo strip.)

4. While the various units, and indeed the soldiers in the individual squads, were distinguished by insignia, the intent here seems to be that (in the daytime) orders should be conveyed by flags with varying insignia. The subsequent sentence, "Settle doubts with flags and pennons," would recapitulate the same theme, although differently implemented. Huo Yin-chang cites a *Kuan-tzu* passage from the "Military Methods" chapter enumerating various insignia, such as the sun and moon, and their correlated actions (CS, pp. 139–140). An enigmatic passage is also found in the *Wu-tzu* in reply to Marquis Wu's question as to whether there is a Tao for advancing and halting: "Do not confront 'Heaven's Furnace' or 'Dragon's Head.' Heaven's Furnace is the mouth of a deep valley. Dragon's Head is the base of a high mountain. You should keep the Green Dragon banner on the left, White Tiger on the right, Vermilion Bird in the front, Mysterious Military to the rear, and Twinkler above from where military affairs will be controlled" ("Controlling the Army"). Further discussion will be found in the commentary.

5. Following CL, pp. 103–104. The chariots would thus have their unit and the rank of their commander clearly distinguished. Other interpretations stress the ranking aspect and would understand the sentence as referring to promotions and demotions based upon the individual's achievements with respect to the responsibilities of his office or command. See, for example, TCT, p. 52.

6. By the Warring States period, systematized taxation based upon local administrative units had already appeared and become widespread. Conscription for military service was localized and founded upon the principle of mutual recognition: Men who would be familiar to each other should be grouped together to form squads and then companies bound by the mutual guarantee system. A hierarchal organization based upon ascending political units could thus be constructed, with the leaders for each level (theoretically) being derived from the appropriate administrative personnel. These methods are prominently seen throughout the military writings and particularly in the *Wei Liao-tzu*, which discusses military organization, responsibilities, and pen-

alties extensively. (See, for example, "Orders for the Squads of Five" and "Orders for Binding the Squads of Five.") However, the most succinct expression is found in Chapter 30, "Agricultural Implements," of the *Six Secret Teachings:* "The units of five found in the fields and villages will provide the tallies and good faith that bind the men together. The villages have officials and the offices have chiefs who can lead the army. The villages have walls surrounding them which are not crossed; they provide the basis for the division into platoons."

7. Huo Yin-chang would understand the sentence as "To unify their weapons first order their marching." (See CS, p. 141; and also SP, p. 75.) Others take it as translated (for example, TCT, p. 52; and SY, p. 86).

8. The SY editors strangely understand *yen chieh* as referring to the army encamping (SY, p. 86; cf. CS, p. 141; and CL, p. 105).

9. Emending "move" to "hunt." However, most commentators emend it to "tread" or "trample" or explain it as some sort of probing action. (See CS, p. 141; CL, pp. 105–106; and SP, p. 75.) "Elongated formation" might also be translated as "rope formation," a formation name otherwise unknown. (Cf. CL, p. 106.) Clearly the emphasis is upon being extensive and encompassing rather than sharp and penetrating, but limited in coverage. (In Chapter 17, "Ten Questions," Sun Pin advises employing a horizontal deployment and also a way to counter an extended, horizontal deployment, but the circumstances of application are different.)

10. "Constraining and contravening" is taken as a formation name by some commentators (for example, CS, p. 141; and SY, p. 87). The first character, *chiao,* is subject to widely differing interpretations ranging from "interacting with" or "joining with" (as in combat) to "correcting." In the latter instance, which seems unlikely, the idea would be to correct the formations to prevent disorder. (See TCT, p. 52; also SP, p. 76.)

11. Again this may refer to a defined but unknown formation. (See CS, p. 141.) TCT suggests that the missing character is "strategic power" and that "danger" indicates superiority. Therefore, the army should be deployed to realize its dangerous (superior) power. (See TCT, pp. 52–53.)

12. Commentators again suggest it designates a particular formation (SP, p. 76; SY, p. 87; CL, p. 107). The deployment obviously consists of linked, mobile formations.

13. The enemy's "fierce beak" refers to its main thrust, which is designed to bite into the army. "Closing envelopment" is thought by some to be a named formation—or, more correctly, deployment—that has as its objective shutting off the enemy's access routes. (See, for example, CS, p. 142; SY, p. 87; and SP, p. 76.)

14. Otherwise understood as the army drawing near to the enemy from outside (CL, p. 107) or surrounding the enemy (TCT, p. 53; HW, p. 118).

15. For "fierce combat" some commentators would understand "noisy combat." (See, for example, CS, p. 142. The *Six Secret Teachings* discusses deliberately raising a great clamor when the fronts are about to clash, behav-

ior that was prominently identified with Chinese troops in the Korean War. For example, Chapter 37, "Movement and Rest," states, "When the battle is joined, beat the drums, set up a clamor, and have your men all rise up together. The enemy's general will surely be afraid and his army will be terrified." Based upon "Responding to Change" in the *Wu-tzu,* Wu Ch'i particularly emphasized the practice and perhaps gave it theoretical respectability.) "Alternated rows" refers to deploying the long and short weapons in mutual support, a tactical principle that appears in Chapter 4, "Ch'en Ch'i Inquires about Fortifications," of the *Military Methods;* and Chapter 2, "Obligations of the Son of Heaven," of the *Ssu-ma Fa,* which states, "When the five types of weapons are not intermixed, it will not be advantageous. Long weapons are for protection; short weapons are for defending." In Chapter 3 their unified use is touted: "Now each of the five weapons has its appropriate use: The long protect the short, the short rescue the long. When they are used in turn the battle can be sustained. When they are employed all at once, the army will be strong."

16. Or possibly "heavy forces." Note that the *Ssu-ma Fa* advises, "In general, in warfare: If you advance somewhat into the enemy's territory with a light force it is dangerous. If you advance with a heavy force deep into the enemy's territory you will accomplish nothing. If you advance with a light force deep into enemy territory you will be defeated. If you advance with a heavy force somewhat into the enemy's territory you can fight successfully. Thus in warfare the light and heavy are mutually related" ("Strict Positions"). Sun Pin does not speak about "light" and "heavy" forces (or troops) apart from this chapter, but of dense and dispersed formations, as in Chapter 29, "The Dense and Diffuse."

17. In Chapter 29 Sun Pin speaks of the dense and diffuse mutually opposing each other, whereas here the light is applied to the dispersed. (Note that "dispersed" is *san,* for which it is reserved throughout our books.) The thoughts are somewhat contradictory, assuming "light" and "dispersed" are not totally different. Chang (CL, p. 108), based upon a passage from Sun-tzu's *Art of War,* suggests "dispersed" refers to fighting on home territory, where the soldiers tend to think of their families and lack commitment. In Chapter 11, "Nine Terrains," Sun-tzu states, "When the feudal lords fight in their own territory, it is 'dispersive terrain. . . . For this reason on dispersive terrain do not engage the enemy. . . . When the troops have penetrated deeply, they will be unified, but where only shallowly, they will be inclined to scatter. . . . For this reason on dispersive terrain I unify their will." However, the context does not support an interpretation that would seem more applicable to Chapter 13, "Expanding *Ch'i.*"

18. *Hsing ch'eng,* translated as "mobile walls," refers to heavy, apparently immense semimobile wall sections designed to blunt the assault of siege troops attempting to scale or otherwise top the permanent fortifications. These sections, perhaps mounted on wagon undercarriages, could be moved to threatened areas to immediately provide greater height and thereby negate

the potential of mobile ladders and towers, denying a foothold for the swarming soldiers. However, here the use is apparently offensive rather than defensive, leading to speculation that it was somehow mounted on a carriage and moved toward the heights targeted for attack. (See CS, p. 142; and SP, p. 76. For a brief overview of siege technology in the Warring States period, see "Siege Engines," pp. 409–451. *Hsing ch'eng* is briefly discussed on pp. 423–424.)

19. This sentence merits note because virtually all the military thinkers advised against attacking heights (even though the T'ai Kung did point out that an army encamped on the heights could also become isolated). For example, Sun-tzu said, "Do not approach high mountains; do not confront those who have hills behind them" ("Military Combat"). Note that Sun-tzu frequently used the analogy of water and other objects rushing downhill to express his concept of *shih*, "strategic configuration of power." In "Treasures of Terrain," Sun Pin himself noted that armies confronting hills are doomed to defeat. While there was not any special advantage in fighting downhill, such as in ancient Greece, where the onrush of the phalanx made downward movement decisive, in general wielding weapons uphill, braving the onslaught of arrows exploiting a naturally greater range, combined with the greater expenditure of energy necessary to advance, made it inadvisable. (Depending upon how the sentence marked by note 21 is interpreted, these two sentences may form a correlated pair.)

20. The main question here is the identities of the attacking and retreating forces. "Martial retreat," which is literally translated, suggests a withdrawal in good order, maintaining a fighting action throughout. However, some commentators understand the sentence as referring to mounting an attack on a retreating enemy. (For example, see CL, p. 109; and CS, pp. 142–143. Compare SP, p. 77; SY, p. 87; and TCT, p. 53–54.) Attacking a retreating force would probably be expressed in more direct terms, such as *kung t'ui*, "attack a withdrawing (force)."

21. The term *approach* (*lin*) entails a connotation of condescension, of superior toward inferior, as the ruler approaching the common people or (appropriately here) superior power and inferior power. If understood as occupying a higher position, it would contrast with the earlier situation, "When you confront heights and deploy, employ a piercing formation." However, some commentators would understand the sentence from a different perspective, such as, "When the enemy's strategic power exceeds ours (or is deployed above us), when deploying to approach them employ a flanking attack on the wings." (See CL, p. 109–110; SP, p. 77, which doubts the missing character is "higher"; and CS, p. 143.) .

22. The SY editors suggest that *fan chan*, translated here as "ordinary warfare," refers to "water warfare" (SY, p. 88).

23. As discussed, ravines provided the means for outnumbered forces to attack an enemy. However, the unstated assumption is that the "few" have already assumed and improved their positions along the sides and at the heights

before the enemy enters. In Sun Pin's situation here, the enemy occupies the ravines; clearly entering it would be foolish, while keeping the enemy bottled up would produce only a temporary standoff. The obvious course is to follow Sun-tzu's dictum to entice the enemy into motion, destabilize him, and then attack as desired. Accordingly, in Chapter 17 Sun Pin advises just this course of action in two situations, including one where the enemy has similarly occupied the mountains: "To strike them, you must force them to move from some of the passes they have taken . . . . . and then they will be endangered. Attack positions that they must rescue. Force them to leave their strongholds in order to analyze their tactical thinking (and then) set up ambushes and establish support forces. Strike their masses when they are in movement. This is the Tao for striking those concealed in strongholds."

24. *Yang* pennants, understood as visible pennants, are thought by some commentators to be properly glossed as "false" pennants. (See, for example, CS, p. 143.) While this reading is possible, it unnecessarily emends the fundamental character. (However, Sun-tzu said, "If there are many visible obstacles in the heavy grass, it is to make us suspicious" ["Maneuvering the Army"].) Clearly amidst tall grass the army's route had to be indicated; multiplying the pennants would accomplish it. (Note that in this same chapter, Sun Pin states, "When the road is thorny and heavily overgrown, use a zigzag advance.") Multiplying the pennants would also cause doubt in the enemy, as observed elsewhere in the chapter, and perhaps prevent being attacked. As several military thinkers observed, tall grass would conceal an enemy preparing an ambush (and also present a risk of incendiary attack if reasonably dry). Accordingly, Sun-tzu advised, "When on the flanks the army encounters ravines and defiles, wetlands with reeds and tall grass, mountain forests, or areas with heavy, entangled undergrowth, you must thoroughly search them because they are places where an ambush or spies would be concealed" ("Maneuvering the Army"). Trapping an enemy in tall grass would of course reduce its mobility but not make attacking it particularly easy. (Note HW, p. 120, which emphasizes the question of clear passage; and TCT, p. 57, which speaks about "opening a road" through the vegetation.)

25. Probably to prevent the soldiers from becoming too lax in their attitude once victory had been secured, while also visibly manifesting the state's might to its enemies. Note that "Middle Strategy" of the *Three Strategies* discusses the issue of victorious armies, and recall the difficulties faced by T'ien Chi portrayed in the Historical Introduction.

26. Or possibly along the ridges to the right. (Cf. KO, p. 218; and SP, pp. 74, 82.) The key character in this sentence is *chü*, GSR 642g, which refers to the "right wing of an army." Some take it as deploying in general (for example, SY, p. 88; and TCT, p. 54) or as establishing a hidden encirclement in the mountains to snare the enemy. (See CL, pp. 111–112; and CS, pp. 143, 147.) The *Ssu-ma Fa* states, "In general, as for warfare: Keep the wind to

your back, the mountains behind you, heights on the right, and defiles on the left" ("Employing Masses"). In the *Art of War* Sun-tzu had similarly advised,: "On level plains deploy on easy terrain with the right flank positioned with high ground to the rear, fatal terrain to the fore, and tenable terrain to the rear. This is the way to deploy on plains." In the same chapter ("Maneuvering the Army") he added, "Where there are hills and embankments you must occupy the *yang* side, keeping them to the right rear. This is to the army's advantage and [exploits the natural] assistance of the terrain."

27. TCT would understand as, "To rest when exhausted employ the Awl Formation" (TCT, p. 54). Since the Awl Formation requires elite, spirited troops, this seems unlikely.

28. Chang (CL, p. 112) takes *tsa kuan* as a formation name. However, it appears to refer to intermixing various elements or commands, such as long and short weapons groups with bowmen and others.

29. Other interpretations for *pi,* translated as "entangled" (from the primary meaning of "overgrown"), include "concealed measures" (SP, p. 78; SY, p. 88; CS, p. 144). Chang thinks it refers to the orientation of the weapons, from a single front turned toward the enemy to facing outward (CL, p. 113). Note that in "Certain Escape" and "Urgent Battles," the T'ai Kung advised using the four-sided martial assault formation to escape in critical situations, which would create an outer orientation in four directions based upon the square. Note that the T'ai Kung also discussed exploiting natural (or entangled) cover: "Deep grass and dense growth are the means by which to effect a concealed escape," and "mountain forests and dense growth are the means by which to come and go silently" ("The Unorthodox Army").

30. The translation assumes that *pien* ends the original sentence rather than beginning the next one. (See CL, p. 113, and SY, p. 88, for differences.) Diverting rivers to inundate major cities as well as army camps was already recorded in the *Tso Chuan,* as discussed in our *History of Warfare in China.*

31. Apparently passes alone would be inadequate to pass through the camp's divisions; additional, task specific authorizations would be required. (See CL, pp. 113–114; note HW, p. 120, for a radically different reading.) Chapter 15, "Orders for Segmenting and Blocking Off Terrain," in the *Wei Liao-tzu,* outlines measures for dividing the camp into secure quadrants and imposing such strict discipline that unauthorized venturing into other sections would be punishable by death.

32. Another view is that the sentence discusses selecting men, the "Death Warriors," who can forcefully penetrate the enemy's interior. (Cf. SP, p. 78; CL, p. 114; and PH, p. 234.) However, *k'ou* ("raider forces" or "marauders") is a pejorative term usually referring to those who plunder one's own state, not the state's elite warriors engaged in a bold penetration of the enemy. It appears in the last part of Chapter 5, "Responding to Change," of the *Wu-tzu.* However, Wu Ch'i's solution, to wait until the plunderers are

burdened and worried, is radically different. The term translated as "Death Warriors," *kuan shih*, is literally "warriors with inner coffins." Clearly *kuan*, "inner coffin," should be understood as *ts'ai*, "talented" or "elite," but the image of warriors dragging along their coffins is a famous one in the West.

33. Or possibly "close-ordered chariot arrays" rather than "long weapons and chariots." (Cf. SP, p. 78; CL, p. 114; SY, p. 89; and TCT, p. 55.)

34. Being forced to engage a superior enemy in battle was a critical problem for ancient generals and thus extensively discussed in the various military writings. (For example, see "Urgent Battles" in the *Six Secret Teachings*, where the danger of being surrounded is discussed.) Within this context Sun Pin advised fully exploiting the firepower provided by missile weapons, including the newly developed crossbows, in conjunction with shock weapons and thereby engaging the enemy in sufficient force to prevent being surrounded and decimated. In Chapter 17 Sun Pin offers several other solutions, including the radically different one of dividing into three and employing "death warriors." Naturally the easiest solution would be to exploit the natural advantages of terrain, such as ravines, if the army can attain them. (Chang believes the "intermixture" refers to forgoing the village-based organizational system. However, this is an extremely suspect interpretation due to the nature of military organization and battlefield deployment [CL, p. 115].)

35. While the "Cloud Formation" does not appear in any other chapters of the *Military Methods*, earlier in this chapter Sun Pin indicated that the "Cloud Formation" is suitable for arrow warfare. (However, in Chapter 16, "Ten Deployments," the Wild Geese Formation is touted as being for "quick archery response" in the view of some commentators.) The sentence has been translated with reference to two chapters in the *Six Secret Teachings*, "Crow and Cloud Formation in the Mountains" and "Crow and Cloud Formation in the Marshes." In the latter the T'ai Kung discusses situations in which one is greatly outnumbered, advises employing the mentioned formations, and describes their strength in effecting and exploiting the unorthodox as follows: "What is referred to as the Crow and Cloud Formation is like the crows dispersing and the clouds forming together. Their changes and transformations are endless." Obviously the analogy with the ever-evolving clouds was not an idle one.

36. Other interpretations include "turbulent winds that shake (the enemy's) formations," "turbulent winds that stir up dust," and even winds that describe the rapid speed in which the "shaking formation" moves forward. (Cf. SY, p. 89; TCT, p. 55; SP, p. 79; CS, pp. 144–145; CL, pp. 115–116.) The "shaking formation" derives its name from its awesomeness, such that it causes the enemy to tremble. The term *shaking* appears in Chapter 45, "Strong Enemy," in the *Six Secret Teachings*, with reference to "shaking invaders"—strong, fierce, numerous assault forces—and Chapter 31, "The Army's Equipment," with reference to "shaking fear." Sun Pin was perhaps

more cognizant of weather factors, although all the writers mentioned the seasons of Heaven and most discussed the difficulties posed by water and rain and advised attacking an enemy worn down by them. (For an example of the effects of rain, see "Gongs and Drums" in the *Six Secret Teachings*.) Moreover, he seems to have been more aware of the disquieting effect of strong winds on morale and armies in general, not to mention the increased difficulty of communicating commands and orders. However, other writers also considered the wind a factor and offered appropriate advice. For example, the T'ai Kung said, "High winds and heavy rain are the means by which to strike the front and seize the rear" ("The Unorthodox Army"; see also "The Army's Indications"). Wu-tzu advised keeping the wind to one's back, apparently considering the wind's effects to be critical: "When about to engage in combat determine the wind's direction. If favorable, yell and follow it. If contrary, assume a solid formation and await the enemy" ("Controlling the Army"). Many of the writings use the analogy of wind to compare the army's swiftness, which has possible echoes here. (For example, the *Three Strategies* states, "In battle they are like the wind arising; their attack is like the release of a pent-up river" ["Superior Strategy"].) Finally, all the tacticians naturally worried about the wind's direction where there was a potential for incendiary attack, and weather officers (termed *astrologers*) with responsibility for predicting and determining the winds were among the command staff included in "The King's Wings" in the *Six Secret Teachings*.

37. Secrecy and deception were watchwords of most of the military writers, frequently remarked upon by Sun-tzu and the T'ai Kung.

38. Clearly the idea is to conceal the army's power in the fastness of the mountains, like a dragon hidden, yet lurking in a pool. Some commentators interpret "dragon" as referring to mountain ridges. (Cf. SY, p. 90; and CL, pp. 116.)

39. "Perverse actions"—actions unexpectedly contrary to normal expectation—are also mentioned in this chapter. In both cases the intent is to deceive the enemy into miscalculating strength or intentions and thereby moving in a desired direction. An enemy crossing a river provided a particularly easy target; therefore, luring it to such a crossing point would prove to be good tactics. For example, Sun-tzu said, "After crossing rivers you must distance yourself from them. If the enemy is fording a river to advance, do not confront them in the water. When half their forces have crossed, it will be advantageous to strike them" ("Maneuvering the Army"). Among the conditions that the T'ai Kung and Wu Ch'i identified as exploitable opportunities, an army crossing a river ranked about highest. The T'ai Kung said, "When they are fording rivers they can be attacked," while Wu Ch'i (just as Sun-tzu) defined the moment more precisely: "When fording rivers and only half of them have crossed, they can be attacked" ("Military Vanguard" and "Evaluating the Enemy," respectively.)

40. The word translated as "unfathomable" means "dark, obscure, or ignorant." Views differ about who is to be kept ignorant, especially as Sun-tzu himself had advised keeping the soldiers ignorant of his plans and intentions. (See "Nine Terrains." Sun Pin evinces a similar thought in "Killing Officers.") It is also suggested that the term refers to "silent warfare," rather unlikely in this context, or "unannounced" engagements. (Cf. SP, p. 79; CS, p. 145; and CL, pp. 116–117.)

41. Note that the T'ai Kung said, "Deep moats, high ramparts, and large reserves of supplies are the means by which to sustain your position for a long time" ("The Unorthodox Army"). Sun Pin previously indicated the utility of ditches and defiles for defense in "The Questions of King Wei."

42. Since pennants and flags were the means for conveying orders and identifying units, estimations of strength, directions of movement, and deployments into formations would be predicated upon observing them. Thus, any unusual activity or dispersal would arouse suspicion, forcing an attempt to determine whether it might be a deliberate ruse or an actual deployment. Doubt would also be among the most common exploitable situations, thoroughly discussed by all the military writers.

43. The whirlwind formation also appears in "The Questions of King Wei."

44. Chang suggests the sentence means, "After you have repressed the enemy, shifting the army is the way to prepare for a strong enemy" (CL, pp. 117–118). As translated the sentence coheres with Sun family doctrines that emphasize mobility and avoidance of direct confrontations with equal or superior forces.

45. "Confined road," following the commentators. As Huo Yin-chang notes, the "Floating Marsh Formation" appears in the *T'ai Pai-yin ching*, although no details are known (CS, p. 145). "Floating Marsh" may also be interpreted more simply as a highly mobile, floating formation or the conveyance of troops across a marsh to mount a flanking attack, which would certainly be a surprising tactic given the problems posed by water obstacles. (Cf. CL, p. 118; SP, p. 80; and SY, p. 90.) Hsü and Wei offer the interesting interpretation that the sentence refers to mounting a light, floating frontal resistance and then attacking from the flanks in a pitched battle, understanding *sui* as "fervent" or "intense" rather than "incendiary" or "road" (HW, p. 122). Other views unrealistically suggest it indicates assembling forces in wet, marshy areas to prepare to engage in incendiary warfare. (See TCT, pp. 55, 58; and SY, p. 90. Both suggest that *sui*, which has "flame" as its primary meaning, refers to incendiary warfare rather than a "confined road" as glossed.)

46. Considerable discussion is occasioned by the first two characters in this sentence, *shan huo*. The translation follows the original gloss of "slowness," although it is not entirely satisfactory. (See CC, p. 230.) Chang provides an

extensive note that is generally followed by other editions, which, based upon understanding *shan* to mean a single layered tunic, in sum yields a translation for the first part of, "By not wearing armor nor helmets and acting in an irregular manner, show the enemy that one is unprepared and entice them to come forth to attack" (CL, pp. 118–119). "Frequent avoidance," found in our translation, does not appear to require further interpretation; however, it is variously understood as "indecisive movements" and "circling back and forth." (Cf. CS, p. 146; SY, p. 90; SP, p. 80; and TCT, p. 56.) As usual, Hsü and Wei have a radically different interpretation based upon slightly different glosses: "Like a golden cicada casting off its shell, have your main strength secretly withdraw in order to entice the enemy" (HW, pp. 122, 125).

47. "Fiery strength" being our translation for "comet."

48. Sun-tzu warned, "If there are many visible obstacles in the heavy grass, it is to make us suspicious" ("Maneuvering the Army").

49. This would seem to reflect Sun Pin's strategy at the battle of Kuei-ling as well. Creating an impression of command incompetence would be the surest way to nurture arrogance and carelessness in an enemy.

50. *Chung hai*, "heavy injury," is also interpreted as making the punishments heavy so as to ensure disciplined troops (TCT, p. 56).

51. Most of the commentators take the first part in an unnecessarily restrictive sense, as "patrol until light." (See, for example, CL, p. 113; and SP, p. 80.)

52. It would be tempting to interpret this sentence as referring to prepositioning provisions, as Ch'i certainly did within its own state prior to the battles of Kuei-ling and Ma-ling. However, once having crossed an enemy's border, the army had to seek out whatever local sources it might find, plundering and foraging as much as possible while resorting to its own reserves only when necessary. Some of the commentators would understand this sentence as referring to keeping one's provisions and supplies dispersed to prevent serious losses in the event of an enemy assault or incendiary attack. (Cf. TCT, p. 56; SP, pp. 80–81; CS, p. 147; and CL, p. 120.)

53. "Reckless withdrawals," preserving the primary meaning of *hu* rather than emending (unnecessarily) to *ku*, "thus" or "with reason," as suggested by Chang (CL, p. 120). "Reckless" should be understood in the sense of "apparently" reckless, meaning energetic, frenetic, and rapid.

54. In general, we are willing to accept one emendation but are less persuaded by a series whereby the original charter is identified as a second one and then the second is interpreted as a third or fourth. Except in unusual circumstances, such chains, even allowing for the dialectical differences that result from assuming the book originated in Ch'i, should perhaps be viewed with considerable skepticism.

55. Since the fragments are identical to portions of sentences contained in the fully reconstructed chapter, they are not repeated in a fragments section.

56. The term *kuan*, translated as "offices," appears early with primary meanings of "office," "official," and "function." (See GSR 157a.) Most of the commentators follow Chang Che-min's imaginative analysis in identifying central themes for each of the five sections and correlating them with different aspects of the military's functions (essentially mission, command, and control). Chang centers his interpretation upon "Establish offices as appropriate to the body," citing Hsün-tzu's use of the term *kuan* as referring to sensory organs. (The original passage is found in "Discussion of Heaven.") This is further underpinned by Sun-tzu's inclusion of the term *kuan* right at the beginning of the *Art of War*, where he states that the "laws," one of the five essential factors in the Tao of the military, encompasses *kuan*, among other items. (The meaning of the Sun-tzu sentence is open to interpretation. Although there are six characters, and commentators sometimes parse them as six distinct items, they would seem to be best understood as two-character combinations, thus obscuring the exact meaning of *kuan* itself. For example, our translation in the *Art of War* reads, "The laws [for military organization and discipline] encompass organizations and regulations, the Tao of command, and the management of logistics." "Tao of command," *kuan Tao*, taken separately might be best understood as "command" [or organs of command] and the Tao.) Extending this primary insight, Chang suggests that the first paragraph speaks about establishing military organs as appropriate to the human body; the second section (covering the sentences from note 8 to note 22), their function in combat requirements; the third (from note 22 to note 30), their manifestation in movement directed toward various targets; the fourth (a brief section from note 30 to note 34), weapons, soldiers, and deployments(!); and the fifth (a lengthy one including all the concrete tactics from note 34 to the end), their function in military applications. (See CL, pp. 101–102.) While ingenious and certainly worthy of approbation for attempting to understand the text on its own terms, this interpretation seems less than appropriate or encompassing. Perhaps a more realistic view would be to understand *kuan* as basically entailing much of what would be subsumed by the twentieth-century concept of command and control. Command requires officers and a command structure; control of the troops requires education, training, discipline, and responsiveness—all nurtured through the senses. Thus, discussions of pennants and drums are well encompassed within the main theme; the other material may be seen as supplemental. (It should be remembered that chapter titles in ancient writings were frequently appended for tangential reasons, often referring to just a couple of significant characters or a single topic in the first few sentences.) For other views, including that *kuan* refers to "qualifications," see HW, p. 117; SP, pp. 74, 81; TCT, p. 56; and SY, p. 81.

57. "Obligations of the Son of Heaven." Also see the examples in note 4.

58. "Military Instructions, I."

59. Ibid.

60. "Military Instructions, II," the *Wei Liao-tzu.*

61. "Military Combat."

62. "The Tao of the General."

63. "Strict Positions."

64. "Responding to Change."

65. "Martial Plans," the *Wei Liao-tzu.* A passage in the *Ssu-ma Fa* indicates that the flags were also directed by the drums: "In general, as for the drums: There are drums directing the deployment of the flags and pennants; drums for advancing the chariots; drums for the horses; drums for directing the infantry; drums for the different types of troops; drums for the head; and drums for the feet. All seven should be properly prepared and ordered" ("Strict Positions").

66. "Controlling the Army," the *Wu-tzu.* Note that in Chapter 56 of the *Six Secret Teachings,* one of the qualifications for strong warriors is the ability to "quickly furl up the flags and pennants."

67. For example, in "Forest Warfare," "The Crow and Cloud Formations in the Mountains," and "Divided Valleys" in the *Six Secret Teachings,* the T'ai Kung states that the flags and pennants should be set out on high ground.

68. "The Army's Indications."

69. "Military Combat." This important dictum is also quoted in Book III of the *Questions and Replies.*

70. "The Tao of the General." Feigned retreats, although difficult to successfully execute, posed great danger to troops that might abandon their own order to rush forward and capitalize on the enemy's apparent weakness, only to be ambushed and enveloped. This "feigned retreat" thus numbered among the more successful unorthodox tactics, and the history books record their disproportionate impact. The *Questions and Replies* contains an illuminating passage: "Whenever the soldiers retreat with their flags confused and disordered, the sounds of the large and small drums not responding to each other, and their orders shouted out in a clamor, this is true defeat, not unorthodox strategy. If the flags are ordered, the drums respond to each other, and the commands and orders seem unified, then even though they may be retreating and running, it is not a defeat and must be a case of unorthodox strategy. The *Art of War* says: 'Do not pursue feigned retreats.' It also says: 'Although capable display incapability.' These all refer to the unorthodox" (Book I). The *Questions and Replies* preserves several interesting discussions about flags and their manipulation in its three books. The early *Ssu-ma Fa* makes an even more radical suggestion—abandon flags altogether to trick the enemy: "In general, as for the Tao of Warfare . . . . . If you are contending for a strategic position, abandon your flags as if in flight, and when the enemy attacks turn around to mount a counterattack" ("Employing Masses").

71. "Evaluating the Enemy."

72. "The Cavalry in Battle," the *Six Secret Teachings*.
73. Ibid.
74. "Movement and Rest," the *Six Secret Teachings*.
75. "Planning for the Army," the *Six Secret Teachings*.
76. "Preparation of Strategic Power."

Chapter 15

1. "Restraint in making impositions" is added in accord with the 1985 edition. (See SP, pp. 87–88.)

2. Sun Pin employs "not what X is urgent about" in Chapter 3 as well. Such occurrences suggest that at least the first part of this chapter, although otherwise marked by some dissimilarities, also preserves his original thoughts.

3. Both the original and revised arrangements of the text include this sentence as part of the chapter's main body; however, it is (as the fragments that follow) clearly from the hands of later writers since it summarizes the achievements of Kings Wei and Hsüan. (Cf. CL, pp. 121–123.)

4. Probably in 325 B.C.

5. Probably in 314 B.C., although some cite 316 B.C. Others suggest this might refer to the famous Battle of the Fire-oxen, in which T'ien Tan revived a nearly vanquished state. (See CL, pp. 125–126; CS, p. 156; and SY, p. 95.) The Battle of the Fire-oxen is extensively analyzed in our *History of Warfare in China: The Ancient Period*.

6. Probably in 301 B.C.

7. The battle's date and the identity of Fan Kao are unknown, but the conflict had to occur before 286 B.C., when the state of Sung was extinguished by Ch'i.

8. Possibly when Ch'i defeated Ch'u in 301 B.C., as just mentioned. (See CS, pp. 156–157.)

9. Although as presently reconstructed the chapter contains numerous late fragments, the main body probably still stems from Sun Pin himself. The inclusion of material clearly subsequent to his life need not be grounds for dismissing the entire chapter as a late fabrication. (For a discussion, see CS, pp. 157–159.)

10. The chronology of Mencius's life and how many visits he may have made to Ch'i have been debated for centuries. For an overview, see D. C. Lau's consideration in Appendix I of *Mencius* (Hong Kong: Chinese University Press, 1984), vol. 2, pp. 309–315.

11. In "Fullness and Emptiness" in the *Six Secret Teachings,* the T'ai Kung is recorded as persuading King Wen to minimize the exactions imposed upon the people. His description of Emperor Yao's personal practices is illuminating in this regard: "What he allotted to himself was extremely meager, the taxes and services he required of the people extremely few. Thus the myriad

peoples were prosperous and happy and did not have the appearance of suffering from hunger and cold. The hundred surnames revered their ruler as if he were the sun and moon and gave their emotional allegiance as if he were their father and mother."

12. See CS, p. 155. "Dispensing provisions" (also known as "kindness" or "beneficence") was the most common way to quickly gain adherents among the people and a technique often employed by powerful individuals to subvert the ruler's authority and control. Such patronage has many echoes in contemporary political practice throughout Asia, including Japan. Note that the *Three Strategies* cites an earlier pronouncement: "A state about to mobilize its army concentrates first on making its beneficence ample. A state about to attack and seize another concentrates on first nurturing the people" ("Superior Strategy").

13. This was not solely a policy of the military thinkers but also a cornerstone of Legalist thinking and one of Lord Shang's most effective policies in strengthening the state of Ch'in. See also note 20.

14. In consequence they became specialists in the tactics and technology of defensive warfare. The *Mo-tzu*, attributed to Mo-tzu himself but the product of many hands, contains extensive materials on siege warfare. (This topic has been extensively explored by Robin D.S. Yates, whose writings may be consulted for further explication.)

15. As discussed in the Historical Introduction. Additional notes will be found under Chapter 27, "Male and Female Cities."

16. For a discussion of the term *Taoist* (which we employ as a rubric for individuals embracing the philosophical perspective and attitude of Lao-tzu and Chuang-tzu), see N. Sivin, "On the Word Taoism as a Source of Perplexity, with Special Reference to the Relations of Science and Religion in Traditional China," *History of Religions*, no. 17 (1978):303–330.

17. Han Fei-tzu integrated many diverse insights to create his synthetic political philosophy, in which the ruler would be marked by deliberate "inaction" or "tranquility." Certain chapters of the *Tao Te Ching* (as tangentially discussed in our previous works and the Notes to Chapter 30, "Unorthodox and Orthodox") discuss employing the military. For an interesting perspective on Taoism and military affairs, see Christopher C. Rand, "Chinese Military Thought and Philosophical Taoism," *MS* 34 (1979–1980):171–218.

18. For example, as described in the passage cited in note 11. For an extensive discussion of the emphasis placed upon nurturing the people as a basis for military power, see the introductions to the various *Seven Military Classics*.

19. Sun-tzu emphasized the great expenditure and adverse impact on the state in Chapter 2, "Waging War," of the *Art of War*.

20. The *Wei Liao-tzu* particularly emphasized the need to attract immigrants, no doubt because the book was composed just when warfare in the Warring States period was escalating to require massive numbers of

infantrymen. For example, "Military Discussions" states, "One who is enlightened about prohibitions, pardons, opening (the path to life), and stopping up will attract displaced people and bring unworked lands under cultivation. When the land is broad and under cultivation, the state will be wealthy; when the people are numerous and well-ordered, the state will be governed. When the state is wealthy and well-governed, although the people do not remove the blocks from the chariots nor expose their armor, their awesomeness instills order on All under Heaven."

21. The *Wei Liao-tzu* preserves extensive material probably reflecting the organizational system imposed in Ch'in, as well as some of Ch'in's regulations and associated punishments for those in military service.

Chapter 16

1. Although Chang Chen-che (CL, p. 134) presents a reasonable argument for GSR 231c being a loan for *chuan* (GSR 231a), "to monopolize power" or "to concentrate" (as in dictatorial control), yielding, "The square deployment is for solidifying control," the chapter's translation retains the original meaning of the character, "to cut" or "to sever." (Cf. CC, p. 248; SY, p. 100; TCT, p. 65; and HW, p. 133. "Cut" is chosen here in contrast with "sever" which is employed for the Awl formation.) The square and circular formations were fundamental, part of every soldier's basic training. For example, the *Wu-tzu* states, "Now men constantly perish from their inabilities and are defeated by the unfamiliar. Thus among the methods for using the military, training and causing them to be alert are first. One man who has been trained in warfare can instruct ten men. Ten men who have studied warfare can train one hundred. One hundred such men can train one thousand. One thousand, ten thousand; and ten thousand who have been trained in warfare can train the entire body of the Three Armies. Have them deploy in circular formations, then change to square ones. Have them sit, then get up; move, then halt. Have them move to the left, then the right; forward and to the rear. Have them divide and combine, unite and disperse. When all these changes are familiar, provide them with weapons" ("Controlling the Army"). (In the next chapter Sun Pin discusses the tactics for attacking a circular formation.)

2. Following the commentators who understand GSR 231a with a wood radical in the sense of "assembling," "bringing together," or "unifying." (See, for example, KO, p. 236; SY, p. 100; TCT, p. 65; and HW, p. 133.) Chang (CL, p. 134) takes it as a loan for *chuan* (GSR 231c), "to turn," which would produce, "The circular deployment is for turning." Most commentators tend to believe that the circular formation is designed for defensive purposes, citing, for example, Sun-tzu's statement that "when the troops have penetrated deeply they will be unified, but where only shallowly, they will [be inclined to] scatter" ("Nine Terrains"; see, for example, CC, p. 249). On the contrary, Chang avers that they err in limiting the function of the square for-

mation to assaults and round formations to defense and cites these sentences from the late *Huang-ti wen Hsüan-nü ping-fa:* "When the enemy deploys in an irregular (curved or bent) formation, you should use a circular deployment to attack them. The circular deployment is a deployment (of the category) of earth. When the enemy deploys in a straight formation, you should use a square deployment to attack them. The square deployment is a deployment (of the category) of metal" (CL, p. 134). Chang's example notwithstanding, the circular formation—which appears throughout the military writings—is normally praised for providing an army with the ability to equally respond in all directions, much like the function of the eight-sided array discussed in the *Questions and Replies*. For example, Sun-tzu said, "In turmoil and confusion their deployment is circular and they cannot be defeated" (Chapter 5, "Strategic Military Power"). Clearly, common practice when under pressure was to assume a circular deployment. Moreover, its employment appears to have been primarily defensive, designed for static deployment rather than ongoing movement. In Chapter 14 Sun Pin has already stated it is appropriate for ravines, a terrain where one would expect to be pressed into a defensive position by an enemy all about. Chang's thought that it is appropriate for turning is also suspect since the circular formation would presumably have its forces oriented toward the exterior. Whether consisting of discrete blocks or simply concentric rings (an issue that bears serious investigation), it would hardly be ideal for turning (in comparison to a square or other linear-based array).

3. The term translated as "diffuse" stands in opposition to that for "concentrated" or "having a high density." Apparently it refers to deploying in very open ranks and therefore can also be translated as either "sparse" or "dispersed." However, since we have employed "dispersed" for another term (*san*) in the earlier chapters and the *Seven Military Classics*, in order to maintain a distinction in translation "diffuse" is employed for *shu* (GSR 90b) here and in Chapter 29. The last character in the sentence, which reappears slightly below, is unclear, although it appears to be a "mouth over a dog" and therefore a variant of *fei*, meaning "to bark." This stimulates some commentators to believe "the sparse deployment is for bluffing," making noise without having any substance behind it. (See, for example, CC, p. 251; and TCT, p. 65.) Our translation of "rapid (flexible) response" is based upon taking this marginally legible character as perhaps *t'u*, "sudden," and the description that follows of the formation's characteristics and functions. Among other possibilities, the SY editors think that it is *fu*, a kind of rabbit net, and that the sentence refers to employing numerous methods to lure the enemy into a trap (SY, p. 100). Chang (CL, p. 131) takes it as "turn about"; Kanaya (KO, p. 231) understands it as "breaking free."

4. Its very density prevents the enemy from amassing enough troops to overwhelm any position. (Chapter 29 discusses the dynamics of dense deployments.)

5. The Awl Formation, by arraying men in a narrow front but to great depth, is designed to penetrate an enemy's lines and thereby wrest a quick vic-

tory. In Chapter 3, "The Questions of King Wei," Sun Pin said, "The Awl Formation is the means by which to penetrate solid (formations) and destroy elite (units)." In Chapter 14 he stated, "To facilitate exhausting the enemy, use the Awl Formation," and "to realize a sharp-edged deployment, use the Awl Formation."

6. The deployment's name invokes the image of wild geese flying in an extended "V" formation. It is traditionally understood as orienting the point toward the enemy, in contradistinction to the hooked formation, which probably opened around the enemy. (However, an inverted orientation is depicted in the late *Wu-pei-chih* and is also appended to the modern PH translation, p. 240.) The translation accords with the early commentators, taking *chieh she* as "engage in (bow) shooting," much like "blades engaging," "clashing." (Cf, CC, p. 250; and SY, p. 100.) However, Chang Chen-che's gloss of *chieh* as "nimble" or "swift," while unlikely, is not impossible and would yield "is for quick archery response" (CL, p. 135). Doubts might also be raised about the traditional understanding for the orientation: To establish an effective killing zone through which an enemy must advance, the archers should be deployed in a semicircle or V, with the zone falling at the focus of the curve, opposite to orienting the V with the point toward the enemy. The explanation that follows stresses the defensive aspects of this deployment, whereas in Chapter 3 Sun Pin has already stated, "The Wild Geese Formation is the means by which to abruptly assault the (enemy's) flanks and respond to [changes]." This would be best accomplished by an open V enveloping the enemy's flanks, requiring the enemy's center forces to move forward to attack the V's midpoint. Orienting the point toward the enemy would effectively prevent advances designed to turn the flanks. (Somewhat contradictorily, in Chapter 14 Sun Pin noted that the Cloud Formation should be used to engage in arrow warfare.)

7. The problem with this sentence is simply that the referents are unclear. Does the hooked formation facilitate making such changes for one's own forces, or does it force the enemy to change its targets and alter its plans? In the absence of an explicit *shih*, "to cause," the former should perhaps be assumed, as translated. However, see Chang (CL, p. 135), who would have "alter the enemy's plans." This would certainly accord with the thrust of the other sentences whereby one does something to an enemy. Unfortunately, the deployment's description is incomplete, precluding any conclusion.

8. "Rising," from the core meaning "to rise up" or "to rear" (i.e., as a horse rearing up), whether with or without the feather radical. (See GSR 730a.) Sun-tzu frequently spoke about keeping the people ignorant, about commanding them without letting them understand the basis or plans. However, "causing doubts in the masses" must be directed toward the enemy's troops rather than one's own because doubt—according to virtually every writer—would immediately doom any military enterprise to failure and inevi-

table defeat. Similarly, causing doubt is designed to make it difficult for the enemy to execute its battle plan, to thwart its assumptions and estimations. Note that Chang rather oddly glosses *ku*, "reason," as "solid." Since *ku*, "solid," appears just below, it seems particularly inappropriate. (See CL, p. 136.)

9. Following Chang's suggestion that *chang*, "to extend," should be understood as the same character with a water radical, meaning "to overflow" or "to flood." (See CL, p. 136. Although unmentioned, "flooding the solid" parallels the previous sentence and also accords with the general thrust of the tactics being discussed.) Most commentators understand *chang* in the sense of "strengthening," something that moats are usually designed to accomplish. (See, for example, SY, p. 101; and TCT, p. 66. The latter, unable to decide which reading to use, gives both, one after the other, in its modern translation on p. 69.) However, armies tended to select terrain for defensive features such as rivers and then improve their positions, deploying to fully exploit any water obstacles. Within the chapter's context it seems likely Sun Pin is discussing aggressive actions rather than the augmenting of static defenses. Diverting streams, redirecting heavy rain running off mountains upon those foolish enough to encamp in low-lying areas (despite Sun-tzu's warning in Chapter 9, "Maneuvering the Army"), and destroying dams were all measures employed by ancient armies to inundate enemy forces, particularly those occupying well-fortified positions. Accordingly, Sun Pin here seems to be advocating employing such techniques to inundate the solidly entrenched just like using incendiary dispositions to seize enemy encampments. However, the expansions that follow speak more about engaging in naval engagements—assaults and defense against assaults—than employing water itself as a weapon to attack static forces. (See the commentary for further discussion.)

10. An uncommon term appears here, *chü chün*, which literally means "occupying army." From other descriptions of the formation's tactics, it is apparent that the term means not simply an "occupying" or "holding" force but rather a form of subtracted or ready reserve that encompasses units designed for responding flexibly, executing unorthodox tactics, and providing aid wherever needed. This would fully accord with Sun Pin's general line of thought to retain significant forces in reserve, as discussed in Chapter 7, "Eight Formations." Most commentators understand *chü chün* as either a reserve or command force and occasionally a flexible response force. (For the former, see SY, p. 101; for the latter, TCT, p. 66. Hsü and Wei interpret it as the main strength [HW, p. 136].)

11. "Rapid response," identical with the chief virtue of the diffuse deployment.

12. In Chapter 4 Sun Pin described a fortified deployment with an empty middle, thus naturally concentrating the forces at the sides. Note that in Chapter 17 Sun Pin discusses a method for attacking the square formation.

13. In "Eight Formations" Sun Pin advised reserving two-thirds of the available forces in the rear "to consolidate the gains" (through unorthodox tactics).

14. As discussed in the commentary to Chapter 14, in general multiplying the number of flags and pennants—as advised slightly farther down in the paragraph—was a standard tactic for confusing the enemy (as well as imposing clarity in the commands directed to one's own troops). An enemy confronted by an apparently chaotic, almost indistinguishable mass would be unable to identify and target particular units or determine the direction or absence of motion. The chief problem addressed by the square deployment is the paucity of men. This is to be compensated for by similarly multiplying the number of flags and pennants, befuddling the enemy and causing it to overestimate troop strength. Similarly, prominently displaying weapons on the outside of the formation would confirm this impression (and explain the subsequent sentence, "Sharpen your blades to act as your flanks"). However, while the main point is clear, the second part of the sentence is probably corrupted, and the common understanding of *shih* as "to show" or "to display" still leaves grammatical problems. (Cf. CC, p. 253.)

15. This might also be understood in terms of the sword analogy previously raised in Chapter 9, "Preparation of Strategic Power"—the edge of the forces must be sharpened to effect a strong defense.

16. "Hold and defend," following Kanaya (KO, p. 234), seems to be a reasonable choice for an otherwise intractable character, as well as being an appropriate correlate for "attack." Some commentators gloss it as another form of "attack" (cf. CL, pp. 137–138; and SY, p. 102); others suggest it is a cognate for *i*, "resolute" (CC, p. 254; TCT, p. 66).

17. "Developing weaknesses" because *shuai* entails an aspect of decline from a former status rather than the more static terms generally seen in the military writings, such as "weakness" or "voids." Some commentators equate it with being tired or exhausted, without justification. (See, for example, SY, p. 102.)

18. Reading as a continuous sentence rather than broken into two clauses. (See CC, p. 242; and CL, p. 138.) The term *hang shou* is somewhat problematic; for unknown reasons some editions understand it as indicating that the formation (or rows) are clearly divided. (See, for example, SY, p. 103.)

19. Supplementing three missing characters based on the context.

20. The *Ssu-ma Fa* also advises having the soldiers sit or squat whenever overwhelmed by fear, suggesting this was probably a fundamental technique for stabilizing units (Book 4, "Strict Positions").

21. "Sound" no doubt encompasses verbal commands as well as the drums and gongs—the intent being to ensure that they would be inescapably noticed by the troops, who might otherwise be too agitated to pay attention to signals conveyed by command flags. As Wu-tzu said, "When the ear has

been awestruck by sound it cannot but be clear" ("The Tao of the General"). The CC editors cite a *Kuan-tzu* passage that states, "The first is called drums. Drums are the means by which they undertake action, by which they rise up, by which they advance. The second is called metal. Metal is the means by which they sit, by which they retreat, by which they cease activities" ("Military Methods"; see CC, pp. 135–136, which also notes additional *Wei Liao-tzu* passages).

22. The idea being that the density and integrity of the concentrated deployment should be preserved at all times because these are its strength. Segmenting troops to pursue and attack enemy forces—which may simply be a lure or enticement in any event—cannot be countenanced. Furthermore, when the enemy advances, he should not be obstructed at any point away from the main formation because the concentrated deployment is designed to withstand attack. It is therefore preferable for the enemy to come forth rather than for the army to move and become disordered in making the attack. Another, although much less likely, reading for the sentence would be within the preceding context of men suffering from fear and becoming disordered rather than as referring to the enemy advancing or retreating. Thus, anyone who deserts would not be pursued; anyone rejoining the ranks would not be stopped.

23. "Circuitous route" (contrary to various interpretations of the commentators) based on Sun-tzu's use of the term in Chapter 7 of the *Art of War*. Chang understands it similarly but strangely explains *jui*, the term translated as "elite troops," as "direct approach." Since *jui* has already appeared in the chapter in the sense of "elite forces," this seems unlikely. (See CL, p. 136.)

24. "Insult" them by treating them with disregard, mounting harassing attacks with outnumbered but compact—and therefore effective—forces.

25. The commentators offer various explanations for the analogy depicting closeness ranging from "feather down" to a closely woven bamboo basket. (For example, see SY, p. 103; and CL, p. 138.) Comparing an army with a mountain is a common image, found for example in the *Wei Liao-tzu* and the *Art of War*.

26. Or "may be compared with a sword," depending upon the understanding of *pi* as a loan for "to cause" or "to compare." (Cf. CL, p. 139; SY, p. 103; and TCT, p. 67.) Sun Pin already employed the image of the sword in Chapter 9, "Preparation of Strategic Power," to characterize the nature of military formations.

27. The "foundation" refers to the body of the sword, the backbone and substance of the blade, and the necessary haft (which he previously identified with tactical reserves in Chapter 9).

28. It is unclear how many characters are missing here. The translation follows the suggested bridge found in TCT, p. 69.

29. The commentators make much of the term *ho* (translated here as "conjoined") as referring to the wings, but what is conjoined is naturally the wings, which no doubt was the origin of the term being used in this fashion. (See, for example, SY, p. 104.)

30. The sentence perhaps continues "no Heaven or Earth." The soldiers must be oblivious to fear, unconcerned about their orientation and where the direction of safety (or flight) lies. In "Martial Plans" in the *Wei Liao-tzu,* the military and its commanding general are described in similar terms: "Now the commanding general is not governed by Heaven above, controlled by Earth below, nor governed by men in the middle. Thus weapons are evil implements. Conflict is a contrary virtue. The post of general is an office of death. Thus only when it cannot be avoided does one employ them. There is no Heaven above, no Earth below, no ruler to the rear, and no enemy in the front. The unified army of one man is like the wolf and tiger, like the wind and rain, like thunder and lightning. Shaking and mysterious, All under Heaven are terrified by it." See also note 32.

31. The sentence contains two pairs of doubled characters that obviously approximate sounds; however, whether they are of men and chariots in movement or the thundering of drums is a matter of speculation. (Cf. TCT, p. 67; SY, p. 105; CC, p. 259; and CL, p. 140.)

32. This image of swift awesomeness is found in several of the military writings. For example, in "Certain Escape," the *Six Secret Teachings,* the T'ai Kung states, "Make your fires and drums numerous, and attack as if coming out of the very ground or dropping from Heaven above."

33. Chang (CL, p. 141) breaks the sentence after "five paces" and therefore understands this measure as referring to the width of the outer ditch to be constructed.

34. The idea being that the piles should provide reasonably uniform flames and heat, thereby preventing the enemy from exploiting a "cooler" portion of the perimeter and readily breaching it.

35. Exactly what they are to construct is not completely clear, but it is probably some sort of spike or palisade system, if not a *chevaux-de-frise.* However, some editions (for example, SY, p. 105) think the sentence describes the torches that the soldiers are to employ.

36. The SY editors think "light and sharp," understood more as "quick and alert," refers to the actions of the soldiers in lighting the fires (SY, p. 105).

37. Retaining the original character *ch'i* ("breath," "vapor," or "pneuma") rather than *chi* ("already") found in most editions in accord with Chang's note (CL, p. 141).

38. This passage, the first of two on incendiary warfare, obviously advances methods for static defense against incendiary attack. Constructing an additional ring of ditches or moats will provide a fire break against an enemy sim-

ply igniting the underbrush around the encampment. Close but equidistant piles of firewood were probably designed to provide ready materials for starting backfires wherever needed or igniting outwardly directed blazes against attacking forces. (See the commentary for additional discussion.) Constructing some sort of linked fence or rail system—perhaps portable—would provide a palisadelike obstacle for rapid deployment to slow any assault troops advancing to exploit an incendiary attack. The missing portion probably discussed further measures to be taken when the enemy enjoyed favorable winds since the extant fragment concludes by advising a retreat in the face of an enemy onslaught should the situation prove untenable.

39. The SY editors gloss *yen*, translated here in its basic meaning of "overflowing" or "abundant," as "low flat terrain" (SY, p. 106). *Hsia*, understood here as "downwind," might also simply mean "below"; however, the wind is obviously the critical factor in igniting fires for assault purposes.

40. Most of the commentators understand this in strangely inverted order—"assist your strategic power with it (i.e., the incendiary attack)." (See, for example, SY, pp. 106–107. In contrast, KO, p. 238, has it correctly.)

41. This series of eight items includes several for whom the meaning of the characters, if not the characters themselves, are essentially unknown (at least in this context). Our translation represents best guesses based upon common loans coupled with the most reasonable glosses of the few commentators, such as Chang Chen-che (CL, p. 143), who venture opinions. Two or three are simply ignored by most editions as impenetrable. Rather than a lengthy, detailed analysis, we would like to note the following: "Repelling poles"—suggested by commentaries cited by Chang, although he concludes that *k'ai*, the character in the text, should be understood as *chevaux-de-frise*. (*K'ai* may be an error for *kun*, meaning "stick" or "pole," since they are essentially the same except for the two components on the right side being inverted. Some editions [such as TCT, p. 68] think "hooks" imply "repelling poles.") "Bamboo raft" is also suggested by some editions (TCT, p. 68). "Cypress wood," long known for its insect and water resistance, would have been prepared in advance for repairs and construction of boats, temporary dikes, etc. Another possibility is the character's meaning of "spreading grasses," for masses of such material could function to stem leaks and bind mud. "Pestle"—used for pounding earth (such as for dikes and defensive walls), but it might equally well be a loan for "hoe," "spade," or even "pitchfork." (Another, although remote, possibility is "mixed mud and straw," which would be indirectly suggested by GSR 465g, apparently a possible loan for "sourfruit." However, Chang takes it as a pestle, while TCT suggests it is a loan for "small boats" [TCT, p. 68].) "Light boats"—following TCT, p. 68. (Virtually no one else is willing to hazard a guess about this character's meaning.) "Oars" and "sails" seem clear enough, while the intermediate word (somewhat unclear in the bamboo strips) probably refers to a type of basket

or woven container for moving earth. Clearly tools and basic materials for both onboard and land use against inundation are encompassed by this listing rather than just weapons for naval engagements. For comparison, Chapter 30, "Agricultural Implements," of the *Six Secret Teachings*, discusses the expeditious military application of normal implements, listing a number of tools similar to these. While it also contains a chapter entitled "The Army's Equipment," nothing comparable is found therein or in any of the other *Seven Military Classics*.

42. Chang believes the character identified as *sui*, "to follow" in the original compilations, should be *chu*, "to expel," in which case the sentence would read, "When (the enemy) advances you must expel (or repel) them." However, an examination of the strip's photograph is inconclusive, while *sui* appears again below (where the strip character is much clearer). (See CL, p. 143.)

43. There is absolutely no question that this sentence describes tactics for a river engagement. However, the exact meaning of *fang*, "square," is somewhat unclear. In the translation it has been taken as a loan for "side" or "flank" as this is common in the *Military Methods*. However, it could mean something like "in all directions" and therefore "everywhere," or possibly it is being used in the less common sense of "side by side," such as two boats lashed together (seen in the *Book of Odes* and followed by the SY and TCT editions). Since the forces are not specified, it could also refer to a flanking attack on the enemy's boats effected by archers on the riverbanks. (Cf. CC, p. 263; and SY, p. 107.)

44. (With reference to note 42), here there is no question that the character is *sui*.

45. How many strips, if any, are missing after these words is unclear. However, the next strip in the reconstructed edition runs from here to the word *fords* near the end of the passage.

46. This sentence echoes a passage found in Chapter 1, the *Art of War.*

47. Or possibly pitchforks (see CL, p. 144). However, Kanaya (KO, p. 239) takes the term *hua/wu* as possibly designating a type of chariot. This would result in a symmetrical sentence: "The infantry have *hua* (chariots), the chariots have defensive (infantry)." Spades would be more appropriate to tactics for defending against water inundation.

48. Chang thinks "defensive infantry"—which in the Warring States period had become common to protect the chariots—should be understood as "drivers and infantry" (CL, p. 144).

49. "Show the people that the infantry is coming," which appears on the last strip, seems rather disjointed, suggesting—even though it clearly concludes the passage—that immediately previous material is missing.

50. Book II, the *Questions and Replies*. Square and round formations are mentioned in "The Unorthodox Army" in the *Six Secret Teachings;* "Con-

trolling the Army" in the *Wu-tzu;* and "Military Discussions" and "Orders for Restraining the Troops" in the *Wei Liao-tzu.*

51. Also see Li Ching's discussion in Book II of *Questions and Replies.*

52. For a discussion of the mythical(?) battles between the Yellow Emperor and his enemies, including an analysis of his tactics and the demythologization of the engagements in terms of tribal conflict, see our *History of Warfare in China: The Ancient Period.*

53. Other variant lists also appear, especially in commentaries and nonmilitary writings, always without characterizing the nature or form of the deployments.

54. See, for example, Chapter 42, "Incendiary Warfare."

55. Although Sun-tzu compares the military to water and obviously studied the dynamics of fluid flow, he strongly cautioned against deploying anywhere near any form or body of water, such as in Chapter 9, "Maneuvering the Army." While the T'ai Kung of the *Six Secret Teachings* also cautioned against becoming enmired on marshy terrain, he also advocated taking advantage of the propensity not to strongly defend water obstacles in Chapter 34, "Certain Escape." However, the remaining *Seven Military Classics* are otherwise silent. For a brief discussion of the history of chariot warfare in the wet states of Wu and Yüeh, see the introduction to our translation of Sun-tzu's *Art of War.*

Chapter 17

1. For a discussion of the terms *guest* and *host,* see Chapter 19, "Distinction Between Guest and Host." (The guest is the invader or aggressor; the host is the state being invaded or the defender in the engagement. Obviously there is some latitude for contradiction in the use of these terms, depending upon the roles being played.)

2. In antiquity armies employed drums for conveying commands as well as designating the requisite marching beat. Different-sized drums were assigned to various units, normally based upon strength, although drums could also be distinguished by unit type (such as chariots, cavalry, or infantry). For example, "Orders for Restraining the Troops" in the *Wei Liao-tzu* states, "Beat the drum once and the left foot steps forward; beat it again and the right foot advances. If for each step there is one beat, this is the pace beat. If for ten steps there is one beat, this is the quickstep beat. If the sound is unbroken, this is the racing beat. The *shang* note is that of the general's drum. The *chiao* note is that of a regimental commander's drum. The small drum is that of a company commander. When the three drums sound together the generals, regimental commanders, and company commanders are all of one mind." As previously noted, the conjoined sounding of all the drums would also have a startling, disorienting effect on an enemy. Furthermore, unified beating would confirm that the army had attained the unity necessary to wrest vic-

tory, particularly in disadvantageous circumstances. As the *Ssu-ma Fa* states, "When the Three Armies are united as one man they will conquer. In general, as for the drums: There are drums directing the deployment of the flags and pennants; drums for advancing the chariots; drums for the horses; drums for directing the infantry; drums for the different types of troops; drums for the head; and drums for the feet. All seven should be properly prepared and ordered" ("Strict Positions"). Because of their great importance in controlling the troops, all the military writings discuss them. (While the character translated as "four" normally designates a team of four horses, and therefore may be interpreted as standing for chariots [or even cavalry by some commentators], it seems clear that the basic meaning of "four," for which it is frequently a loan, is intended. However, compare CL, p. 149, n. 6.)

3. As this situation seems closely related to the deployments discussed in Chapter 16 (where it also states that the "circular is for unifying"), "diffuse" has been supplied for the missing character. (Note HW, p. 143; compare CL, p. 150.) Unfortunately, neither chapter describes the circular formation's composition.

4. There are several possibilities for the missing character, such as "prepare for" or "harass" them. "Fragment" coheres best with the thrust of Sun Pin's (and Sun-tzu's) thought, especially as found in the immediately preceding chapters. (See, for example, the tactics for aquatic deployments in Chapter 16.) (Cf. HW, p. 143; CL, p. 150.)

5. This clause might also be understood as, "Observe and depart from them." (See, for example, KO, p. 247.)

6. Most editions understand this clause as simply "attacking their rear" (for example, SY, p. 111; and TCT, p. 75. However, see HW, p. 145; and KO, p. 247).

7. As several commentators note, the "sharp" deployment is perhaps the "Awl Formation" discussed in the previous chapter.

8. Thereby deploying a horizontal array (possibly in the hook formation?) to apparently oppose the vertical, menacing Awl Formation. In Chapter 29 Sun Pin advances the principle that the dense (which would characterize men in the Awl Formation) should not be opposed by the dense but by the diffuse, while in Chapter 14 he said, "Defend against and surround (the enemy) with an entangled, flowing formation," and "to hunt down the (enemy's) army, use an elongated formation." Note that the T'ai Kung said, "Stretching out distant formations to deceive and entice the enemy are the means by which to destroy the enemy's army and capture its general" ("The Unorthodox Army").

9. Supplying our best guess for the missing characters.

10. "Death warriors"—warriors committed to fighting to the death—were a highly valued component of ancient military thought. Many strategists, such as the famous general Wu Ch'i, advocated selecting and grouping

men of exceptional strength and motivation, as in the first book of the *Wu-tzu:* "The ruler of a strong state must evaluate his people. Among the people those who have courage and strength should be assembled into one unit. Those who take pleasure in advancing into battle and exerting their strength to manifest their loyalty and courage should be assembled into another unit. Those who can climb high and traverse far, who are nimble and fleet should be assembled into a unit. Officials of the king who have lost their positions and want to show their merit to their ruler should be assembled into a unit. Those who abandoned their cities or left their defensive positions and want to eradicate the disgrace should be assembled into a unit. These five will constitute the army's disciplined, elite troops. With three thousand such men, from within one can strike out and break any encirclement or from without break into any city and slaughter the defenders" ("Planning for the State"; a similar discussion, "Selecting Warriors," is found in the *Six Secret Teachings*). When elite units were not available or extraordinary measures were called for, thinkers such as Sun-tzu advocated temporarily creating them by casting the soldiers onto "fatal ground," forcing them to engage in a death struggle merely to survive, just as certain insertion tactics in the Vietnam conflict did. (For example, see Chapter 11, "Nine Terrains.") In "The Questions of King Wei" Sun Pin already stated, "Selecting the troops and strong officers is the means by which to break through enemy formations and capture their generals." Thereafter, the principle of employing selected men to wrest breakthroughs or contend with fierce enemies is repeated in "Selecting the Troops," "Preparation of Strategic Power," and "Offices, I."

11. Chang Chen-che would interpret the phrase as "ten or a hundred times our (forces)." (See CL, p. 150, n. 16.) "Hundred" seems very unlikely because even the most foolhardy would not attack in such circumstances and such discrepancies in force levels were never discussed. The SY editors take the comparisons as referring solely to the disparity in chariots and cavalry, but this is not what the text states. (See SY, p. 112; KO, p. 248, has it correctly.) In the Warring States period, chariots, while still of major importance, had been largely displaced in operational terms by infantry, while cavalry was just appearing, rarely in numbers exceeding 7,000–10,000. The number of infantry combatants would be the focal concern, against which disproportionate component strength—in comparison with the enemy and terrain requirements—would be weighed.

12. Most editions ignore the "grass radical" on the strip character *pao* when reproducing the text, and even Chang (CL, p. 151), who transcribes it correctly, understands it as a loan for *pao,* "to preserve," "to protect," or a "stronghold." While the former is often a loan for the latter, and in fact "preserve" does not appear without the grass radical in the *Military Methods'* bamboo strips (and only twice in the extant *Art of War*), in this case the meaning of *pao* (with the grass, GSR 1057f) is very appropriate to the discus-

sion—"luxuriant foliage," "to cover," or "to conceal." Accordingly, the translation reads, "You should conceal yourselves in the ravines and take the defiles as your base." The characters *pao* and *tai* resonate with each other because *tai* (GSR 315a), meaning "base" (of a flower or fruit), also appears in the bamboo strips with a grass radical (which is dropped by most editions in their printed text). *Tai* (without the grass radical—GSR 315a, "belt" or "sash") causes the commentators considerable difficulty and is usually glossed as something like "to occupy and control" rather than "stretched out through (like a belt)." (See, for example, SY, p. 112.)

13. "Easy terrain," which obviously refers to ground easily traversed, is a technical classification in Sun-tzu's thought and thereafter, referring in particular to terrain accessible to chariots. It is to be distinguished from otherwise open but difficult terrain, such as areas marked by ravines, undulations, and water hazards—all of which may pose little obstacle for infantry forces but spell doom for chariots (with their lack of ball bearings, spring suspensions, and pivotable axles). For a discussion, see the appendix and discussions in the *Seven Military Classics;* and "Battle Chariots" in the *Six Secret Teachings.*

14. Mounting an attack against forces already solidly emplaced in ravines was considered difficult to impossible by all the military writers and viewed as certain to result in heavy casualties at best, total annihilation at worst. Consequently, outnumbered forces generally took advantage of them, while armies that held them were also thought to be cowardly and afraid: "One who occupies ravines lacks the mind to do battle" ("Tactical Balance of Power in Attacks," the *Wei Liao-tzu*). Even Sun Pin advised resorting to the occupation of ravines in situations where it is imperative to avoid battle (see Chapter 3). The T'ai Kung is also recorded as saying, "Holding defiles and narrows is the means by which to be solidly entrenched" (Chapter 23, "The Unorthodox Army," the *Six Secret Teachings*). In Chapter 14 Sun Pin already suggested indirect methods for dealing with an enemy "bottled up" in a ravine: "Release the mouth in order [to entice] them farther away."

15. "Vacuities" is a technical term probably originated by Sun-tzu, who stressed exploiting the enemy's weaknesses and attacking positions bereft of men. For example, in the seminal chapter "Vacuity and Substance" he stated, "To effect an unhampered advance strike their vacuities," and "the army's disposition of force avoids the substantial and strikes the vacuous."

16. In "Nine Terrains" Sun-tzu defined "contentious terrain" as follows: "If when we occupy it, it will be advantageous to us while if they occupy it, it will be advantageous to them, it is 'contentious terrain.'"

17. The parsing and meaning of the characters in this sentence are a matter of disagreement. The translation follows the original reconstruction, with which Chang also concurs. (See CC, p. 269; and CL, p. 151.) Some editions continue the sentence through to have the officers and army acting brutally

and disdainfully toward the feudal lords, interpreting *nieh* in the sense of "causing to bow under." (See, for example, SY, p. 114; and TCT, p. 73.)

18. The meaning of *tai* is variously interpreted here, ranging from "prepare against" them to "contend with" them. (See, for example, CL, pp. 151–152.)

19. The translation of "sit about" preserves the primary meaning of *tso*, "to sit," in contrast to the troops rising up, prepared to engage in battle. In this sense it echoes passages in the preceding chapters, particularly the thought that when the men are afraid, common practice would be to have them sit. Some commentators to the various military writings insist that "sit" in this sense means "kneel"; however, the term "to kneel" is also found, so a distinction should be preserved. Obviously the troops could either be sitting on the ground or squatting down.

20. Following CL, p. 147, in reading *sui*, "accord," rather than *tuo*, "indolent," for the strip character. However, others (such as TCT, p. 72) adhere to the original transcription and therefore understand the sentence as "in order to make their thoughts arrogant and their ambitions (or will) indolent." (Making people arrogant while undermining their ambitions seems psychologically contradictory.) According with the probable intentions, desires, or ambitions of others was frequently recommended for reducing an enemy's wariness and alertness and thereby creating an opportunity to mount a surprise attack. (The T'ai Kung even advocated such measures in the civil realm, constituting perhaps the first principles of psychological warfare. See "Civil Offensive" in the *Six Secret Teachings*.) The CC commentators note that the *T'ung Tien* preserves the following (mythical) dialogue between Sun-tzu (Sun Wu) and Wu Ch'i, two great strategists of different but close generations: "At the end of the Spring and Autumn period Wu-tzu asked Sun Wu: 'If the enemy is courageous and unafraid, arrogant and unworried, while his soldiers are numerous and strong, how should we plan against them?' Wu said: 'Be yielding in your posture in order to accord with their thoughts. Don't let them become conscious of your plans in order to increase their indolence and laxity. In accord with their (troop) movements secret your forces in ambush and await them. Don't look at their advancing lines nor pay heed to their retreating ones, but strike the middle. Even though they are numerous they can be taken. The Tao for attacking the arrogant precludes engaging their front in combat.'" (See CC, pp. 279–280; or CL, p. 152. Cited from *chüan* 152 of the *T'ung Tien*.)

21. Doubt dooms every military enterprise, as the ancient strategists repeatedly stressed. This paragraph essentially summarizes principles propounded in Sun-tzu's *Art of War* and subsequently found in many other writings and Sun Pin's earlier chapters as well.

22. The SY editors understand the attack as having destroyed their "haughty and martial" spirit, causing them to be racing away in such haste

that they would not be able to see each other (SY, p. 115. However, compare TCT, p. 76).

23. This sentence basically repeats the four character clause discussed in note 12 but with "passes" substituted for "defiles."

24. An open, flat basket such as those used for carrying dirt or concrete even today at Asian construction sites is perhaps intended. Rather than being spherical with high sides, it is shaped more like a shovel. The term for "disposition of forces," *hsing*, frequently employed by Sun-tzu, suddenly appears here rather than *chen*, "deployment," or *hsing*, "formation."

25. Chapter 1, "Initial Estimations." It should be noted that Sun-tzu's views in the *Art of War* are not unequivocal: At one point he advises avoiding much larger forces; at another he indicates that relative numbers are unimportant. For a discussion, see the analytic introduction to our translation of the *Art of War*.

26. Chapter 6, "Vacuity and Substance."

27. Chapter 7, "Military Combat." Remarkably, the *Ssu-ma Fa* expressed similar principles: "If you divide your forces and attack in turn, a small force can withstand a large mass" (Book V, "Employing Masses"). Wu Ch'i spoke about segmenting into five operational forces to confuse the enemy and thwart his defenses (Chapter 5, "Responding to Change"). The T'ai Kung stated, "Dividing your troops into four and splitting them into five are the means by which to attack their circular formations and destroy their square ones" ("The Unorthodox Army").

28. Chapter 6, "Vacuity and Substance."

29. Several possibilities are found in Chapter 1, "Initial Estimations": "Although you are capable, display incapability to them. When committed to employing your forces, feign inactivity. When your objective is nearby, make it appear as if distant; when far away, create the illusion of being nearby."

30. Earlier Sun-tzu said, "Display profits to entice them" ("Initial Estimations"), and "one who excels at moving the enemy deploys in a configuration to which the enemy must respond. He offers something which the enemy must seize. With profit he moves them, with the foundation he awaits them" ("Strategic Military Power"). Sun Pin of course used this strategy in the two famous battles of Ma-ling and Kuei-ling.

31. For example, "The Unorthodox Army" in the *Six Secret Teachings* states, "Valleys with streams and treacherous ravines are the means by which to stop chariots and defend against cavalry. Narrow passes and mountain forests are the means by which a few can attack a large force."

32. Chapter 10, "Configurations of Terrain."

33. Chapter 5, "Responding to Change," the *Wu-tzu*.

34. Chapter 36, "Approaching the Border," the *Six Secret Teachings*.

35. Chapter 55, "Equivalent Forces." In subsequent passages he cuts the chariot's effectiveness in half when on difficult terrain.

36. Chapter 58, "Battle Chariots." The eight conditions leading to victory largely turn upon exploiting weaknesses and debilitation in the enemy rather

than any inherent strength of the chariots. Apparently by the time this chapter was written—presumably late in the Warring States period—it was felt that the chariots could not simply create victory from nothing, could not conquer just because of their power and mobility.

37. Chapter 59, "The Cavalry in Battle." The cavalry's chief virtue was seen to be its mobility; when this was affected by the terrain or actions of the enemy, the cavalry could be trapped and destroyed.

38. Chapter 60, "The Infantry in Battle."

Chapter 18

1. About the only comment the modern analysts will venture relates to the title, *lüeh chia,* which they would understand as "seizing mailed troops." While one of the minor meanings of *lüeh* (GSR 766v) is "to seize" or "to grab," the primary meaning is "to regulate," "to plan," or "to lay out." It appears that Sun Pin intended the title to refer to a discussion about regulating and ordering mailed or armored troops in combat, not "seizing" the enemy's armored troops. However, see CL, p. 155.

2. Note that Wu Ch'i said, "Now if there is a murderous villain hidden in the woods, even though one thousand men pursue him they all look around like owls and glance about like wolves. Why? They are afraid that violence will erupt and harm them personally. Thus one man oblivious to life and death can frighten one thousand" ("Stimulating the Officers").

3. Chang has "deaths" for ranks (CL, p. 154).

4. "Rolling up one's armor" in the sense of leaving it behind is found both in Sun-tzu's *Art of War* ("Military Combat") and Sun Pin's own writing.

5. The style of characters found on this chapter's bamboo strips is the same as the previous two chapters (CC, pp. 285–286). Consequently, some of the fragments may actually belong to the earlier discussions. However, for unknown reasons the damage is unusually severe.

Chapter 19

1. The meaning of the first two sentences turns upon the interpretation of the character *fen,* "to divide." In the first sentence it is taken as "distinction," referring to the distinction between the guest and host, although some commentators would understand it as indicating a quantitative measure; others, as "position." In the second, based upon the chapter's discussions, it refers to comparative share rather than absolute numbers, as will be discussed in the commentary. (Cf. CL, 159; and SY, p. 120. Note that "host" is literally "master of men.")

2. While this is essentially a definition, its comparative nature should be noted: A "host" that has arrayed its defensive lines, when compelled to shift them to a new position in order to counter the "guest" who, declining to deploy in the unexpected location, chooses terrain more to its advantage becomes essentially a "guest." Sun Pin's thought encompasses this possibility,

but only the later writers make it explicit. (For further discussion, see the commentary.)

3. If they retreat, they face capital punishment, yet they dare to do so rather than advance into an impossible situation. (Note the severe penalties imposed on soldiers and their families for defeat and failure preserved in the *Wei Liao-tzu*.) However, the commentators take it more generally, as simply indicating that they are courageous in contending with great dangers to retreat rather than face vanquishment. (Cf. CL, p. 160; and SY, p. 120.) The CC edition notes that a similar passage appears in Chia I's famous essay, "The Excesses of Ch'in" (CC, pp. 291–292). Teng (TCT, p. 80) thinks the passage refers to the general, citing the following sentences from Sun-tzu's *Art of War:* "If the Tao of Warfare indicates certain victory, even though the ruler has instructed that combat should be avoided, if you must engage in battle it is permissible. If the Tao of Warfare indicates you will not be victorious, even though the ruler instructs you to engage in battle, not fighting is permissible. Thus a general who does not advance to seek fame, nor (fail to retreat) to avoid (being charged with the capital) offense of retreating, but seeks only to preserve the people and gain advantage for the ruler is the state's treasure" ("Configurations of Terrain").

4. "Strategic configuration of power" is discussed in the commentary.

5. Supplementing the parallel phrases that have been lost.

6. Note the size of the forces stated here—several hundred thousand. As discussed in the Historical Introduction, the entire troop count of the strongest states barely approached such numbers in Sun Pin's time, while campaign armies generally required 100,000 or so.

7. Various interpretations have been offered for this sentence, which clearly has several characters missing at the end because the strip is broken off. The idea appears to be that, even though the state's farmers produce a surplus of food, it is inadequate for the army's needs. However, another possibility is that the army is large and the farmers prosperous—the picture of the ideal Warring States power. (Cf. SY, p. 121; and TCT, p. 82.)

8. The standing forces being maintained in peacetime within the state are an excessive burden on the economy; however, when in the field they are found to be inadequate because the numbers employed are too few. Thus, Sun Pin appears to say, a proper relationship between standing forces and campaign forces should be maintained. However, the strip is broken after the next partial sentence, and the tactics proposed (in the sentence marked by note 10) seem somewhat inappropriate, suggesting important material has been lost that would illuminate the thrust of this fragment.

9. Following CC, p. 292. Most commentators take the partially legible character as "continue," but Chang suggests it is "cut" (CL, p. 160). This would result in interpreting the sentences as, "If they are cut off a thousand by a thousand, ten thousand will remain."

10. Assuming that *i* ("remain") should be understood in the sense of "send out." Possibly it is written for *ch'ien*, "to dispatch," which is similar in appearance. (Cf. SY, p. 121.)

11. Guessing that the missing character is "hand." However, compare TCT (p. 82), which understands the clause as "like killing a sacrificial lamb."

12. "Smallest amounts" by extension from the probable characters *tzu chu*, which refer to extremely small weights used for precious metals in antiquity. Although they are not particularly legible, all the commentators agree upon them. (See CC, p. 293; and CL, p. 160.) They also appear in a similar passage in the *Huai Nan-tzu*.

13. The missing characters are supplemented from an identical sentence in the next chapter in accord with prevailing opinion. (See CC, p. 294; and CL, p. 161.) Whether this is appropriate or not is somewhat questionable because "know about each other" does not cohere well with the sense of the sentence at that point and some reference to strength of numbers being inadequate might also be expected. "Deep moats and high fortifications," or sometimes, "Make the moats deep and fortifications high," is a stock phrase found in many of the military writings, including the *Art of War* ("Vacuity and Substance") and the *Six Secret Teachings* ("Approaching the Border").

14. The CC edition (pp. 294–295) notes that a similar passage appears in Hsün-tzu's chapter "Discussion of the Military."

15. Sun-tzu said, "The victorious army first realizes the conditions for victory and then seeks to engage in battle. The vanquished army fights first and then seeks victory" ("Military Disposition," the *Art of War*). Note that Sun Pin also said, "To know beforehand whether one will be victorious or not victorious is termed 'knowing the Tao'" (Chapter 4, "Ch'en Chi Inquires About Fortifications").

16. Understanding *ch'u* as going forth out of the state and *ju* as reentering the state's borders after a campaign. However, some of the commentators interpret the terms as advancing into battle and retreating or even as going forth and then crossing the border into the enemy's state. (Cf. SY, p. 124. "Advancing" is less likely because *chin* is normally employed for going forth into battle.)

17. The character translated as "completely," thought to be *pi* (which also means "exhausted"), apparently is somewhat uncertain but generally accepted. (See, for example, CC, p. 297; KO, p. 257; SY, p. 125; and TCT, p. 80.) Chang (CL, pp. 161–162) suggests it is *i*, meaning "different" or "to differ" rather than "exhausted," which appropriately parallels "make them tired." This would yield, "If the Three Armies' warriors can be forced to differ with each other and lose their determination, victory can be attained and maintained." (The problem of consolidating and profiting from battlefield victories was remarked upon by several of the military writers, as already noted.)

18. In discussing the power of the well-trained and well-prepared army, the *Wei Liao-tzu* characterizes its effects in terms of an ability to disrupt the enemy in similar terms: "A heavy army is like the mountains, like the forests, like the rivers and great streams. A light force is like a roaring fire; like earthen walls it presses upon them, like clouds it covers them. It causes the enemy's troops to be unable to disperse and those that are dispersed to be unable to reassemble. The left is unable to rescue the right, the right unable to rescue the left" ("Military Discussions").

19. As discussed in the immediately preceding chapters, soldiers were generally instructed to sit when anxious or afraid, so the primary meaning of the character is appropriate here. However, it might also be understood as referring to the army "standing down," not initiating any action.

20. The supplemented portion seems obvious as frequent battles were thought by virtually every military writer, including Sun Pin, to debilitate the army and inevitably lead to defeat. For example, Sun Pin said that the army's "injury lies in frequent battles" (Chapter 5, "Selecting the Troops").

21. Book II.

22. "Strict Positions."

23. For example: "To travel a thousand *li* without becoming fatigued, traverse unoccupied terrain" ("Vacuity and Substance").

24. See the discussions in the introductory material for Sun-tzu's *Art of War* in either the *Seven Military Classics* edition or the single-volume *Art of War*. Also see the translation and discussion for Chapter 9, "Preparation of Strategic Power," of the *Military Methods*.

25. "Selecting the Troops." In Chapter 9, "Preparation of Strategic Power," Sun Pin already noted, "Now the Tao of the army is fourfold: formations, strategic power, changes, and strategic imbalance of power."

26. "Lunar Warfare."

27. "Eight Formations."

28. "Initial Estimations."

29. "The Questions of King Wei."

30. "Strategic Military Power."

31. For example, see "Responding to Change."

32. Chapter 9, "Maneuvering the Army."

33. See especially Chapter 6, "Vacuity and Substance."

Chapter 20

1. The left portion of the character is not clear. While it probably means "strong," it might also be another character for "nimble." The commentators generally take it as "strong." (See, for example, CC, p. 301; and CL, p. 164.)

2. The vital importance of exploiting natural obstacles and avoiding being compelled to occupy or travel through them is again evident in Sun Pin's thought.

3. The commentators generally take "expanding and contracting" as referring solely to advancing and retreating, attacking and defending. The terms are essentially self-explanatory, referring equally well to increasing or decreasing the size of the army (such as by adding or reducing the number of troops mobilized or deployed) or altering the spatial disposition of forces in an engagement to make the army's deployment more or less extensive.

4. Chang implies, with some justification, that the character read as "army" may be a similar one meaning "storehouse" (CL, p. 162).

5. From the people.

6. "Doubtful" and "suspicious" are alternatives suggested for "rancorous." (See CC, p. 303; CL, p. 164; and SY, p. 127.)

7. This is generally understood by the commentators as referring to the army not being forced into ravines when moving to the left or right. (See, for example, SY, p. 128.) While this is certainly implied by the preceding sentence, it is not invariably necessary.

8. Supplementing the missing characters from the parallel phrases that follow. SY suggests they should be "impose their awesomeness" upon the enemy. However, this seems rather doubtful because the immediately preceding sentences discuss measures that cannot be imposed upon one's own army by the enemy and would therefore require changing perspective and breaking the parallelism. (See SY, p. 128.)

9. These four clauses essentially paraphrase sentences and embody tactical principles found in Sun-tzu's *Art of War*. In "Military Combat" he states, "If you abandon (literally, 'roll up') your armor to race forward day and night without encamping, covering two days normal distance at a time, marching forward a hundred *li* to contend for gain, the Three Armies' generals will be captured." In "Vacuity and Substance" he states, "If the enemy is rested you can tire him; if he is well fed you can make him hungry; if he is at rest you can move him. Go forth to positions to which he must race. Race forward where he does not expect it."

10. Understanding *po* in its primary meaning, "thin," "to thin out," rather than as a loan for "to press," as the commentators do. (Cf. CL, p. 165.)

11. Thereby realizing Sun-tzu's fundamental principle: "In general, whoever occupies the battleground first and awaits the enemy will be at ease; whoever occupies the battleground afterward and must race to the conflict will be fatigued. Thus one who excels at warfare compels men and is not compelled by other men" ("Vacuity and Substance").

12. See note 18 to Chapter 19.

Chapter 21

1. The names consist of two characters each. "Awesomely Strong" could equally well be translated as "Awesome and Strong," but the former seems more appropriate as a designation of type. "Fearfully Suspicious" follows

Chang's suggestions (CL, pp. 167–168). There are numerous possibilities for each combination.

2. "Disrupt," by extension from "to move," a probable loan for *t'ung*. (See CL, p. 168.) *T'ung* has two basic meanings—"to advance" and "to draw out." (KO, p. 271, follows the latter.) Other possibilities include the idea of attacking or penetrating the enemy, but in the context of attacks only being advised when the enemy goes forth (in the next sentence), *t'ung* perhaps encompasses the idea of agitating and disrupting them.

3. This term is appended to the bottom of the bamboo slip, as if a title or encapsulating summary (similarly for "five brutalities").

4. The essential question here is the range of activities encompassed by "normality," the activities that make up usual behavior. Chang Chen-che, whose commentaries are generally illuminating, believes it refers to the army's pillaging and plundering to obtain provisions (CL, pp. 168–169). This reflects Sun-tzu's emphasis on reducing the state's expenditures by foraging in enemy territory. In "Waging War," where he outlines the dire economic effects of military campaigns on the home state, he concludes, "The wise general will concentrate on securing provisions from the enemy. One bushel of the enemy's foodstuffs is worth twenty of ours; one picul of fodder is worth twenty of ours." However, the question of foraging seems more likely to pertain to the second instance of being respectful, and additionally to the fourth, than to the initial one. Obviously some strong, defining action is demanded of the invaders when first entering enemy territory; otherwise, they will be rebuffed, the enemy will be able to marshall its forces for a successful defense, and defeat will loom. It makes little sense for the army to advance and then simply do nothing, except in cases where the threat of overwhelming force might cower the defenders into voluntarily submitting. Therefore, the broadest sense of normality, as encompassing all the actions appropriate to a campaign army, should be understood here.

5. This would seem to echo Sun-tzu's statement: "One who excels in employing the military does not conscript the people twice or transport provisions a third time. If you obtain your equipment from within the state and rely on seizing provisions from the enemy, then the army's foodstuffs will be sufficient" ("Waging War"). However, compare Chang, who understands the term *liang* ("millet") here (but not in previous contexts) as referring to acquiring *additional* foodstuffs, not the basics (CL, p. 169; and TCT, p. 88). Most editions emend the character "millet" to "provisions" (also *liang*). However, the original editors seem to have it right—rather than food for the men, fodder for the animals. (See SY, p. 132.) Early on the army's food supply for the men might still be adequate, but the horses and draft animals would normally forage in the captured countryside. Since "four respects" result in a lack of food, "two respects" would probably result in a loss of access to the surrounding fields for grazing the animals.

6. Compare Chang's unique interpretation (CL, p. 169).

7. Compare CL, p. 169; and KO, p. 272.

8. Chang interprets the character *hua*, normally meaning "flowery" or "glorious," as "unorthodox," "perverse" (CL, p. 169; cf. SY, p. 132). However, while his chain of equivalents is intriguing, the interpretation seems far-fetched. When the army has entered enemy territory, if it engages in battle or acts "brutally" twice, it displays its awesome martial power. This makes it glorious; the soldiers can strut, and the populace will be awestruck. Moreover, they can freely forage and plunder, obtaining the fodder denied if they are respectful twice.

9. The defenders are terrified; however, the limit to such brutality has now been reached.

10. Literally, "the troops and officers see manifest deceit." (Previously the character "see" has been glossed by the commentators as "to know"; therefore, "to know deceit" might also be possible. We understand the character for "see" as an indicator of the passive and therefore translate as "have been deceived.") While the sentence seems clear enough, Chang Chen-che suggests that it is the people who are deceived (presumably by the troops? Possibly because they were naively thought to be saviors rather than oppressors, according to ancient [and already outmoded] theories of justification? See CL, p. 169). While some commentaries and modern translations agree with our version, Hsü and Wei think it means the army will be deceived—that is, be tricked and fall into ambushes (HW, p. 165).

11. As Sun-tzu indicated in "Waging War." Also see the commentary that follows regarding excessive battles.

12. "Planning for the State."

13. For example, in the famous opening lines of the *Art of War* Sun-tzu said, "Warfare is the greatest affair of state, the basis of life and death, the Tao to survival or extinction." And in "Incendiary Attacks" he stated, "If it is not advantageous, do not move. If objectives cannot be attained, do not employ the army. Unless endangered do not engage in warfare." As for extirpating the evil, the *Ssu-ma Fa* apparently preserves a list of the situations or offenses of the various feudal lords that would move the king to mount a punitive campaign. (See Book I, "Benevolence the Foundation.")

14. "Waging War."

15. And also the *Wu-tzu*, while Mencius—the famous Confucian standard-bearer—also flourished in this period.

16. See, for example, Chapter 40, "Occupying Enemy Territory," in the *Six Secret Teachings;* and "Military Instructions, II" in the *Wei Liao-tzu.*

17. Sun-tzu said, "When you plunder a district, divide the wealth among your troops. When you enlarge your territory, divide the profits" (Chapter 7, "Military Combat").

18. "Planning for the State."

19. Are battles and brutality synonymous? Does brutality refer to killing the innocent and pillaging the land or simply to forcefully applying military power and strictly imposing martial law upon conquered areas? In the context of rather draconian legal codes, cruelty would have to be dramatic to even be noticed.

Chapter 22

1. Unfortunately, the bottom two-thirds of this strip have been lost, and there are not any parallel passages to provide a basis for supplementing the lost portion. Some modern editions take the premise found in this sentence fragment as continuing on through the next strip, but this seems unlikely. (See, for example, PH, p. 248.)

2. The principle being that it is foolish to attempt to bolster an inherent weakness in order to confront the enemy's strength. Armies should focus upon attacking weaknesses with strength and thereby gaining an easy victory.

3. An army that is quickly "bent over."

4. "Insulted" army is a term frequently encountered in the early military writings, meaning essentially that it was manhandled by the enemy.

5. Chang Chen-che (CL, p. 172) glosses *tso* as "broken," but this is contextually extreme. The army's advance is suppressed, frustrated, because its weapons are ineffective against the enemy. This is the opposite of the previous situation. (By extension, the army is defeated.)

6. An army conquering its enemy may be just a question of localized strategic power; for a state to prevail in the long run, another concept and set of values are required. (Note also HW, pp. 170–172; while their translation can hardly be recommended, their observations are insightful.)

7. Understanding *hui* in the basic sense of "meeting" or "assembling" rather than as the CC commentators, "opportunity," which is a derived meaning. (See CC, p. 314.) "Dividing and assembling," expressed by various terms, is a critical concept in most military writings, including the *Art of War* and Sun Pin's *Tactics*.

8. The parallel passage from the *I-Chou-shu* is discussed in the commentary.

9. The concept of time is discussed in the commentary.

10. Sun-tzu spoke about the importance of being formless in order to befuddle the enemy and about perceiving victory and defeat before engaging in battle. Similarly, the T'ai Kung is recording as saying, "One who excels in warfare does not await the deployment of forces. One who excels at eliminating the misfortunes of the people manages them before they appear. Conquering the enemy means being victorious over the formless. The superior fighter does not engage in battle. Thus one who fights and attains victory in

front of naked blades is not a good general. One who makes preparations after (the battle) has been lost is not a Superior Sage!" ("The Army's Strategic Power").

11. The sentence begins with "the army"; however, only the commander has the power to make such judgments. Note the parallel passages discussed in the commentary.

12. "Dragon," which has been left in the translation since it obviously imagizes surpassing strength and power, is usually glossed as "favored," as in someone being favored by a ruler. As this seems contextually inappropriate, we have retained "dragon." The character translated as "lustful"—"greedy" or "covetous"—occasions some emendations by the commentators. However, it seems the original meaning is intended, for in the ancient East desire was clearly recognized as being the great motivational force. Without it neither armies nor their commanders would strive forward. The key, therefore, is constraining the desire for conquest and glory and acting correctly, just as in the previous chapter, where "brutality" and "respect" are contrasted. (Compare CL, p. 174, where Chang suggests the translation should read as if the connectives were understood as "able to." Therefore, for example, "while weak able to be strong." TCT, p. 91, concludes that "dragon" should be understood as "arrogance.")

13. As the CC commentators note, in this passage Sun Pin seems to be describing two different "Taos," so that if one implements one "Tao," a certain result will be obtained; a different one will result from the other. However, in the nearly identical passage in the *Six Secret Teachings* (provided in the commentary below), the T'ai Kung speaks about the Tao arising and stopping in these distinctly different behavioral patterns. (See CC, pp. 317–318.)

14. See note 15.

15. The title of the chapter is *wang p'ei*, which literally means "(what) the king Wears (at his waist)." This actually refers to the jade pendants symbolizing Virtue normally worn by the king and high-ranking officials, but this symbol is taken as the occasion for a diatribe on upright, solicitous behavior.

16. "Frequent regret" is understood from the similar passage in Sun Pin.

17. Obviously the nature of their relationship merits serious study. This part of the *Military Methods* is not uniformly thought to be from Sun Pin's hand, and the likelihood of borrowing from commonly available sources, including other military works, is possible. For further discussion of the development of the *Six Secret Teachings*, see the introduction to our translation in the *Seven Military Classics*.

18. Chapter 5, "Clear Indications."

19. Chapter 7, "Preserving the State's Territory."

20. Chapter 27, "The Unorthodox Army."

21. Chapter 7, "Preserving the State's Territory."

22. Chapter 2, "Evaluating the Enemy," the *Six Secret Teachings*.

23. The *Three Strategies* appears to have been heavily influenced by this school of thought. (For a discussion, see the introduction to our translation in the *Seven Military Classics*.) The question of what is, or is not, Taoist has stimulated considerable debate in recent years. Unfortunately, even minimal commentary falls beyond the scope of this book.

24. Chapter 76 of the traditional recension, Chapter 19 of the recently recovered tomb text. Our translation differs somewhat from standard versions, emphasizing consistency with the terms found in the military writings. For comparison, any of numerous, readily available translations may be consulted. Note: The word *break* in the third to last line follows the tomb text.

25. Chapter 78 of the traditional recension, Chapter 20 of the tomb text.

26. Chapter 8, "Martial Plans."

27. "Superior Strategy," *Huang Shih-kung*.

28. Ibid.

29. Chapter 11, "Nine Terrains."

30. Chapter 4, "The Tao of the General."

Chapter 23

1. The sentence form "cannot but," which is equivalent to "must," emphasizes that there is no alternative to being righteous.

2. Throughout the military writings "awesomeness" was commonly believed to be the ultimate basis of personal authority; therefore, it had to be assiduously cultivated. The foundation of awesomeness was primarily identified with the authority to inflict punishments but also with the twin handles of rewards and punishments. For example, the T'ai Kung said, "The general creates awesomeness by executing the great" ("The General's Awesomeness"). See also "Superior Strategy" in the *Three Strategies*.

3. This is not simply tautological. When the army lacks achievement, the officers and soldiers will not be rewarded. Therefore, because they do not experience the fruits of victory, the usual concrete incentives may lose their stimulus value, and the soldiers will lack the motivation necessary to fight.

4. The next chapter discusses the general's Virtue in operational terms, thereby indicating that Sun Pin is using the term in its broadest meaning of "personal power" rather than simply behavioral adherence to basic ethical codes. The latter is essentially encompassed in *te*, Virtue. However, the transcendent dimensions are more important: According to the thought of his era, the man of *Te* embodies the Tao. (See note 3 to Chapter 5.)

5. Although the commentators tend to run the last fragments together, based upon the formulaic sentences above, at least one or two full bamboo strips must be missing. While there is no indication of what they might contain, the body analogy might be expected to encompass the heart; otherwise, the use of "belly" would be odd.

6. While both short lists include "wisdom," their texts also frequently discuss generals and command in terms of knowledge, including knowledge of the Tao, tactics, terrain, rewards and punishments, and maneuvering. The other writings also emphasized knowledge over courage, as will be seen in the commentary to the chapters that follow.

7. Sharing hardship with the troops is frequently advocated as the best way to bind them to a commander. Wu Ch'i was famous for embodying this principle; his actions have been preserved in his biography, and several of the military writings praise him as an exemplar. For a general statement of so-called appropriate actions, see "Combat Awesomeness" in the *Wei Liao-tzu;* and "Encouraging the Army" in the *Six Secret Teachings.* For Wu Ch'i's famous behavior, see the introduction to our translation of the *Wu-tzu.*

8. For example, in "Martial Plans" in the *Wei Liao-tzu,* the commanding general's role is discussed, and Wu Ch'i is cited: "When Wu Ch'i approached the field of battle his attendants offered their swords. Wu Ch'i said: 'The general takes sole control of the flags and drums, that is all. Approaching hardship he decides what is doubtful, controls the troops, and directs their blades. Such is the work of a general. Bearing a single sword, that is not a general's affair.'"

9. Sun Pin also emphasized loyalty. See Chapter 5, "Selecting the Troops," including the associated commentary and note 8.

10. Chapter 3, "Planning Offensives." Some twenty or more characteristics are mentioned at different points in the *Art of War.* For an extensive discussion, see the introduction to our translation of the *Art of War.* For a typical comparative analysis of Sun-tzu and Sun Pin, see HW, pp. 174–177.

11. Chapter 1, "Initial Estimations."

12. Chapter 4, "The Tao of the General."

13. Chapter 2, "Military Discussions."

14. "Superior Strategy."

Chapter 24

1. "Clumps of earth" follows CL, p. 177, which explains *chieh* as a loan for "clump." Most commentators take *chieh,* "mustard plant," as some sort of grass, as "insignificant things," so essentially they all agree that the troops will be treated as if their lives count for nothing, as is appropriate even for the most "caring" commander. (See SY, p. 141; and TCT, p. 195.) The translation would then be "employs them like earth and grass."

2. This is somewhat ambiguous because it might refer either to orders being directed to the general, essentially depriving him of command initiative, or to the army in general, thereby circumventing and undermining his authority.

3. Norms of behavior, appropriate standards.

4. Understanding *wei* as referring to being affected or "tied" by the individual's status rather than focusing upon the actions themselves. "External threats" basically follows CL, p. 178. Reading the strips as continuous rather than broken after "affected by," in agreement with Chang, is based upon Heir-apparent continuity in the reproductions. However, for broken readings see SY, p. 142; and TCT, p. 97.

5. These sentences are also found in Book II, *Questions and Replies*. Furthermore, Sun-tzu said, "If you impose punishments on the troops before they have become attached, they will not be submissive. If they are not submissive, they will be difficult to employ. If you do not impose punishments after the troops have become attached, they cannot be used. Thus if you command them with the civil and unify them through the martial, this is what is referred to as 'being certain to take them'" ("Maneuvering the Army").

6. Chapter 64 of the traditional recension. We have retained the primary meaning of "defeat" for *pai* to emphasize the original meaning with respect to the military writings. Normally, with reference solely to the *Tao Te Ching*, it might be translated as "overturned" or "frustrated." (See CC, p. 324; and TCT, p. 95.)

7. "The Tao of the General."

8. "Planning Offensives." The chapter also discusses the ways in which a ruler adversely interferes with the army's mission by meddling in its command.

9. "Middle Strategy," citing the *Army's Strategic Power* (possibly a book title).

10. Chapter 21, "Appointing the General."

11. Defeat meant execution for the commanding general. However, there were also numerous regulations governing the loss of a squad member, the defeat of a unit, and desertion—examples of which are preserved throughout the *Wei Liao-tzu*.

12. Chapter 19, "A Discussion of Generals," the *Six Secret Teachings*.

13. For example, "The General's Awesomeness" in the *Six Secret Teachings*, states, "The general creates awesomeness by executing the great and becomes enlightened by rewarding the small. Prohibitions are made effective and laws implemented by careful scrutiny in the use of punishments. Therefore if by executing one man the entire army will quake, kill him. If by rewarding one man the masses will be pleased, reward him. In executing value the great; in rewarding value the small. When you kill the powerful and honored, this is punishment that reaches the pinnacle. When rewards extend down to the cowherds, grooms, and stable men, these are rewards penetrating downward to the lowest. When punishments reach the pinnacle and rewards penetrate to the lowest, then your awesomeness has been effected."

14. Chapter 2, "Obligations of the Son of Heaven."

Chapter 25

1. "Light" in the sense of not "weighty"—that is, he acts without proper consideration, changes his opinion too quickly, and does not proceed in an orderly fashion.

2. Supplementing based upon CC, p. 326.

3. He "brings about chaos by himself," such as by changing deployments at the wrong time.

4. Means and methods for evaluating men were much discussed in ancient China, and many men were famous for their ability to recognize an unknown talent. For a discussion, see Ralph Sawyer, *Knowing Men* (Taipei: Kaofeng, 1979).

5. The tests are as follows: "First, question them and observe the details of their reply. Second, verbally confound and perplex them and observe how they change. Third, discuss things which you have secretly learned to observe their sincerity. Fourth, clearly and explicitly question them to observe their virtue. Fifth, appoint them to positions of financial responsibility to observe their honesty. Sixth, test them with beautiful women to observe their uprightness. Seventh, confront them with difficulties to observe their courage. Eighth, get them drunk to observe their deportment. When all eight have been fully explored, then the Worthy and unworthy can be distinguished" ("Selecting Generals"). In recent decades an informal test was conducted by engaging the individual in a game of majong, or Chinese chess.

6. Courage inevitably is one of the general's defining virtues. However, as Wu Ch'i noted, it is not enough: "Courage is but one of a general's many characteristics for the courageous will rashly join battle with the enemy. To rashly join battle with an enemy without knowing the advantages and disadvantages is not acceptable" ("The Tao of the General").

7. Chapter 19, "A Discussion of Generals."

8. Chapter 8, "Nine Changes."

9. Chapter 9, "The Tao of the General."

10. "Superior Strategy."

Chapter 26

1. The commentators variously understand the "means for going and coming" as an inability to deploy the troops, a choosing of inappropriate actions (such as going when he should come), or a deficient military organization. (See CC, p. 330; CL, p. 182; SY, p. 145; and TCT, p. 99.) While none can be excluded with total certainty, the military writers were acutely aware of the need for routes of passage, for secure roads for advancing and retreating, so as to avoid being forced onto enmiring or constricted terrain, which seems a more probable interpretation here.

2. Chang understands the clauses in a connected sense—that is, because the general has gathered together useless troops, he seems to have resources but does not really have any usable ones (CL, p. 182). As the *Ssu-ma Fa* states, "In general, in warfare: It is not forming a battle array that is difficult; it is reaching the point that the men can be ordered into formation that is hard. It is not attaining the ability to order them into formation that is difficult; it is reaching the point of being able to employ them that is hard. It is not knowing what to do that is difficult; it is putting it into effect that is hard" ("Strict Positions").

3. The sentence is translated with reference to the general, but it might equally well describe conditions within his command, as many of the subsequent items do.

4. The absolute need for unity was stressed by most of the military writers. For example, the T'ai Kung said, "In general, as for the Tao of the military, nothing surpasses unity. The unified can come alone, can depart alone" ("The Tao of the Military"). The *Wei Liao-tzu* further notes, "One who is unified will be victorious; one who is beset by dissension will be defeated" ("Army Orders, I").

5. "Old army" was variously defined, but in general it referred to an army that had campaigned long, was largely exhausted, and lacked faith in its commanders and purpose. For example, the *Huang Shih-kung* essentially provides a definition in this paragraph: "Now the one who unifies the army and wields its strategic power is the general. The ones that bring about conquest and defeat the enemy are the masses. Thus a disordered general cannot be employed to preserve an army, while a rebellious mass cannot be used to attack an enemy. If this sort of general attacks a city it cannot be taken, while if this type of army lays siege to a town it will not fall. If both are unsuccessful then the officers' strength will be exhausted. If it is exhausted then the general will be alone and the masses will be rebellious. If they try to hold defensive positions they will not be secure, while if they engage in battle they will turn and run. They are referred to as an old army" ("Superior Strategy").

6. It was generally thought that soldiers would be more inclined to think about home when they had only just left their state's borders rather than when immersed deeply into enemy territory. For example, Sun-tzu said, "When the troops have penetrated deeply, they will be unified, but where only shallowly, they will be inclined to scatter" ("Nine Terrains").

7. It is also possible that this sentence means, "if their weapons have become dull," assuming a slight emendation for "desert." Sun-tzu and others spoke about weapons growing dull after prolonged campaigns.

8. Being suddenly frightened—as opposed to being constantly fearful—appears in many texts; examples will be found in the commentary.

9. "One who occupies ravines lacks the mind to do battle" ("Tactical Balance of Power in Attacks," the *Wei Liao-tzu*). Concentrating solely upon employing ravines is symptomatic of an army that finds itself in difficult straits and has no alternative to static emplacement. While Sun-tzu advocated the

tactical exploitation of terrain such as ravines, here the army is obviously dispirited and exhausted. Ravines are not being aggressively exploited as one of many possibilities but chosen out of hopelessness.

10. See the commentary for parallel passages.

11. "Combat Awesomeness" in the *Wei Liao-tzu* states, "One who excels at employing the army is able to seize men and not be seized by others. This seizing is a technique of mind. Orders unify the minds of the masses. When the masses are not understood, the orders will have to be changed frequently. When they are changed frequently, then even though orders are issued the masses will not have faith in them. Thus the rule for giving commands is that small errors should not be changed, minor doubts should not be publicized. Thus when those above do not issue doubtful orders, the masses will not listen to two different versions. When actions do not have questionable aspects, the multitude will not have divided intentions. There has never been an instance where the people did not believe the mind of their leader and were able to attain their strength. It has never been the case that one was unable to realize their strength and yet attain their deaths in battle."

12. The T'ai Kung cites "not regarding their generals seriously" ("The Army's Indications") as one sign of weakness, while Wu Ch'i noted that the officers despising their commanding general could be easily exploited. (See the commentary.)

13. The commentators unnecessarily offer various explanations for the word "lucky," failing to notice its use in such texts as the *Wei Liao-tzu*, which speaks about armies "being lucky not to be defeated" ("Tactical Balance of Power in Attacks"). When the troops have experienced good luck, they are inclined to leave things to chance rather than act diligently. (Cf. CL, p. 183.) Moreover, "victory through luck" is a symptom of fundamental flaws in the army's organization, discipline, planning, or command—a situation certain to end in disaster.

14. Wu Ch'i said, "It is said that the greatest harm that can befall an army's employment stems from hesitation, while the disasters that strike the Three Armies are born in doubt" ("Controlling the Army").

15. See the commentary for parallel passages.

16. While Sun-tzu and others advocated attacking the enemy when his *ch'i* declined, they all emphasized aggressively manipulating the enemy to attain this condition. Naturally, if, for example, the enemy becomes dispirited, the day has passed, the road has been long, or it is raining, the troops' debilitated *ch'i* state can be exploited. However, one who passively waits for this to occur has lost the initiative and is therefore doomed.

17. The logic is similar to that for the thirteenth instance: Ambushes are but a single tactical tool in the commander's arsenal and are not to be exclusively relied upon, especially out of an inability to take other action.

18. The sentence apparently describes a situation in which the advance soldiers and those in the rear guard are not coordinated and have somehow become intermixed and disordered. This might be due to the rear guard's

overenthusiasm; fear in the front ranks, causing them to lag behind their targeted positions; or poor deployment methods. (Sun Pin discusses the requisite organization and flexible employment of forces for the front and rear in several chapters, including "Preparation of Strategic Power," "Nature of the Army," and "Unorthodox and Orthodox.") However, another explanation offered is that the troops in the front have inappropriate weapons, the weapons for the rear guard. (See, for example, TCT, p. 101; and CL, p. 184.) The need for assigning different weapons groups to the front and rear also appears in several chapters of the *Military Methods,* including "Ch'en Chi Inquires About Fortifications."

19. This is precisely the result Sun-tzu aims to achieve by being formless and unknowable: "The location where we will engage the enemy must not become known to them. If it is not known, then the positions which they must prepare to defend will be numerous. If the positions the enemy prepares to defend are numerous, then the forces we engage will be few. Thus if they prepare to defend the front, to the rear there will be few men. If they defend the rear, in the front there will be few. If they prepare to defend the left flank, then on the right there will be few men. If there is no position left undefended, then there will not be any place with more than a few" ("Vacuity and Substance"). In Sun Pin's instance, the general goes to even greater extremes and creates actual voids.

20. That is, *chün, shih,* and *ping.* For a brief history of these terms, consult the appendix on military organization in the *Seven Military Classics.*

21. Chapter 4, "Strict Positions."

22. Chapter 3, "Determining Rank."

23. Chapter 4, "The Tao of the General."

24. Chapter 2, "Evaluating the Enemy."

Chapter 27

1. A title supplied by the editors based upon the chapter's contents. The characters refer to the "male of birds" and "female of birds." As employed by Sun Pin, male is synonymous with strong and martial; female, with weak and passive.

2. The term for "mound" or "hillock" is modified by the character *fu,* which is generally taken to mean "small earthen mountains" rather than its original meaning of "attached" or "appended to." (See CL, p. 186; and compare CC, pp. 337–338, which would gloss it as "larger mounds.") Kanaya suggests that they are about two stories high (KO, p. 294).

3. In antiquity, as today, illness rapidly decimated even the most stalwart armies, particularly when they foraged on unfamiliar ground, encountered deliberately poisoned wells, and were forced to suddenly adapt to hostile environments. Armies often melted away over time; therefore, discovering and

preserving uncontaminated water sources have been a focal concern for three thousand years. As Sun-tzu said, "Now the army likes heights and abhors low areas, esteems the sunny and disdains the shady. It nourishes life and occupies the substantial. An army that avoids the hundred illnesses is said to be certain of victory" (Chapter 9, "Maneuvering the Army"). Sun Pin previously mentioned "water of life" in Chapter 8, "Treasures of Terrain," where further notes may be found. Here the principle is simply that swiftly flowing water is less likely to be contaminated, while "water that doesn't flow is deadly."

4. The principle of attacking those whose *ch'i* has been debilitated is closely identified with Sun-tzu and has previously been discussed in Sun Pin's Chapter 13, "Expanding *Ch'i*." The T'ai Kung also states, "Taking advantage of their exhaustion and encamping at dusk are the means by which ten can attack a hundred" (Chapter 27, "The Unorthodox Army"). The missing portion is supplied from the parallel passage below.

5. "Vacuous" (*hsü*) is a technical term deriving from the *Art of War*, where Sun-tzu stressed the importance of creating "vacuities" by manipulating the enemy's forces and then striking them. (See especially Chapter 6, "Vacuity and Substance." Also see note 15 to Chapter 17 of the *Military Methods*.) Although the term is the same as in T'ai Kung's Chapter 42, "Empty Fortifications," the latter focuses upon evaluating cities that might actually be "empty" rather than vacuous because of strategic weakness, disorganized defenders, or insufficient supplies. Concrete examples of the latter may also be seen in Chapter 6, "Tactical Balance of Power in Attacks," the *Wei Liao-tzu,* where such cities are termed "empty and void."

6. Sun Pin offers no explanation for his characterization. Presumably the ground is unable to sustain life, all the vegetation having been burned, the trees destroyed, and the water fouled. Natural cover being absent, troops encamped upon it would be easy targets for missile weapons. Conversely, if the area were large, the attackers would also have to cross open terrain and could equally be slaughtered during their advance. Burned-over terrain would have the single advantage of denying materials for an incendiary attack, precluding the need to employ backfires or prepare "safe ground," as discussed in "Incendiary Warfare" in the *Six Secret Teachings*. Note that the commentators unnecessarily make the initial characters overly complex. "Thoroughly incinerated" is perfectly clear. (Cf. CL, p. 187; and CC, p. 340.)

7. *Fan*, translated here as "stagnant," actually means "overflowing" or "overflowed," as in a flood. By itself it would have to be understood as water found in marshes or other over-the-banks situations or possibly even small derivative streams and backwashes, but coupled with the earlier stipulation of "flowing water," the term clearly indicates stagnant water, such as in pools or marshes. See also CL, p. 187, which, among others, glosses it as "contaminated."

8. The CS edition, for example, does not include the last fifteen chapters. However, numerous analytical articles stress the categorization of cities as a key advance marking Sun Pin's era and thought.

9. Chapter 9, "Planning Offensives."

10. Ibid.

11. "Four Changes."

12. See Chapter 8, "Martial Plans."

13. Chapter 4, "Combat Awesomeness."

14. Chapter 22, "Military Instructions, II."

15. Chapter 5, "Tactical Balance of Power in Attacks."

16. For the complementary side, see Chapter 6, "Tactical Balance of Power in Defense."

17. Chapter 5, "Responding to Change," the *Wu-tzu*. The importance of occupying heights is stressed by virtually every military writer and also emphasized in many philosophical works, particularly those with chapters focusing on military affairs. For example, as the commentators note, "Configuration of Terrain" in the *Huai Nan-tzu* states, "As for configurations of terrain . . . . . the heights constitute life, below constitutes death. Mounds and hillocks are male, gorges and valleys are female." Of course no issue is ever so simple, as the T'ai Kung pointed out: "Whenever the Three Armies occupy the heights of a mountain, they are trapped on high by the enemy. When they hold the land below the mountain, they are imprisoned by the forces above them" (Chapter 47, "Crow and Cloud Formation in the Mountains," the *Six Secret Teachings*).

18. Chapter 27, "The Unorthodox Army," the *Six Secret Teachings*.

19. Chapter 7, "Military Combat." Sun Pin's Chapter 8, "Treasures of Terrain," also advises against confronting hills.

20. Chapter 9, "Maneuvering the Army." Note that the *Ssu-ma Fa* advocates the following principles: "Keep the wind to your back, the mountains behind you, heights on the right, and defiles on the left" (Chapter 5, "Employing Masses").

21. The advantages of height were previously outlined by Sun Pin in Chapter 8, "Treasures of Terrain."

Chapter 28

1. The character translated as "army" is *ping*, which normally means "weapons." Sun Pin generally but not exclusively uses the term *chün* throughout the text.

2. As the relative imbalance is unclear, and part of the character *jang* also obliterated, this represents our best guess at the meaning. The commentators fail to hazard any possibilities.

3. As will be discussed in the commentary, Sun Pin and Sun-tzu both advised against confronting strengths because it would be inherently self-defeating.

4. That is, what is most important to him, what he values above all else.

5. The character for "seizing," *ch'ü,* might have been expected here, but the text has *chuo,* meaning "to grasp." Thus the translation of "nine graspings."

6. For example, with reference to establishing cities and their defense, see Chapter 2, "Military Discussions," in the *Wei Liao-tzu.*

7. Chapter 4, "Military Disposition," the *Art of War.*

8. See "Controlling the Army" in the *Wu-tzu.* In the same chapter Wu Ch'i identifies situations wherein an enemy can be easily defeated, such as when it has raced to a battlefield. The *Ssu-ma Fa,* reflecting the concerns of an earlier age and more limited warfare, also stresses measure and control throughout its chapters.

9. Chapter 20, "Those Who Excel." As already discussed in the Historical Introduction, these were the tactical principles Sun Pin employed against Wei's armies at Kuei-ling and Ma-ling.

10. Chapter 7, "Military Combat." However, note that the *Wei Liao-tzu,* apparently to create the necessary excitement and commitment in the troops, actually advocates making a forced march just before engaging in battle: "When the Three Armies have assumed formation, they should advance for a day and on the next day make a forced march to complete a total of three days distance. Beyond three days' distance they should be like unblocking the source of a river" (Chapter 8, "Martial Plans").

11. Chapter 22, "The Army's Losses."

12. Chapter 12, "Incendiary Attacks."

13. Chapter 3, "Planning Offensives."

Chapter 29

1. The character for "diffuse" is *shu,* basically meaning "distant" or "separate." It previously appeared in Chapter 15, where note 3 explains why it is translated as "diffuse" rather than "dispersed" (which would more conveniently express some of the thoughts in this chapter). Note that Chang (CL, pp. 191–192) thinks the meaning is different in this chapter because of inconsistencies in tactical analyses pertaining to the formation, although it seems unlikely. In this chapter the "diffuse" is correlated with the "dense," whereas in the former it was discussed and employed with respect to the "concentrated." Most commentators assume the terms are identical but without any basis for doing so. The translation preserves the distinction.

2. "Vacuous," a technical term, was previously encountered in Chapter 27. See note 5 to that chapter.

3. As Chang points out, "shortcuts" and "roads" do not readily transform into each other, and the pair is rather odd. (Cf. CL, p. 192.) He suggests that "shortcuts" be understood as hasty, direct movement as contrasted with normal roads, which imply ordered, measured travel. However, another possibility would be Sun-tzu's concept of the direct and indirect, with shortcuts be-

ing the most direct routes (setting aside any questions about roughness, passability, width, and constriction) and roads standing for the indirect.

4. Presumably forcing them into lassitude and weakness. Several of the military classics warn against allowing the troops to become too rested, too at ease, for they will then lose their spirit and energy, not to mention conditioning. For example, in "Employing Masses" Wu-tzu said, "If you move first it will be easy to become exhausted. If you move after the enemy, the men may become afraid. If you rest, the men may become lax; if you do not rest, they may also become exhausted. Yet if you rest very long, on the contrary, they may also become afraid." However, normally Sun-tzu, among others, would advise making the rested tired. Making the tired more tired would also be effective.

5. This is standard doctrine for Sun Pin and Sun-tzu, as is discussed in the commentary.

6. For a general discussion of Sun-tzu's pairs, see the introduction to our translation of the *Art of War*, pp. 130–132.

7. Chapter 7, "Military Combat." This passage is also commented upon and expanded by Li Ching in *Questions and Replies*.

8. See CL, p. 192.

9. At least one edition understands this formula rather differently, interpreting it along the lines of, "If one should be X, then X." (See PH, p. 254.)

10. Dissipating an enemy's tactical advantage by forcing it toward the complementary condition is an orthodox approach, an aspect of manipulating the enemy. Naturally, dispersing oneself is more directly attainable.

11. Chapter 40, the *Tao Te Ching*. The concept also appears in the first sentence of Sun Pin's next chapter, "Unorthodox and Orthodox," indicating the concept's importance in his thought.

12. Chapter 8, "Martial Plans." Note that Sun-tzu (and Sun Pin) sometimes advised employing this principle to exploit flaws in temperament, such as, "If they are angry, perturb them; be deferential to foster their arrogance" (Chapter 1, "Initial Estimations"). Also compare the views in "Superior Strategy" of the *Three Strategies*.

13. Book IV, "Strict Positions."

14. If "shortcut" is interpreted as the direct, presumably it means forcing the enemy to choose or be compelled onto "direct" paths that will totally exhaust it by the nature of the terrain and condition of the path. If it is interpreted as "hasty" (even though the term *urgent* appears separately), then it indicates forcing the enemy to extreme haste. Neither explanation is particularly satisfactory.

15. Chapter 7, "Military Combat."

16. The implied indirect reference, the enemy, is then manipulated in relatively common—and therefore comparatively orthodox—fashion. Grammatically, the connective *ku* is understood as "then," as Chang suggests. (See

CL, p. 192; and the modern translation of PH, pp. 254–255.) SY suggests it stands for *shih,* "to cause" or "to make." Therefore, the sentences would read, "If the enemy is dense, cause him to disperse" (SY, p. 158; and TCT, p. 107).

Chapter 30

1. Chapter 40 of the traditional *Tao Te Ching* states, "Reversal is the movement of the Tao." Similar expressions are found in many pre-Ch'in writings, including the *Huai Nan-tzu* and *Kuan-tzu.* (See CL, pp. 195–196.) Observations of natural cyclic phenomena are also encountered in several of the military writers, including Sun-tzu, where they underlie tactical conceptualizations. (See the commentary and note 20.) While we concur with those who think the two missing characters might be *yin* and *yang,* as translated, some suggest the concrete embodiment of *yin* and *yang's* waxing and waning, the "sun and moon," is intended. (See, for example, CC, p. 353; PH, p. 255; and SY, p. 160.) However, note that Sun-tzu said, "Heaven encompasses *yin* and *yang,* cold and heat, and the constraints of the seasons" ("Initial Estimations").

2. The five phases, sometimes also called the five elements, represent categories of activity that can subsume all phenomena in the universe. The five—water, fire, wood, metal, and earth—can be interrelated through production and conquest cycles. For example, water conquers fire, fire conquers metal, metal conquers wood, wood conquers earth, and earth conquers water. Several production cycles exist, including one in which wood produces fire, fire produces earth, earth produces metal, metal produces water, and water produces wood. For a discussion, see *History of Scientific Thought.* Given the existence of these "elemental categories" and their inherent interrelationships, once battlefield phenomena are characterized in terms of them, the relationships can be exploited to provide conquest measures. Unfortunately, this is a subject beyond the scope of a note or single article. Some editions yield somewhat different translations. While we could also elide slightly to read, "conquering and not conquering," others would (unacceptably) have "conquering and being conquered." (See, for example, PH, p. 255.) Obviously there can be things that something conquers, others that it cannot, yet still not be conquerable by the latter.

3. "Form," *hsing,* also translatable as "disposition" in a military context, and *shih,* "strategic power," are two concepts extensively developed in the *Art of War.* For a discussion, see the introduction to our single-volume *Art of War.* This sentence might also be phrased as, "Being excessive, being insufficient."

4. See the commentary for a discussion of the underlying philosophy. Various commentators point out that a similar phrase appears in *Kuan-tzu.* (See CC, p. 354; and CL, p. 196.)

5. With the aspects of the myriad things that will produce victory, with their particular strengths.

6. Sun-tzu said, "In accord with the enemy's disposition we impose measures on the masses that produce victory, but the masses are unable to fathom them. Men all know the disposition by which we attain victory, but no one knows the configuration through which we control the victory. Thus a victorious battle strategy is not repeated, the configurations of response to the enemy are inexhaustible" (Chapter 6, "Vacuity and Substance").

7. Anciently, writing was done on specially prepared bamboo strips, such as Sun Pin's tomb text was preserved upon, and silk rolls because paper did not really appear until late in the Han dynasty. The states of Ch'u and Yüeh were noted bamboo growing areas and therefore centers for the production of such strips. Noting that even all their strips would be inadequate to record the myriad forms of conquest is equivalent to stating the latter are virtually infinite. This reference to the bamboo of Ch'u and Yüeh is cited, as noted in the Historical Introduction, by some as evidence that Sun Pin accompanied T'ien Chi in his exile in Ch'u.

8. That is, in accord with their particular strengths.

9. Some commentators would understand the sentence as, "They grasp victory just like using water to conquer fire." (See CL, p. 198.)

10. According to Chang, the character remnants suggest "three unities" or, by extension, "three forces," which coheres with Sun Pin's tendency to divide the army into three operational forces (CL, pp. 199–200). Some commentators would take *fen*, found at the end of the last sentence and the beginning of this one (translated as "differentiate," "differentiation"), in a more restricted sense, "to divide," and just as narrowly for *shu*, "numbers," translated here as "techniques" (for which it frequently stands). Thus they would have, "That the unorthodox and orthodox are inexhaustible is due to division. Divide (your forces) according to unorthodox numbers." They cite Sun-tzu's statement as the basis: "In general, commanding a large number is like commanding a few. It is a question of dividing up the numbers. Fighting with a large number is like fighting with a few. It is a question of configuration and designation" ("Strategic Military Power"; see CC, p. 360; and CL, p. 199, among others). While this reading is possible, there is no need to restrict the meaning to "numbers" because the techniques for the unorthodox are manifold, while forces can be configured both in large and small groups according to their nature and function within a battlefield context. The late *Questions and Replies* contains several discussions of this nature.

11. The SY edition suggests the missing portion expresses the idea or examples of employing forces of the same type and composition to attempt to wrest victory, only to fail, with the present fragment presenting the necessary conclusion (SY, p. 164).

12. All pairs commonly found in the military writings, including Sun Pin's earlier chapters.

13. Some texts include a marker indicating that the last character from the previous sentence, *ch'i*, "unorthodox," is repeated at the start of this one, while others do not. (For example, the CC edition does not; the CL edition does.) We have translated on the assumption that it is not reduplicated; however, if it were duplicated, the sentence would be rather different: "When the unorthodox is initiated and becomes (thereby) the orthodox, what has not yet been initiated is the unorthodox." Furthermore, there is some question as to the meaning of *fa* in this sentence—rather than "initiated" (such as an attack being "launched"), some would understand it as, "When (actions) are discovered they are orthodox; when not yet discovered they are unorthodox." (See SY, p. 164; and TCT, p. 111.) However, this overlooks the fact that many apparently orthodox attacks are "discovered," but they conceal unorthodox tactics. It is not "discovery" or visibility that defines the characteristics, but their employment. For an incisive analysis, see Li Ching's discussion in Book I, *Questions and Replies*.

14. KO (p. 309, n. 18) suggests this sentence might be understood in terms of Sun-tzu's statement: "For this reason attaining one hundred victories in one hundred battles is not the pinnacle of excellence. Subjugating the enemy's army without fighting is the pinnacle of excellence" ("Planning Offensives"). Moreover, in "Lunar Warfare" Sun Pin said, "If in ten battles someone is victorious ten times, the general excels, but it leads to misfortune." (The character glossed as "misfortune" in the text is actually "surpass," as in "surpassing victory.") Kanaya implies that a surplus of the unorthodox will thereby lead to misfortune in conquest, and Chang (CL, p. 200) arrives at a similar conclusion but through considering that what exceeds is somehow not complete. However, given the strong, consistent thrust of the entire chapter, their interpretations seem unfounded. For comparison, note the discussion about the unorthodox and "surplus" in Book I, *Questions and Replies*.

15. The body analogy is frequently used by the military writers to explain the necessity for unified command, comparing the commander to the head and the four limbs to the forces or armies. The distraction of a finger is also found in *Mencius*, VIA12.

16. Presumably when advancing and retreating, respectively.

17. Meng Pen, who was famous for his great personal courage, was previously encountered in Chapter 9.

18. The missing character seems to be "enemy," but some suggest it should be "living" men. For example, HW's modern translation envisions the measures listed as including clearing the battlefield of the dead so that the soldiers will not perceive death, only life (HW, p. 213). However, note that

the *Ssu-ma Fa* states, "When men have minds set on victory, all they see is the enemy. When men have minds filled with fear, all they see is their fear" ("Strict Positions").

19. Sun-tzu uses the power of flowing water as an analogy to explain the nature of strategic power in a famous sentence: "The strategic configuration of power is visible in the onrush of pent-up water tumbling stones along" ("Strategic Military Power"). Much later in the *Questions and Replies* Li Ching employs another water image, indicating its popularity among the military thinkers: "According to Sun-tzu, employing spies is an inferior measure. I once prepared a discussion and at the end stated: 'Water can float a boat, but it can also overturn the boat. Some use spies to be successful; others, relying on spies, are overturned and defeated.'"

20. Note a similar use in the *Six Secret Teachings:* "The Sage takes his signs from the movements of Heaven and Earth; who knows his principles? He accords with the Tao of *yin* and *yang* and follows their seasonal activity. He follows the cycles of fullness and emptiness of Heaven and Earth, taking them as his constant. All things have life and death in accord with the form of Heaven and Earth" ("The Army's Strategic Power").

21. For a discussion, see the introduction to our single volume *Art of War*, especially pp. 147–150.

22. The possibilities are, of course, unlimited. For a discussion of their intricacies, see the *Questions and Replies*.

23. For a brief discussion of the cavalry's history in China, see Appendix B to the *Seven Military Classics*. The generally accepted date for the introduction of cavalry forces in China, 307 B.C.—which may have to be rethought in the light of recent textual discoveries, such as the *Military Methods*—falls at the end or just after Sun Pin's life. As would be expected, although the cavalry is mentioned in the early chapters and apparently used at the battles of Kuei-ling and Ma-ling, the *Military Methods* still does not contain any tactics explicitly designed to exploit its greater speed and mobility.

24. For a discussion, see the introduction to our single-volume *Art of War*.

25. Chapter 5, "Strategic Military Power."

Bibliography of Selected Chinese and Japanese Works

Insofar as full bibliographic information for all Western-language works cited in the Historical Introduction and commentaries is provided in the Notes, and an extensive bibliography of Western books and articles may be found in the *Seven Military Classics of Ancient China*, only original sources in Chinese and Japanese are provided here. For the convenience of readers conversant with these languages, they are cited in traditonal character format. (Secondary works, having already been fully indicated in romanized form in the Notes, are not repeated.)

柏楊, 柏楊版資治通鑑, 遠流出版公司, 台北, vol. 1, 1991.

張震澤, 孫臏兵法校理, 明文書局, 台北, 1985.

陳鼓應, 老子註譯及評介, 中華書局, 香港, 1987.

鄭賢斌, 白話中國兵法, 成都出版社, 成都, 1992.

中國兵書集成編委會, 孫臏兵法, 中國兵書集成, 解放軍出版社, 遼沈書社, 北京 / 沈陽, vol. 1., 1987.

徐培根, 魏汝霖, 孫臏兵法註釋, 黎明文化事業股份有限公司, 台北, 1976.

霍印章, 孫臏兵法淺說, 解放軍出版社, 北京, 1986.

金谷治, 孫臏兵法, 東方書店, 東京, 1976.

李均明, 孫臏兵法譯注, 兵家寶鑑, 河北人民出版社, 河北, 1991.

劉心健, 孫臏兵法新編注譯, 河南大學出版社, 開封, 1989.

馬持盈, 史記今註, 台灣商務印書館, 台北, 1979.

村山孚, 孫臏兵法, 德間書店, 東京, 1976.

沈陽部隊後勤部 (孫臏傳) 編寫組, 孫臏傳附 (孫臏兵法) 今譯, 遼寧人民出版社, 沈陽, 1978.

沈陽部隊 (孫臏兵法) 注釋組, (孫臏兵法) 注釋, 遼寧人民出版社, 沈陽, 1975.

世界書局編輯部, 新校史記三家注, 世界書局, vols. 1—5, 1975.

瀧川龜太郎, 史記會注考證, 藝文印書館, 台北, 1972.

鄧澤宗, 孫臏兵法注譯, 解放軍出版社, 北京, 1986.

王守謙, 戰國策全譯, 貴州人民出版社, 貴州, 1992.

王雲五, 宋刊本武經七書, 商務印書館, 台北, 3 vol., 1971 (1935).

楊家駱主編:

　戰國策高氏注, 世界書局, 台北, 3 vol., 1967.

　韓非子集解, 世界書局, 台北, 1969.

　淮南子注, 世界書局, 台北, 1969.

　荀子集解, 世界書局, 台北, 1974.

　新校資治通鑑注, 世界書局, 台北, 1972.

　逸周書集訓校釋, 世界書局, 台北, 1980.

　國語韋氏解, 世界書局, 台北, 1968.

　孫子十家注, 世界書局, 台北, 1984.

　老子新考逑略, 老子本義, 世界書局, 台北, 1972.

　管子校正, 商君書解詁定本, 世界書局, 台北, 1973.

　鬼谷子等九種, 世界書局, 台北, 1974.

楊伯峻, 春秋左傳注, 中華書局, 北京, 4 vol., 1990.

楊鍾賢, 郝志達, 文白對照全譯史記, 國際文化出版公司, 北京, 5 vol., 1992.

嚴一萍, 百部叢書集成 (宋本子部), 藝文印書館, 台北, 1965.

Glossary of Names and Selected Chinese Terms

action 動
administration 制, 治, 政
  civil 文制, 文治
  military 軍制, 軍治, 軍政
advance 進
advance force 前鋒
advantage 利
afraid 恐
agents (spies) 間
Ah 阿
All under Heaven 天下
alliances 交
altars (of state) (國) 社
ambush 伏
ancestral temple 廟
An-i 安邑
archers 弓者
armor 甲
army 軍, 師
  Awesomely Strong 威強
  Contrary 逆軍
  Doubly Soft 重柔 (柔)
  Fearfully Suspicious 助忌
  Fierce 暴軍
  Firmly Unbending 剛至
  Hard 剛軍
  Loftily Arrognt 軒驕
  of the Center 中軍
  of the Left 左軍
  of the Right 右軍
  "old" 師老
  Righteous 義軍
  Six (Armies) 六師

  Strong 強軍
  Three (Armies) 三軍
arrayed walls 行城
arrow 矢
Art of War 孫子兵法
artifice 譎
ascension and decline 盛衰, 興衰
assemble 合
  and divide 分合
attack 攻
  incendiary 火攻
  orthodox 正攻
  sudden 突攻, 襲
  unorthodox 奇攻
augury 卜
auspicious 吉
authority (*ch'üan*) 權
awesomeness 威
ax 斧

balance of power 權
Bamboo Annals 竹書紀年
bamboo slips 竹簡
barbarian dress 胡服
barricade 塞
battalion 旅, 廣
battle 戰
battle array 戰陣, 陣, 戰行
bells 鈴
beneficence 惠
benevolence 仁
besiege 圍
bestowals 賚

boats 舟
border 境, 垠
bows 弓
brigade 師
brigand 賊
brutal 暴
byways 俓

calculate 算, 數
caltraps 疾利, 蒺藜
capture 擒
cavalry 騎, 騎兵
  armored 甲騎
  attack 陷騎
  elite 銳騎, 選騎, 鐵騎
  fighting 戰騎
  heavy 重騎
  light 輕騎
  roving 游騎
  swift (fast) 疾騎
certitude 信
Chang I 張儀
Chang Kai 張丐
Chang Liang 張良
change 變
  and transformation 變化
Chan-kuo Ts'e 戰國策
Chao (state of) 趙
  King Wu-ling of 武靈王
  Prince Ch'ao of 朝公子
chaos 亂
character 性
  evaluating 考人, 知人
  exploiting 利, 用
  flaws 隙, 過
chariot 車
  assault 衝車
  attack 攻車
  battle 戰車
  defensive 守車

heavy 重車
light 輕車
Ch'en Chi 陳忌
Cheng 鄭
chevaux-de-frise 行馬
chi 戟
ch'i (pneuma, breath) 氣
Ch'i (state of) 齊
  Duke Huan of 桓公
  King Hsüan of 宣王
  King Min of 湣王
  King Wei of 威王
Chi River 濟水
Ch'i-ch'eng 齊城
chieh (constraint) 節
Ch'ien-fu 黔夫
Ch'ih-ch'iu 茌丘
Chi-hsia 稷下
Ch'ih-yu 蚩尤
Chin (state of) 晉
Ch'in (state of) 秦
Ch'in Shih Huang-ti 秦始皇帝
Chin Wen-kung 晉文公
Chou 周
  Duke of 周公
  dynasty 周朝
  Eastern 東周
  Western 西周
Chou li (Rites of Chou) 周禮
Ch'u (state of) 楚
ch'üan 權
  authority 權
  tactical balance of power 權
Ch'üan 鄟
Ch'un Ch'iu 春秋
Chung Kuo (Central States) 中國
Chung-shan 中山
Chung-shou 種首
Chuo-che 濁澤
circuitous 迂

city 城
civil 文
  and martial 文武
clamorous 譁
clarity (clearness) 明
cloud ladder 雲梯
combat 戰
combat platoon 戰隊
commands 命, 號
commandant 尉
common people 平民, 老百姓
company 卒, 閭
compel others 致人
concentration of force 集力
configuration and designation 形
  名
configuration of power 勢
confront 當, 向
Confucius 孔子
confuse 惑
confusion 亂, 擾亂
conquer 勝, 克, 剋
constant 常
constraints 節
contrary Virtue 逆德
cook fires 竈
counterattack 反攻, 逆擊
courage 勇
court 朝
crack troops 銳兵
credibility 信
criminal 罪者
crossbow 弩

danger 危
death struggle 死戰
deceive 詭, 詐
decline 衰
defeat 敗
defense 守

defiles 阻
deflated (in spirit) 失氣
deploy 陣
deployment 陣
  aquatic 水陣
  concentrated 數陣
  Dark Rising 玄襄之陣
  diffuse 疎陣
  hooked 鈎行之陣
  incendiary 火陣
depths of Earth 九地之下
desert 逃
designation 名
destroy 破
dilatory 慢, 失時
disadvantage 害, 不利
disaster 災, 殃
discipline 兵制, 兵治
disharmony 不和
disordered 擾, 不治
dispirited 失氣, 挫氣, 傷氣
disposition and strategic power 形
  勢
disposition of force 形
disposition of power 勢
ditches 溝, 洫, 瀆
dividing 分
  and reuniting 分合
divination 卜筮
doubt 疑, 狐疑
Dragon's Head 龍頭
drums 鼓

Earth 地
Eastern Hu 東胡
elite force 銳士
embankment 隄防
emblem 章
emoluments 祿
emotionally attached 親附

emotions 情
employing men 用人
employing the military 用兵
empty 空
encampment 營
encircle(d) 圍
enemy 敵
entice 利之, 動之以利, 誘
entrenchments 延
error 過, 失
estimate 計
estimation 計
evaluate 考, 策, 察, 測
　enemy 察敵情, 料敵
　men 考人, 察才
evil implement 凶器
excess (flaw) 過
execute 誅
exterior 表
xternal 外

failure 失, 敗
fatal 死
fathom (the enemy) 相, 測, 占 (敵)
fear 畏
feigned retreats 佯北
Feng-che 逢澤
feudal lords 諸侯
few 少
fields 田, 野
fines 罰
five
　affairs 五事
　colors 五色, 五彩
　flavors 五味
　grains 五穀
　grasses 五草
　killing grounds 五墓殺地
　notes 五音
　phases 五行

soils 五壤
terrains 五地
weapons 五兵
flags 旗
flanks 偏, 旁
flee (run off) 走
flourish and decline 勝衰
flying 飛
　bridge 飛橋
　hook 飛鉤
　river 飛江
　tower 飛樓
foodstuffs 食
foot soldiers 步兵
force
　heavy 重兵
　light 輕兵
ford (rivers) 渡 (水), 濟 (水)
forest 林
formation 陣, 陳
　angular 銳陣
　aquatic 水陣
　assault 衝陣
　Awl 錐行
　battle 戰陣
　circular 圓陣
　close 皮傅
　closing envelopment 闔燧陣
　Cloud 雲陣
　concentrated 數陣
　curved 曲陣
　Dark Rising 玄襄之陣
　dense 積陣
　diffuse 疏陣
　dispersed 散陣
　Eight 八陣
　elongated 索陣
　entangled, flowing 贏渭陣
　female 牝陣
　Fierce Wind 剽風

Floating Marsh 浮沮
hooked 鉤行之陣
incendiary 火陣
male 牡陣
piercing 刲陣
round 圓陣
Six Flowers 六花
solid 固
square 方
straight 直
striking 衝
training 習
Wheel 輪
Whirlwind 敷陣
Wild Geese 鴈行 (雁行)
formless 無形
fortification 城, 壘, 保, 堡
foundation 本
four 四
  Heads and Eight Tails 四頭八尾
  limbs 四肢
  quarters 四方
  seasons 四季
four-sided martial assault formation
  四武衝陣
frightened 驚懼
front 前, 鋒, 逢
frontal assault force 戰鋒隊
Fu Sui 斧遂
full 實
funeral mounds 墳墓

gate 門
general 將, 帥
  commanding 主將
  enlightened 明將
  grand 大將
  ignorant 闇將
  of the army 將軍
  subordinate 副將, 裨將

ghost 鬼
glory 榮
gong 金
gorge 谿
grain 粟
granary 倉
guarantee 保
guest 客

halberd 戈
Han 漢
  dynasty 漢朝
  Kao-tsu 漢高祖
  state of 韓
Han Ch'üeh 韓厥
Han Shu 漢書
handles of state 國柄
Han-tan 邯鄲
hard and strong 剛強
harm 害
harmony 和
hasty 疾, 急
heaven 天
Heavenly
  deployment 天陣
  float 天浮
  offices 天官
Heaven's
  Fissure 天隙
  Furnace 天竈
  Huang 天潢
  Jail 天牢
  Net 天羅
  Pit 天陷
  Well 天井
hegemon 霸
heights 高
heights of heaven 九天之上
helmet 盔, 冑
Heng 橫

hero 雄, 傑
high official (*ta-fu*) 大夫
hillock 丘
holding force 駐隊
honor 貴
Hopei 河北
horse 馬
host 主
Hsia 夏
　dynasty 夏朝
　King Chieh of 桀王
Hsiang-ling 襄陵
Hsi-men Pao 西門豹
Hsin Shu (New Book) 新書
Hsing (form) 形
Hsü-chou 徐州
Hsün Hsi 荀息
Hsün-tzu 荀子
Hsü-tzu 徐子
Huai-nan tzu 淮南子
Huang Shih-kung 黃石公
human effort (affairs) 人事
human emotions 人情
hundred 百
　illnesses 百病
　surnames 百姓

implements 器
impositions 斂
indications 徵
infantry 步兵, 徒
　heavy 重兵
　light 輕兵
insignia 表, 號, 符
instructions 敎, 練
intelligence (military) 敵情
interior 裏
internal 內
invader 寇, 客

jails 囹圄

K'ai-feng 開封
Kao-t'ang 高唐
Kao-yüan 高宛
know yourself 知己
Kuan 管
Kuan Chung 管仲
Kuan-tzu 管子
Kuei Ku-tzu 鬼谷子
Kuei-ling 桂陵
Kung Kung 共工
Kung-sun Han 公孫閈
Kuo-yü 國語
Ku-pen chu-shu 古本竹書

labor service 役
Lao-tzu 老子
law 法
li (rites, forms of etiquette) 禮
li 里
Li Ching 李靖
Li K'o 李克
Li K'uei 李悝
Liang 梁
limbs and joints 肢節
Lin-i 臨沂
Lin-tzu 臨菑 (淄)
Liu Pang 劉邦
long weapons 長兵
Lord of Ch'eng 成侯
Lord of Liang 梁君
Lord Shang 商君 (商鞅)
lost state 亡國
love the people 愛民
Lu 魯

majesty 威
Ma-ling 馬陵

mandate 命
many 多
Marquis Ching 敬侯
Marquis Lieh 烈侯
Marquis T'ien 田侯
Marquis Wen 文侯
Marquis Wu 武侯
marsh 澤
marshy depression 坳澤
martial, the 武
masses 眾
Master of Fate 司命
material resources 資, 財
measure 量
Mencius 孟子
Meng Pen 孟賁
merit 功
method 法
military
  administration 軍制
  campaign 軍旅, 行師, 出軍
  discipline 軍治
  power 軍勢
Military Methods 兵法
Military Pronouncements 軍讖
misfortune 患, 禍
moat 溝, 池, 塹
mobilize the army 起軍, 興兵, 舉
  兵, 起
Mo-tzu 墨子
mountains 山
movement and rest 動止, 動靜
mutual
  change 相變
  conquest 相剋
  production 相生
  protection 相保
  responsibility 相任
Mu-yeh 牧野

name and function 形, 形名
Nan-liang 南梁
narrow passes 隘塞
nonaction 無為
not being knowable 不可知
nurture the people 養人, 養民

oars 楫, 櫓
oath 誓
observation post 斥侯, 長關
observe (the enemy) 觀, 伺 (敵)
occupy 佔, 居, 處
offense 攻
officers 士
officials 吏, 官
omen 兆
opportunity 機
oppose 當, 拒
order 令
ordinance 律
orthodox (*cheng*) 正

Pa 巴
P'ang Chüan 龐涓
P'ang-tzu 朌子
pardon 赦, 舍
party 黨
pennant 旌
people 人, 民
perverse 邪
P'ing-ling 平陵
pipes and whistles 笳笛
plains 原
plan 計, 策
platoon 隊, 倆 (兩) 屬
Po-hai 勃海
Po-wang 博望
power 勢
precipitous 險
preservation 全

press (the enemy) 逼, 壓, 薄, 迫
pretend 僞
probe (the enemy) 刺 (敵), 角之
probing force 跳盪 (隊)
profit 利
prohibitions 禁
protracted fighting 戰久
provisions 糧
punish 罰
punishment 罰
punitive expedition 討
pursue 追

Questions and Replies 問對

raiding force 寇
rain 雨
ramparts 壘
rank 爵
ravine 險
rear 後
rectify 正
regiment 旅, 師
regimental commander 帥
regulations 律, 制
repel 禦
repress 挫
resentment 怨
respect 共(恭)
responsibility 任
rested 佚
retreat 退, 北
rewards 賞
   and punishments 賞罰
righteous 義
rites 禮
river 川, 水
rows and files 行列
ruler 主
   enlightened 明主
   obtuse (ignorant) 無知之主

rules 法
rumor 讒讟

Sage 聖人
salary 祿
sated 飽
scouts (遠) 斥
seasonal occupations 時事, 季事
security 安
segmenting and reuniting 分合
seize 奪
sericulture 蠶, 織桑
sever 絕, 斷
Seven Military Classics 武經七書
shaman 巫
shame 恥
Shang 商
   dynasty 商朝
   king Chou of 紂王
Shao-liang 少梁
shape (*hsing*) 形
Shen, Prince of Wei 太子申
Shen Nung 神農
Shen Pu-hai 申不害
shield 盾
shih (strategic power) 勢
Shih chi 史記
Shih-ch'iu 市丘
short weapons 短兵
Shu 蜀
Shu-lü 蜀祿
Shun (Emperor) 舜
siege 圍
Six Secret Teachings 六韜
soft 軟
soldiers 卒, 兵, 士
solid 固
Son of Heaven 天子
spear 矛
spies (see agents) 間

spirit (morale) 氣
spirits 神
Spring and Autumn 春秋
squad 伍
Ssu-ma Ch'ien 司馬遷
Ssu-ma Fa 司馬法
Ssu-shang 泗上
stalwart 堅
standoff 相拒
steelyard 衡
stimulate (the enemy) 作(敵)
storehouse 庫
strategic
  advantage 利, 地利
  configuration of power (*shih*)()勢
  point 要, 要塞, 要點, 塞, 險
strategy 計, 法, 兵法
strength 力
strike 擊
strong 強
strongholds 固
Su Ch'in 蘇秦
substantial 實
subterfuge 陰謀
subtle 微, 機
subtle change 機
Sun Chen 孫軫
Sung 宋
Sun Pin 孫臏
Sun Wu 孫武
Sun-tzu 孫子
supply wagon 輜
sword 劍, 長劍

ta-fu 大夫
T'ai Kung 太公
T'ai Shan (Mt.) 太山
T'ai-pai-yin ching 太白陰經
T'ai-p'ing yü-lan 太平御覽
take (the enemy) 取(敵)

Ta-liang 大梁
T'ang 湯
T'an-tzu 檀子
Tao (Way) 道
  of men 人道
  of the military 兵道, 軍道
  of warfare 戰道, 兵道
Tao Te Ching 道德經
Taoist 道家
taxes 賦
technique 術
terrain
  accessible 通地
  advantages of 地利
  broken-off 絕地
  classification of 分地
  configurations of 地形
  confined 狹地
  constricted 隘地
  contentious 爭地
  deadly (fatal) 死地
  dispersive 散地
  distant 遠地
  easy 易地
  encircled 圍地
  entrapping 圮地
  expansive 遠地
  exploiting 利地
  fatal (deadly) 死地
  heavy 重地
  isolated 絕地
  light 輕地
  near 近地
  open 通地
  precipitous 險地
  sinking 圮下
  stalemated 支地
  suspended 挂地
  tenable 生地
  traversable 交地

treacherous 險地
wet 濕地
terrified 恐懼
Three Armies 三軍
Three Chin 三晉
Three Miao 三苗
Three Strategies 三略
Ti 狄
T'ien (clan) 田
T'ien Ch'en-ssu 田臣思
T'ien Chi 田忌
T'ien Ho 田和
T'ien P'an 田盼
T'ien Tan 田單
T'ien Ying 田嬰
T'ien-t'ang 天唐
tired 勞, 困
tomb 墓
town 邑
training 練, 習
tranquil 靜
transformation 化
trebuchet 投幾
troops 卒
troubled 憂
trust 信
trustworthy 有信, 可信
Ts'ao 曹
Tso Chuan 左傳
Tsou Chi 鄒忌
Tuan-han Lun 段干綸
T'ung Tien 通典
Tung-yang 東陽
Tunhuang 敦煌

unconquerable 不可勝
unity 一, 專一
unorthodox (*ch'i*) 奇

vacillate 猶豫
vacuity 虛

vacuous 虛
valley 谷
vanguard 踵軍
victory by turn of events 曲勝
village 村
villain 賊
Virtue 德
vital point 機
vulnerable point 空點, 弱點, 虛點
 (地)

wage war 作戰
wall 牆, 城
war 戰
ward off 拒
warfare 戰
 explosive 突戰
 forest 林戰
 mountain 山戰
Warring States 戰國
warrior 士, 武
 armored 甲士
 death 死士
 elite 選士, 銳士
water 水
weak 弱
weapons 兵器
Wei 魏
 King Hui of 惠王
Wei Liao-tzu 尉繚子
well 井
wetlands 沮, 澤, 沛
Wey 衛
withdraw 退, 却
Worthy 賢人
Wu 武
Wu Ch'i 吳起
Wu-tzu 吳子

yang 陽
Yang Chu 楊朱

Yang-tze River 江, 長江
Yao (Emperor) 堯
Yellow Emperor 黃帝
Yellow River 黃河, 河
Yellow Sea 黃海
Yen (state of) 燕
Yi 夷

Yi 羿
yin 陰
yin and *yang* 陰陽
Yin-fu Ching 陰符經
Yü 禹
Yüeh (state of) 越
Yung Gate 雍門

About the Book and Translator

"They eat people and cook the bones, the officers have no thought to turn outside—these are the troops of Sun Pin." This graphic description from the *Shih Chi* depicts the awesome army commanded by Sun Pin, a direct descendent—probably the great-grandson—of the legendary Sun Tzu. Sun Pin studied and adopted many of the ideas found in his predecessor's masterpiece, the *Art of War,* but he also developed his own distinct strategic style suited to the changing conditions of his time, the fourth century B.C.

In addition to translating this "eighth military classic," Ralph Sawyer has prepared insightful chapter-by-chapter commentaries and a vivid general introduction that describes Sun Pin's life and times, analyzes in detail Sun Pin's tactics in important battles, and compares Sun Tzu's strategic thinking with Sun Pin's.

After study at MIT, Harvard, and National Taiwan University plus a brief period of university teaching, Ralph D. Sawyer spent twenty years in international consulting throughout Asia. He is the translator of *The Seven Military Classics of Ancient China* (Westview, 1993) and *Sun Tzu: Art of War* (Westview, 1994).

Index of Strategic and Tactical Principles in the Text of the Military Methods

References are to chapter and page numbers. Entries are sequenced by topic without regard to the following prefatory words: a, the, employ, establish, exploit, and maintain. Throughout, the emphasis has been upon selecting significant, illuminating passages rather than upon comprehensiveness.

For a more complete listing of relevant topics, the General Index—presented in traditional format—should be consulted.

Selected Measures to Implement General Principles

Employ ambushes, 3 (88), 14 (148)

Attack weakness:

command deficiencies (flaws, weaknesses, ignorance), 23 (197), 25 (204), 26 (208-10)

the confined, 8 (114), 14 (147)

the confused, 7 (108), 17 (168)

the debilitated, 26 (208), 27 (214)

the disordered, 17 (167), 26 (208), 32 (240)

the dispirited, 13 (139), 14 (146), 19 (179), 26 (208-9), 27 (217)

the disunited, 13 (139), 17 (169), 26 (208-9)

the doubtful, 14 (148), 17 (167, 169)

the fearful and terrified, 21 (187), 26 (208-9)

the lax, 17 (169), 26 (209), 32 (240)

the slow, 29 (224)

the tired, 1 (82), 26 (208-9), 27 (214)

the vulnerable, exposed, or on fatal terrain, 8 (113), 27 (214), 30 (231)

crossing water obstacles, 8 (113), 14 (148)

when and where unexpected, 3 (89, 91), 17 (169), 32 (240)

when and where unprepared (undefended), 3 (89, 91), 17 (169), 26 (209), 32 (240)

the vacuous and vacuities, 17 (168-9), 26 (210), 27 (214), 29 (224), 30 (230), 32 (240)

the weak, 7 (108)

Concentrate force, 17 (167-8)

Confuse and disrupt the enemy, 3 (89), 9 (117), 14 (148)

Destabilize the enemy into motion, 1 (82), 3 (88, 90), 14 (147), 17 (167, 169, 170), 20 (185), 28 (220)

Employ elite units as decisive force, 3 (91), 5 (101), 7 (108), 14 (147), 16 (171), 17 (168, 170), 18 (176)

Envelop the enemy, 14 (146)

Formations require a strong front, 9 (116), 10 (122), 14 (147), 16 (162)

Employ methods to cause misperception and disorder:

create clamor and noise 21 (187). *See also* Clamor in the General Index.

multiply the flags, 9 (117), 14 (148), 16 (162)

Improve defenses with equipment at hand, 4 (96)

Manipulate the enemy to weaken him by:

changing his strengths to weakness, 20 (185)

deceiving him, 1 (81-2), 3 (88), 14 (148), 17 (169)

enticing him with apparent profit, 1 (82), 3 (88), 7 (109), 14 (147-9), 17 (167, 169), 18 (176), 21 (187)

feigning defeat and chaotic retreat, 17 (167-8)

Commonly Encountered Situations with Selected Tactics

Offensive Measures

General Index